METAL MATRIX COMPOSITES

METAL MATRIX COMPOSITES

Nikhilesh Chawla
Arizona State University, Tempe, AZ

Krishan K. Chawla
University of Alabama at Birmingham, AL

Springer

Library of Congress Cataloging-in-Publication Data

A C.I.P. Catalogue record for this book is available
from the Library of Congress.

ISBN 10 0-387-23306-7 e-ISBN 10 0-387-28567-9
ISBN 13 9780387233062 e-ISBN 13 9780387285672

Printed on acid-free paper.

Printed in the United States of America.

9 8 7 6 5 4 3 2 1 SPIN 11051367

springeronline.com

मनःप्रसादः सौम्यत्वं मौनमात्मविनिग्रहः ।
भावसंशुद्धिरित्येतत्तपो मानसमुच्यते ॥

Serenity, simplicity, gravity, self-control, and purity of thought are the austerities of the mind.

Bhagvad Gita,
Chapter 17, shloka 16

To The Women in Our Lives

Anita, Kanika, and Nivedita

CONTENTS

PREFACE

Metal matrix composites (MMCs) have been around for a long time, but have only been recognized as legitimate engineering composite materials in the second half of the twentieth century. From their humble research beginnings, MMCs have gone from "niche" materials to several high performance applications in aerospace, electronic packaging, automotive, and recreational products. Much of the significant research advances have been published in several journal articles, conference proceedings, and multi-authored volumes. It is surprising that only two texts have been published in this area. Taya and Arsenault (1989) published the first book on MMCs, which was followed by that of Clyne and Withers (1993). Thus, it has been a long time since a textbook on MMCs has been published. During this period there has been a myriad of new and exciting advances and applications in the field. These include continuous fiber reinforced MMCs for cables in power transmission, high temperature superconducting wires, particulate MMCs in civilian aircraft and automotive applications, and high volume fraction, high thermal conductivity substrates for electronic packaging. All these factors make this a very opportune time to bring out the present text.

We have several goals in writing this textbook on MMCs: (i) To provide a thorough and coherent coverage of MMCs that will be useful to the beginning student who wants to learn about this exciting field, as well as the practicing scientist or engineer who may be familiar with some aspects of MMCs, but has not been able to keep with all the exciting new developments in the field; (ii) summarize and consolidate the vast body of work that has been done in the last 50 years or so; (iii) illustrate several examples of MMC applications, with the hope that this will spur further research and innovation, and have an even bigger impact on MMCs for future applications.

Throughout this textbook, the reader will encounter a recurring theme, which is the synergistic relationship among processing, microstructure, and properties of metal matrix composites. In this regard, we begin with a brief definition and introduction of metal matrix composites. This is followed by a chapter each on reinforcements and common matrix materials. A chapter on the very important topic of processing of MMC is then presented. This is followed by a chapter on interfaces in MMCs, their characterization and techniques to obtain interfacial properties. Next there are chapters on monotonic mechanical and physical properties; followed by cyclic fatigue, creep, and wear resistance. We conclude with a chapter on applications of MMCs.

Acknowledgments

NC would like to acknowledge long-term support for research in the MMC arena from the Office of Naval Research (A.K. Vasudevan, Program Manager); and The United States Automotive Partnership, through the Department of Energy (W. Jandeska, *GM*, R. Chernekoff, *Ford*, and J. Lynn, *Daimler-Chrysler*, Program Managers). Several individuals were instrumental in whetting his appetite and enhancing his knowledge of MMCs, through fruitful interactions, collaborations, and discussions: W.H. Hunt, Y.-L. Shen, J.J. Lewandowski, J. LLorca, J.W. Jones, and J.E. Allison. NC would like to thank a few key individuals in the administration at Arizona State University for providing the means and encouragement to carry out this endeavor; in particular his department chair S. Mahajan, Dean P. Crouch, and President M.M. Crow. This book would not have been possible without the outstanding work and inquisitive nature of his current and former students and postdocs: A. Ayyar, G. Crawford, X. Deng, M. Dudek, V.V. Ganesh, M. Kerr, G. Piotrowski, R. Saha, R.S. Sidhu, J.J. Williams, and B. Wunsch. G. Piotrowski was also instrumental in making several figures and diagrams in the text.

KKC would like to acknowledge the grant of a sabbatical leave from the University of Alabama at Birmingham, part of which was spent at Arizona State University (ASU) where the hospitality of Professor S. Mahajan was a great help in the writing of this book. During other parts of the sabbatical, KKC had the pleasure of stimulating discussions with P.D. Portella (BAM, Berlin), M. Elices and J. LLorca (University of Madrid), and G.M. Gladysz (Los Alamos National Laboratory). He is also grateful to the researchers in his group, K. Carlisle, M. Koopman, R. Kulkarni at University of Alabama at Birmingham for their cheerfulness, sincerity, and hard work. He is very thankful to B.A. MacDonald, National Science Foundation, and S. Fishman and A.K. Vasudevan, Office of Naval Research, who over the years have funded his research program. He also thanks A. Mortensen, B.R. Patterson, B.V. Patel, J.M. Rigsbee, and N. Stoloff who were always there to discuss things.

The following individuals were particularly generous in providing us with micrographs and figures of their work: H. Devé, A. Dlouhy, A. Drake, G. Eggeler, J. Eldridge, D. Herling, P. Krajewski, J. Liu, J. Llorca, D. Lloyd, D. Miracle, M. Ochionero, W. Nix, S. Rawal, H. Ruppert, and T. Wang. We are very thankful to Greg Franklin and Carol Day for their patience and understanding during the course of writing this book.

Finally, for the curious reader who might be wondering about the common last names of the authors, we should let the cat out of the bag. This book is collaboration between two colleagues and friends, who happen to be father (KKC) and son (NC). As such, it was a mutually fulfilling and special family endeavor. Furthermore, this book would not have been possible without the support and encouragement of the rest of the Chawla family: Anita Chawla (wife and daughter-in-law), Kanika Chawla (sister and daughter), and Nivedita Chawla (mother and wife). Special thanks to Anita Chawla for her invaluable assistance with formatting and index preparation.

Nik Chawla
Tempe, Arizona

Krish K. Chawla
Birmingham, Alabama

References

Clyne, T.W., and P.J. Withers, (1993) *An Introduction to Metal Matrix Composites*, Cambridge University Press, Cambridge, U.K.

Taya, M., and R.J. Arsenault, (1989) *Metal Matrix Composites, Thermomechanical Behavior*, Pergamon Press, New York.

Chapter 1

INTRODUCTION

Metal matrix composites (MMCs), like all composites, consist of at least two chemically and physically distinct phases, suitably distributed to provide properties not obtainable with either of the individual phases. Generally, there are two phases, e.g., a fibrous or particulate phase, distributed in a metallic matrix. Examples include continuous Al_2O_3 fiber reinforced Al matrix composites used in power transmission lines; Nb-Ti filaments in a copper matrix for superconducting magnets; tungsten carbide (WC)/cobalt (Co) particulate composites used as cutting tool and oil drilling inserts; and SiC particle reinforced Al matrix composites used in aerospace, automotive, and thermal management applications.

A legitimate question that the reader might ask is: Why metal matrix composites? The answer to this question can be subdivided into two parts: (a) advantages with respect to unreinforced metals and (b) advantages with respect to other composites such as polymer matrix composites (PMCs). With respect to metals, MMCs offer the following advantages:

- Major weight savings due to higher strength-to-weight ratio
- Exceptional dimensional stability (compare, for example, SiC_p/Al to Al)
- Higher elevated temperature stability, i.e., creep resistance
- Significantly improved cyclic fatigue characteristics

With respect to PMCs, MMCs offer these distinct advantages:

- Higher strength and stiffness
- Higher service temperatures
- Higher electrical conductivity (grounding, space charging)
- Higher thermal conductivity
- Better transverse properties

- Improved joining characteristics
- Radiation survivability (laser, UV, nuclear, etc.)
- Little or no contamination (no out-gassing or moisture absorption problems)

1.1 TYPES OF MMCs

All metal matrix composites have a metal or a metallic alloy as the matrix. The reinforcement can be metallic or ceramic. In some unusual cases, the composite may consist of a metallic alloy "reinforced" by a fiber reinforced polymer matrix composite (e.g., a sheet of glass fiber reinforced epoxy or aramid fiber reinforced epoxy).

In general, there are three kinds of metal matrix composites (MMCs):

(i) particle reinforced MMCs
(ii) short fiber or whisker reinforced MMCs
(iii) continuous fiber or sheet reinforced MMCs

Figure 1.1 shows, schematically, the three major types of metal matrix composites: Continuous fiber reinforced, short fiber or whisker reinforced, particle reinforced, and laminated or layered composites. The reader can easily visualize that the continuous fiber reinforced composites will be the most anisotropic of all. Table 1.1 provides examples of some important reinforcements used in metal matrix composites as well as their aspect ratios (length/diameter) and diameters.

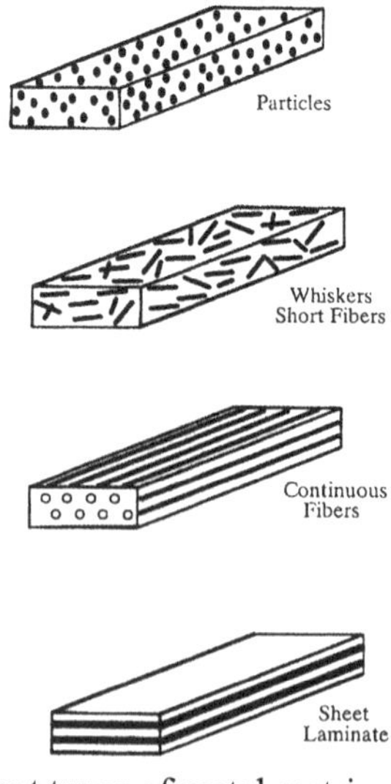

Fig. 1.1 Different types of metal matrix composites.

Table 1.1 Typical Reinforcements Used in Metal Matrix Composites

Type	Aspect Ratio	Diameter, μm	Examples
Particle	1–4	1–25	SiC, Al_2O_3, BN, B_4C, WC
Short fiber or whisker	10 – 10000	1–5	C, SiC, Al_2O_3, $Al_2O_3+SiO_2$
Continuous fiber	>1000	3 – 150	SiC, Al_2O_3, C, B, W, Nb-Ti, Nb_3Sn

Particle or discontinuously reinforced MMCs (the term discontinuously reinforced MMCs is commonly used to indicate metal matrix composites having reinforcements in the form of short fibers, whiskers, or particles) have assumed special importance for the following reasons:

- Particle reinforced composites are inexpensive *vis-à-vis* continuous fiber reinforced composites. Cost is an important and essential parameter, particularly in applications where large volumes are required (e.g., automotive applications).
- Conventional metallurgical processing techniques such as casting or powder metallurgy, followed by conventional secondary processing by rolling, forging, and extrusion can be used.
- Higher use temperatures than the unreinforced metal.
- Enhanced modulus and strength.
- Increased thermal stability.
- Better wear resistance.
- Relatively isotropic properties compared to fiber reinforced composites.

Within the broad category of discontinuously reinforced composites, metal matrix composites made by liquid metal casting are somewhat cheaper to produce than powder metallurgy composites. There are two types of cast metal matrix composites:

- Cast composites having local reinforcement.
- Cast composites in the form of a billet having uniform reinforcement with a wrought alloy matrix. Such composite billets are forged and/or extruded, followed by rolling or other forming operations.

1.2 CHARACTERISTICS OF MMCs

One of the driving forces for metal matrix composites is, of course, enhanced stiffness and strength. There are other characteristics which may be equally valuable. As examples, we can cite the ability to control thermal expansion in applications involving electronic packaging. By adding ceramic reinforcements, one can generally reduce the coefficient of linear thermal expansion of the composite. Electrical and thermal conductivity characteristics may be important in some applications. Clearly, superconductors require superconducting characteristics. The metallic matrix provides a high thermal conductivity medium in case of an accidental quench, in addition to holding the tiny superconducting filaments together. Other important characteristics that may be of immense value include wear resistance (e.g., in WC/Co composites used in cutting tools or oil drilling inserts and SiC_p/Al rotor in brakes). Thus, although one commonly uses the term reinforcement by particle or fibers in the context of metal matrix composites, it worth pointing out that strength enhancement may not be the most important characteristic in many applications. In the chapters that follow we explore these and other unique and important attributes of metal matrix composites.

Chapter 2

REINFORCEMENTS

Reinforcement materials for metal matrix composites can be produced in the form of continuous fibers, short fibers, whiskers, or particles. The parameter that allows us to distinguish between these different forms of reinforcements is called the aspect ratio. Aspect ratio is nothing but the ratio of length to diameter (or thickness) of the fiber, particle, or whisker. Thus, continuous fibers have an aspect ratio approaching infinity while perfectly equiaxed particles have an aspect ratio of around one. Table 2.1 lists some important reinforcement materials available in different forms. Ceramic reinforcements combine high strength and elastic modulus with high temperature capability. Continuous ceramic fibers are also, however, more expensive than ceramic particulate reinforcements.

In this chapter, we describe some general features of reinforcements and then provide a more detailed description of processing, microstructure, and properties of some important ceramic reinforcements that are commonly used in metal matrix composites.

2.1 FIBROUS MATERIALS

One can transform practically any material (polymers, metals, or ceramics) into fibrous form (Chawla, 1998). A fiber can be defined as an elongated material having a more or less uniform diameter or thickness of less than 250 μm and an aspect ratio of more than 100. Note that this is not only an operational definition but also a purely geometrical one that applies to any material.

Fibers have some unique features that stem mainly from their relatively small cross-section and large aspect ratio:

Table 2.1 Some important reinforcements for metal matrix composites.

Continuous Fibers	Al_2O_3, Al_2O_3+SiO_2, B, C, SiC, Si_3N_4, Nb-Ti, Nb_3Sn
Discontinuous Fibers	
(a) Whiskers	SiC, TiB_2, Al_2O_3
(b) Short Fibers	Al_2O_3, SiC, (Al_2O_3+SiO_2), vapor grown carbon fibers
Particles	SiC, Al_2O_3, TiC, B_4C, WC

- High degree of flexibility.
- Higher strength than the bulk material of the same composition.

The long length of fibers also makes it imperative, in most cases, to incorporate them in some continuous medium, i.e., the matrix, to hold them together to make a fiber reinforced composite. It should be emphasized that this, by no means, is the sole purpose of the matrix in a composite.

2.2 FIBER FLEXIBILITY

Flexibility is an important attribute of fine fibers. A high degree of flexibility is an intrinsic characteristic of a material having a small diameter and a low modulus (Dresher, 1969). Flexibility of fibers permits a variety of techniques to be employed for making fiber reinforced composites. Consider a fiber as an elongated elastic beam; then its flexibility will be an inverse function of its elastic modulus, E, and the second moment of area or moment of inertia of its cross-section, I. The elastic modulus of a material is generally independent of its form or size and is generally a material constant for a given chemical composition (assuming a fully dense material). Thus, for a given composition and density, the flexibility of a material is determined by its shape, or more precisely by its diameter. Consider a bending moment, M to the elastic beam, which will bend it to a radius R. The product of the bending moment (M) and the radius of curvature (R) is called *flexural rigidity*. We can use the inverse of this flexural rigidity as a measure of flexibility. From elementary strength of materials, we have the following relationship for a beam bent to a radius R (see the boxed section for a derivation of this relationship):

$$\frac{M}{I} = \frac{E}{R}$$

where E is the Young's modulus of the material and the other terms are defined above. For a beam or a fiber of diameter, d, we have I = $d^4/64$. Substituting this in the above expression and rearranging the expression, we can write:

$$MR = EI = \frac{E\pi d^4}{64}$$

EI (or MR) being the flexural rigidity, we have flexibility equal to 1/MR. Thus,

$$\frac{1}{MR} = \frac{64}{E\pi d^4} \tag{2.1}$$

where d is the equivalent diameter and I is the moment of inertia of the beam (fiber). Equation 2.1 indicates that flexibility, 1/MR, is a very sensitive function of diameter, d. Given a sufficiently small diameter, it is possible to produce, in principle, a flexible fiber whether it is a polymer, metal, or ceramic.

Figure 2.1 shows different forms in which fibers can be used. Figure 2.1(a) shows unidirectionally aligned fibers in an untwisted, continuous tow or roving of fibers. Figure 2.1(b) shows a woven fabric in plane weave pattern, while Fig.2.1(c) shows a knitted fabric. Such architectures of fibers, especially those shown in Figs. 2.1(b) and (c), are possible only with small diameter fibers which possess the required flexibility for weaving.

2.3 CARBON FIBERS

Carbon is a very versatile element. It is very light, has a theoretical density of 2.27 g/cm^3, and exists in a variety of forms. The two most important forms of carbon are diamond and graphite. Among the relatively new forms of carbon are the Buckminster Fullerene, also known as Buckyball, and the drawn out version of Buckyballs, which may be in the form of nanotubes or nanowires. It is the graphitic form of carbon that is important in understanding the carbon fibers. The hexagonal structure of carbon in the graphitic form is shown in Fig. 2.2.

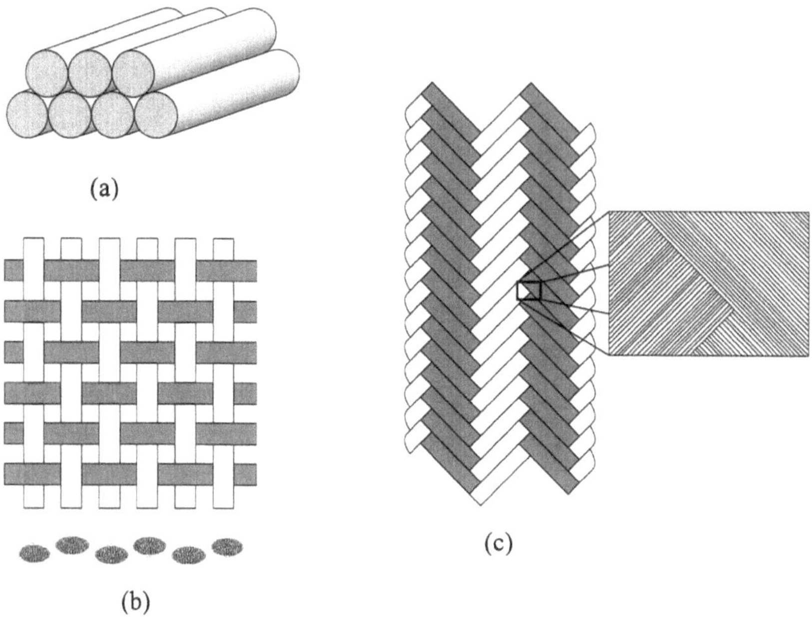

Fig. 2.1 Forms of fibrous materials (a) unidirectional fibers, (b) woven fiber fabric, and (c) fabric made of knitted fibers.

Consider a fiber to be a cylindrical beam under an applied moment, M.

$$M = \int y\sigma \, dA = \int y \frac{\sigma}{y} y \, dA = \frac{\sigma}{y} \int y^2 \, dA = \frac{\sigma I}{y}$$

where σ is the stress in an element a distance y from the neutral axis of the beam. The $\int y^2 \, dA = I$ is called the second moment of area of the beam. Rearranging the above expression, we get:

$$\frac{M}{I} = \frac{\sigma}{Y} \qquad (1)$$

$$\varepsilon = \frac{(R + y)d\theta - Rd\theta}{Rd\theta}$$

$$\varepsilon = \frac{y}{R}$$

Incorporating Hooke's law,

$$\varepsilon = \frac{\sigma}{E}$$

$$\varepsilon = \frac{\sigma}{E} = \frac{y}{R}$$

$$\frac{\sigma}{y} = \frac{E}{R} \qquad (2)$$

From Eqs. (1) and (2), we get,

$$\frac{M}{I} = \frac{E}{R}$$

The term carbon fiber represents a family of fibers. It is one of the most important reinforcement fibers in all kinds of composites. It is worth emphasizing that graphite fiber is a special form of carbon fiber that is obtained after heating to a temperature greater than 2400°C. This process, called *graphitization*, results in a highly oriented, layered crystallographic structure, which, in turn, leads to significantly different chemical and physical properties than non-graphitic forms of carbon. An extreme case of the graphite structure is, of course, a graphite single crystal. Such a single crystal has hexagonal symmetry and, consequently, anisotropic characteristics. In particular, in a carbon fiber, properties such as clastic modulus will be different along the length of fiber and along the transverse or radial direction. The more aligned the basal planes in a carbon fiber, i.e., the more graphitic the structure, the higher the modulus in the direction of the axis and the greater the degree of anisotropy.

2.3.1 Fabrication of Carbon Fibers

Carbon fibers are fabricated by controlled pyrolysis of an organic fiber precursor (Ezekiel and Spain, 1967; Watt and Johnson, 1969; Johnson and Tyson, 1969; Watt, 1970; Diefendorf and Tokarsky, 1975; Singer, 1979). As it turns out, the concept of starting with an organic precursor fiber and

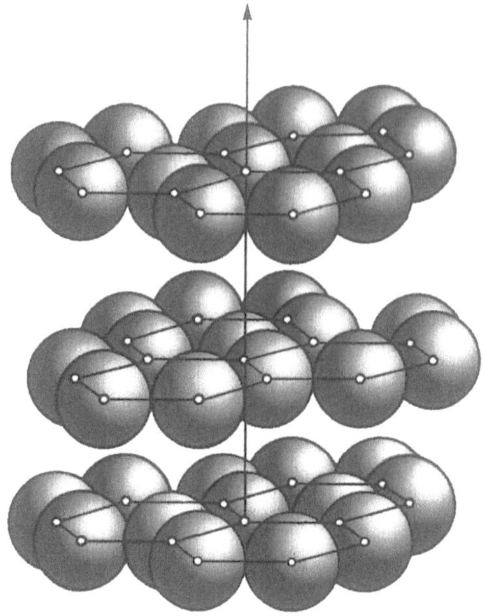

Fig. 2.2 Hexagonal structure of graphite.

converting to an inorganic fiber is quite general, as we shall see for ceramic fibers. Depending on the precursor and the processing route, one can obtain a variety of carbon fibers with different strengths and moduli. Some important types of carbon fibers are High Strength (HS), High Modulus (HM), Intermediate Modulus (IM), Super High Modulus (SHM), etc.

Polyacrylonitrile (PAN) is commonly used as a precursor fiber for making carbon fibers. Pitch based carbon fibers are also important; carbon fibers obtained from pitch can have much superior properties than those obtained from a PAN precursor.

The steps common to all processes of carbon fiber manufacture are as follows:

(a) Fiberization, i.e., extrusion of a polymer melt or solution into a precursor fiber.

(b) Stabilization (oxidation or thermosetting) is done at relatively low temperatures (200-450°C), usually in air. This renders the precursor infusible during subsequent high temperature processing.

(c) Carbonization is carried out in an inert atmosphere (pure N_2 generally) at 1000-2000°C. At the end of this step the fiber has 85-99% carbon content.

(d) Graphitization (optional) is done in Ar or N_2 at a temperature greater than 2500°C. This step increases the carbon content to more than 99% and imparts a very high degree of preferred orientation to the fiber.

2.3.2 PAN-Based Carbon Fibers

The PAN precursor fiber has a flexible polymeric chain structure. A flow diagram showing the various processing steps for PAN-based carbon fiber is shown in Fig. 2.3. The PAN precursor has a flexible polymer chain structure, but it has an all-carbon backbone chain containing polar nitrile groups, Fig. 2.4(a). During the stabilization treatment, the PAN precursor fiber is heated to 220°C *under tension*. During this treatment, the nitrile groups react to

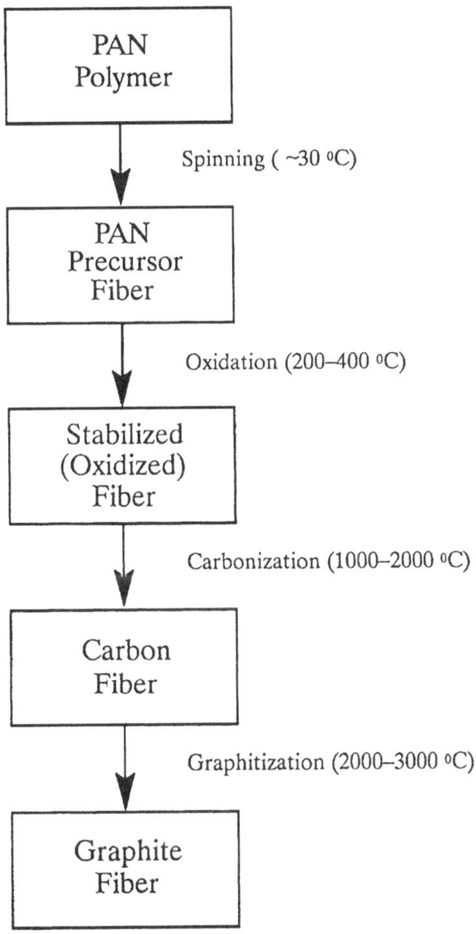

Fig. 2.3 Flow diagram for the fabrication of PAN-based carbon fiber.

form a ladder structure, which is a rigid and thermally stable structure, Fig. 2.4(b). Also, when PAN is heated in air at 220°C, the absorbed oxygen cross-links the chains, the fibers turn black, and a stable ladder structure is formed. This treatment, also done under tension, helps in maintaining the orientation of the ring structure even after unloading. During the carbonization treatment between 1500°C, well-developed hexagonal networks of carbon form and a considerable amount of gases are evolved. This gas evolution is partly responsible for some crack formation in the fiber, resulting in a lower tensile strength.

Commercially produced carbon fibers have a protective surface coating, called a *sizing or size*. The size serves two purposes: (i) ease of handling of the fiber, and (ii) improved adhesion with a polymeric matrix. The uncoated carbon fibers pick up surface charge easily when they come in contact with rubbing surfaces such as rollers, pulleys, guides, spools, etc. Handling such fibers (winding, weaving, or braiding) can cause fiber breakage and tiny fragments of carbon fibers can become airborne and short electrical machinery.

2.3.3 Pitch-Based Carbon Fibers

Pitch is an important and low cost raw material for producing carbon fibers. There are three common sources of pitch:

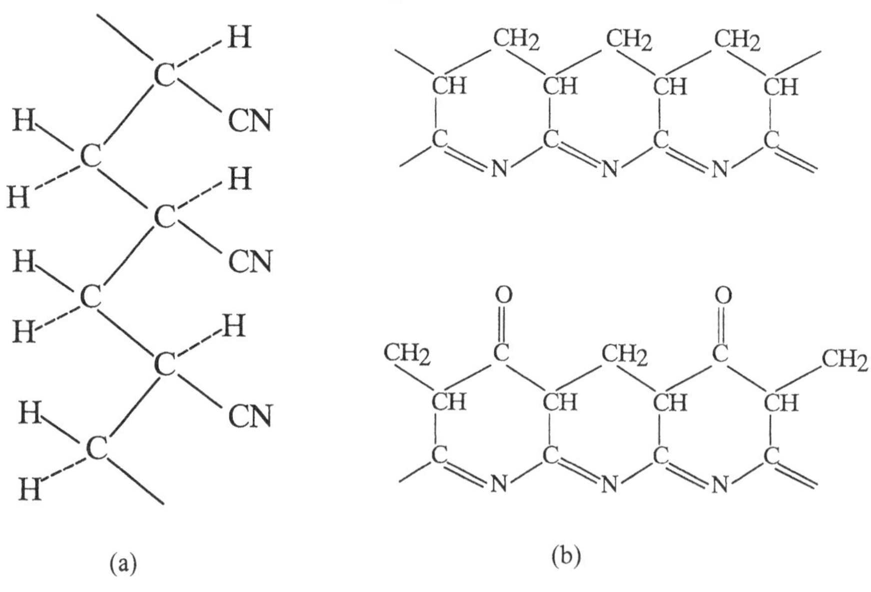

Fig. 2.4 (a) The carbon backbone chain structure of PAN (b) the ladder structure of PAN after stabilization.

• Petroleum asphalt
• Coal tar
• Polyvinyl chloride (PVC)

Pitch, being thermoplastic in nature, is difficult to carbonize without being first stabilized against fusion during pyrolysis. A schematic of the process of making carbon fibers from pitch is shown in Fig. 2.5. It involves the following steps:

• Extrusion or melt spinning into fibrous form
• Stabilization between 250°C and 400°C
• Carbonization
• Graphitization

Spinnability of the pitch and its conversion into a nonfusible state are the most important steps (Diefendorf and Tokarsky, 1975). These characteristics depend on the chemical composition and molecular weight distribution of the pitch. The molecular weight (MW) of the pitch controls the viscosity and the melting range, i.e., MW controls the temperature and speed of spinning. The pitch composition depends on its source. In fact, there is tremendous

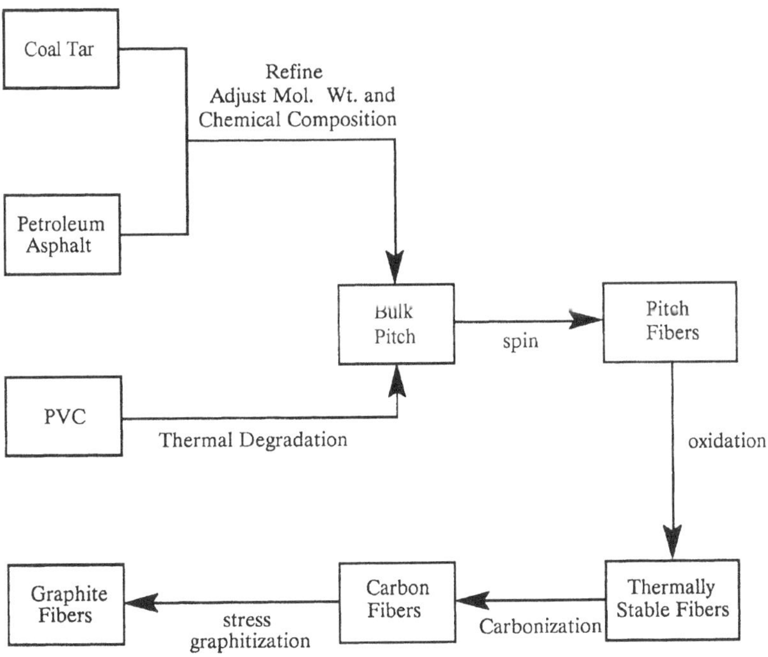

Fig. 2.5 Schematic of the process of making carbon fiber from a pitch.

variability in the composition of a pitch because it can be a mixture of hundreds of different species, varying with the source of crude oil and the processing conditions in the refinery.

Suitability of pitch for conversion to carbon fiber depends on a number of factors. The pitch should have a high carbon content (> 90%), high aromatic content (> 50%), low impurities and molecular weight and molecular weight distribution, viscosity and rheological characteristics consistent with the ease of spinning, followed by an easy conversion to the infusible state (Singer, 1979).

2.3.4 Spinning and Rheology of Mesophase Pitch

Although melt spinning of a pitch to produce a precursor fiber is commercially popular, centrifugal spinning and jet spinning can also be used. Mesophase pitch, a thermoplastic, has a nematic liquid crystal structure, wherein ordered domains of rigid rodlike molecules float in an isotropic matrix. The two-phase precursor pitch is agitated prior to spinning to form a homogeneous mixture and spun into filaments in the temperature range where the viscosity is in the 1-20 Pa.s range. Fibers can be spun at speeds of 3-100 $m.min^{-1}$ with diameters of 10-20 μm and having the same composition as the mesophase pitch. The as-spun mesophase fibers are anisotropic because of the nematic liquid crystal structure. These fibers have large, elongated, anisotropic domains (about 4 μm diameter) aligned along the fiber axis and are thermoplastic in nature. Fibers drawn from isotropic pitch or the ones drawn from a PAN precursor do not exhibit such anisotropic domains.

An oxidation treatment is given next to thermally stabilize the fibers against internal relaxation and render them infusible for subsequent processing steps. During the stabilization treatment, oxygen enters the precursor fibers and provides cross-linking between the pitch molecules of hydrocarbons comprising pitch. There is a slight weight gain in the fibers in this step. The rest of the process is essentially similar to the PAN-based carbon fiber fabrication, except that stress is applied during graphitization. The step of carbonization serves to remove the non-carbon atoms (mostly hydrogen and oxygen) with an attendant loss in weight of the fibers.

2.3.5 Structure and Properties of Carbon Fibers

Table 2.2 presents property data for three different types of PAN-based carbon fibers, while Table 2.3 gives the properties of mesophase pitch-based

Table 2.2 Properties of PAN-based carbon fiber (strand data) (after Riggs, 1985).

Characteristic	High Strength[a]	Super High Strength[b]	High Modulus[c]
Filament diameter, μm	5.5–8.0	5.4–7.0	8.4
Density, g/cm^3	1.75–1.80	1.78–1.81	1.96
Carbon content, wt%	92–95	99–99^+	99^+
Tensile strength, MPa	3100–4500	2400–2550	1865
Tensile modulus, GPa	25–260	360–395	520
Strain at fracture, %	1.3–1.8	0.6–0.7	0.38
Electrical resistivity, μΩm	15–18	9–10	6.5
Thermal conductivity, $W(mK)^{-1}$	8.1–9.3	64–70	120

[a] Thornel T-300, T-500, T-600, T-700; Celion 3000, 6000, 1200; AS2, AS4, AS6, IM6; [b] Thornel T-50, Celion G-50, HMS; [c] Celion GY-70

Table 2.3 Properties of mesophase pitch-based carbon fiber (after Singer, 1981)

Property	Thornel P555	Thornel P755	Thornel P200
Filament diameter, μm	10	10	10
Density, g/cm^3	2.02	2.06	2.15
Carbon content, wt. %	99	99	99^+
Tensile strength, MPa	1895	2070	2240
Tensile modulus, GPa	380	517	690
Strain at fracture, %	0.5	0.4	0.3
Electrical resistivity, μΩm	7.5	4.6	2.5
Thermal conductivity, W/mK	110	185	515

carbon fibers. Note the high density and high modulus of pitch-based fibers compared to PAN-based fibers. The stress applied at very high temperatures during the graphitization step (to a temperature as high as 3000°C) increases the degree of order in carbon fibers. This is also accompanied by a large increase in the elastic modulus of the fiber. The tensile strength of PAN-based fibers when subjected to high temperature treatment, however, decreases, Fig. 2.6 (Watt 1970). This is attributed to the presence of discrete flaws on the fiber surface and within the fiber. Most of the volumetric defects in carbon fibers originate from:

(i) Inorganic inclusions
(ii) Organic inclusions
(iii) Irregular voids from rapid coagulation
(iv) Cylindrical voids precipitated by dissolved gases

These defects get transformed during the high temperature treatment into diverse imperfections. Basal-plane cracks called *Mrozowski* cracks are perhaps the most important type of flaw that limits the tensile strength of carbon fibers. These occur as a result of anisotropic thermal contractions within the ribbon structure on cooling from the high temperature treatment (> 1500°C).

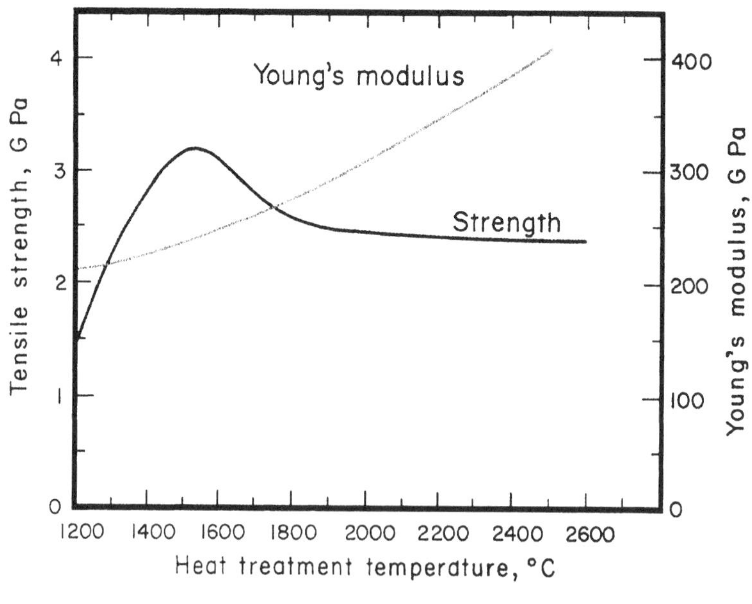

Fig. 2.6 Elastic modulus and tensile strength of carbon fiber as a function of high temperature treatment (after Watt, 1970).

These cracks are generally aligned along the fiber axis. Their presence lowers the tensile strength of the fiber by providing easy crack nucleation sites. The fiber elastic modulus, however, is unaffected because the number of flaws is relatively small, and the elastic strains involved in the modulus measurement are also small. Surface flaws can also limit the tensile strength of the carbonized fibers. Various oxidation treatments are commonly given to carbon fibers, which tend to remove surface defects and, thus, increase the strength of the fiber.

2.3.6 Structure of Carbon Fibers

Many researchers have studied the structure of carbon fibers; for a good summary, see, for example, Peebles, 1995. Here we provide the salient features of the structure of carbon fibers. Scanning electron micrographs of pitch-based carbon fibers showing graphitic, sheet like morphology can be seen in Fig. 2.7 (Kumar et al., 1993). At a submicrometer level, as seen in a transmission electron microscope, the microstructure of carbon fibers is quite heterogeneous. In particular, there is a pronounced irregularity in the packing of graphitic lamellae inward from the fiber surface. The basal planes are much better aligned in the near surface region of the fiber. In very general terms, the graphitic ribbons are oriented more or less parallel to the fiber axis with random interlinking of layers, longitudinally and laterally. An example of such a lamellar structure of structure of PAN-based carbon fibers is shown in Fig. 2.8 (Deurbergue and Oberlin, 1991). This is a high resolution electron micrograph of a longitudinal section of a fiber carbonized at 1330°C. The micrograph shows <002> lattice fringes, which represent the

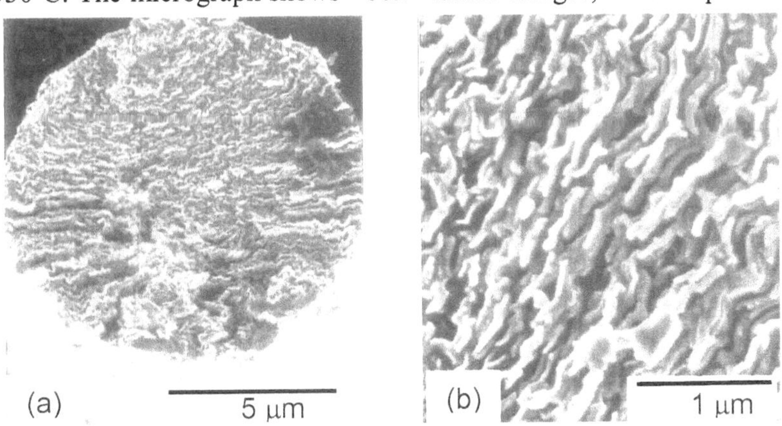

Fig. 2.7 Scanning electron micrograph of pitch-based P-100 fiber at low (a) and high (b) magnification showing graphitic sheet like morphology (Kumar et al., 1993; courtesy of S. Kumar).

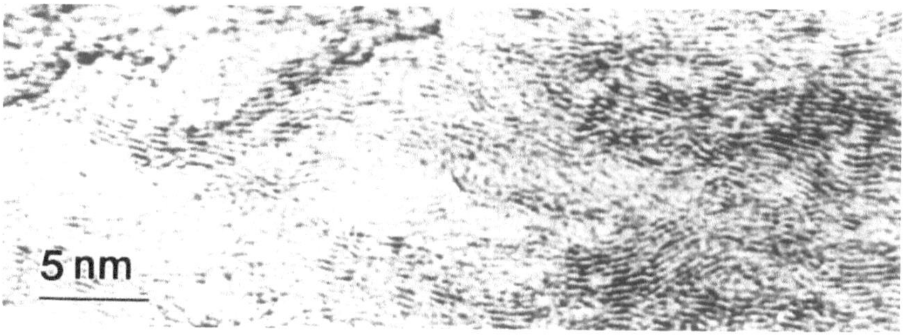

Fig. 2.8 High resolution transmission electron micrograph showing lamellar structure of structure of PAN-based carbon fibers (courtesy of Oberlin).

graphitic planes of the carbon fiber. The degree of alignment of the basal planes increases with the final heat treatment temperature.

2.4 BORON FIBERS

Boron is another elemental fiber, like carbon, that has high strength and high stiffness. It is commonly made by chemical vapor deposition (CVD) on a substrate such as tungsten or carbon, see Fig. 2.9. Some important characteristics of any CVD process of making fibers are:

- Essentially, a coating is deposited on a fibrous substrate. The result is a large diameter sheath with a core, i.e., the fiber itself is a composite.
- Unlike conventional fiber forming processes, the CVD fibers are not formed by stretching or drawing, i.e., there is no lengthening; instead, lateral growth by thickening occurs.
- The final diameter of the fiber may be as much as 10 times that of the starting fiber substrate.

Boron on a tungsten substrate, denoted by B(W) and SiC on a carbon substrate, denoted by SiC(C) are two examples. Boron on a carbon substrate is also possible. B(W) fibers find applications mostly in aerospace and the sporting goods industry. Major drawbacks of most CVD produced fibers are there large diameter (which makes them less flexible) and high cost.

Elemental boron is produced by the reduction of a boron halide by hydrogen:

$$2BX_3\,(g) + 3H_2\,(g) \rightarrow 2B(s) + 6HCl(g)$$

where X denotes a halogen such as Cl, Br, or I. B is deposited on a substrate of tungsten or carbon. Typically, the commercially produced boron fiber can

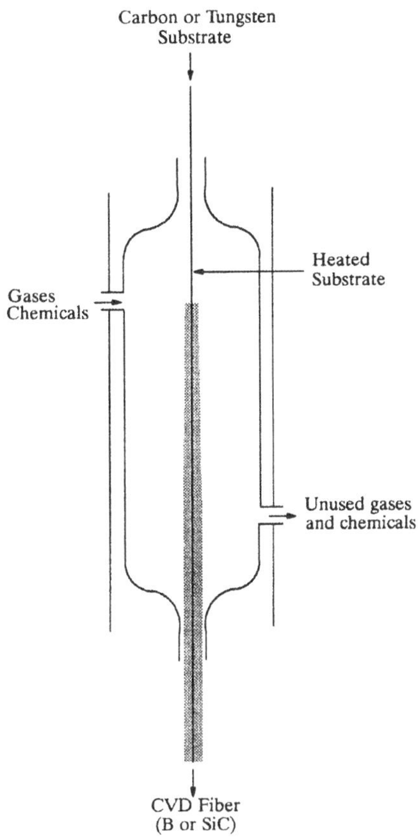

Fig. 2.9 Chemical vapor deposition (CVD) of boron or silicon carbide on a tungsten or carbon substrate.

have a diameter of 75, 100, 140, 200 μm, with 100 μm being the most common. Sometimes a surface coating of SiC or B_4C is applied by CVD on the B fiber, for use in aluminum or titanium matrix composites, to prevent chemical reaction between boron and the metal matrix.

Boron fiber made by deposition on a tungsten substrate can result in complex reaction products at the core/sheath interface. A boron fiber cross section (100 μm diameter) is shown in Fig. 2.10(a) while Fig. 2.10(b) shows, schematically, the various subparts of the cross-section. Depending on the temperature conditions during deposition, the W core may consist of a series of compounds such as W, W_2B, WB, W_2B_5, and WB_4. These tungsten boride phases form by diffusion of B into W. Generally, the fiber core consists only of WB_4 and W_2B_5. On prolonged heating, the core may be converted completely into WB_4. As boron diffuses into the tungsten substrate to form borides, the core expands from its original 12.5 μm (original W wire diameter) to 17.5 μm. The SiC coating shown in Fig. 2.10(b) is a barrier coating used to prevent any adverse reaction between B and the matrix, such as Al or Ti, at high temperatures. The SiC barrier layer is vapor deposited onto the boron using a mixture of hydrogen and methyldichlorosilane.

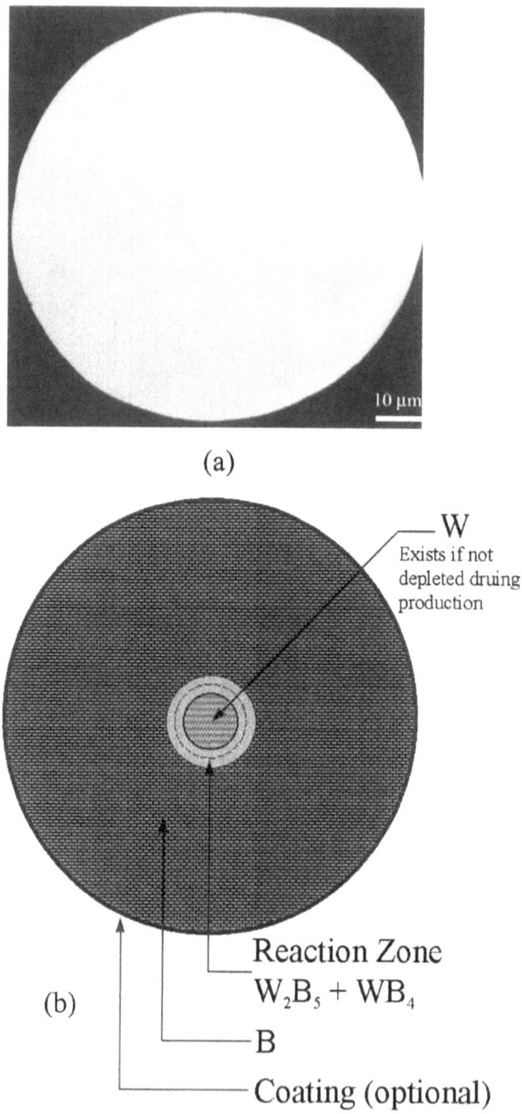

Fig. 2.10 (a) Boron fiber cross-section (100 μm diameter) (b) schematic of boron fiber showing the subparts of the cross section.

As-deposited boron (β rhombohedral) has a nanocrystalline structure. The surface of boron fibers shows a corn-cob structure. This stems from the growth of nodules of boron, each of which starts as an individual nucleus and grows outward in a conical form. Frequently, foreign particles or inclusions are also trapped at the boundaries between nodules.

2.4.1 Residual Stresses

Fibers made via CVD, such as boron fiber, have inherent residual stresses which originate from chemical vapor deposition. Growth stresses in the nodules of boron, stresses introduced due to diffusion of boron into the tungsten core, and stresses generated due to the difference in the coefficient of expansion of deposited boron and tungsten boride core, contribute to the residual stresses. These internal stresses are algebraically added to the applied stresses, and thus can have a considerable influence on the mechanical properties of the fiber. Morphologically, the most conspicuous aspect of these internal stresses is the frequently observed radial cracks in the transverse section of these fibers.

Brittle materials such as boron show a distribution of strengths rather than a single value. Imperfections, such as inclusions or nodular boundaries on the boron fiber surface, lead to stress concentrations. As the brittle material is not capable of deforming plastically in response to these stress concentrations, fracture ensues at one or more such sites. Boron fiber is indeed a very brittle material and cracks commonly originate at preexisting defects located at the boron/core interface or at the surface. The surface defects are due to the nodular surface which results from the growth of boron cones. In particular, when a nodule coarsens due to exaggerated growth around a contaminating particle, a crack can result from this large nodule and weaken the fiber.

2.5 OXIDE FIBERS

Ceramic oxide fibers, continuous and discontinuous, have been commercially available since the 1970s. We describe below some important aspects of processing and microstructure of these fibers.

2.5.1 Alumina Type Oxide Fibers

Alumina can have different allotropic forms such as γ, δ, η, and α; α-alumina being the thermodynamically stable form. Many different alumina-based oxide fibers are available commercially. 3M produces a series of oxide fibers with compositions ranging from pure alumina to mixtures of alumina and silica or alumina and mullite. Sumitomo Chemical Co. produces a fiber that can have a wide composition range: 70-100% Al_2O_3 and 0-30% SiO_2. A short fiber, called staple fiber, of δ-alumina (96%) (tradename Saffil) is available commercially. Single crystal continuous aluminum oxide or sapphire fibers can be produced by drawing from molten alumina. A fiber produced by this method, called Saphikon, has a hexagonal structure with its c-axis parallel to the fiber axis, i.e., the basal plane (0001) is perpendicular to the fiber axis. The diameter is rather large, between 75-250 μm. We describe below the salient features of some of these fabrication methods and the properties of fibers obtained.

Alumina + Silica Fibers and α-Alumina Fiber

A series of alumina + silica fibers via sol-gel route is available commercially. In particular, 3M Co. has developed a series of such fibers called *Nextel* fibers. In this series, *Nextel 610* is a polycrystalline, α-alumina fiber. The sol-gel process of making fibers involves the following steps common to all sol-gel processing:

(a) Formulate sol
(b) Concentrate to form a viscous gel
(c) Spin the precursor fiber
(d) Calcine to obtain the oxide fiber

Specifically, in the case of the 3M process of making Al_2O_3 fiber, the following steps are involved:

- Use an organic basic salt solution as a precursor.
- Drive out (decompose and volatilize) the organics without causing cracking, blistering, or other defects.
- Fire at 1400°C under carefully controlled conditions.
- Apply a low temperature straightening treatment.

A fine-grained α-Al_2O_3 fiber is obtained by seeding the high temperature α-alumina with a very fine hydrous colloidal iron oxide (Wilson and Visser, 2001). The fine iron oxide improves the nucleation rate of α-Al_2O_3, with the result that a high density, ultrafine, homogeneous α-Al_2O_3 fiber is obtained. The rationale for seeding with iron oxide is as follows. Basic salts of aluminum decompose into transition aluminum oxide spinels such as α-Al_2O_3 above 400°C. These transition cubic spinels convert to hexagonal α-Al_2O_3 on heating between 1000 to 1200°C. The problem is that the nucleation rate of pure α-Al_2O_3 is too low and results in large grains. Also, during the transformation to α phase, large shrinkage results in a rather large porosity (Kumagai and Messing, 1985; Suwa et al., 1985). Seeding of alumina with fine particles would appear to be a solution. α-Fe_2O_3 is isostructural with α-Al_2O_3, only 5.5% lattice mismatch (Wilson, 1990). Hydrous colloidal iron oxide sol appears to be an efficient nucleating agent. According to Wilson (1990), without the seeding of iron oxide, the η-alumina to α-alumina transformation occurs at about 1100°C. With 1% Fe_2O_3, the transformation temperature was decreased to 1010°C, while with 4% Fe_2O_3, the transformation temperature came down to 977°C. Concomitantly, the grain size was refined. Nextel 610 fiber has 0.4-0.7% Fe_2O_3. Besides Fe_2O_3, about 0.5 wt. % SiO_2 is added to reduce the final grain size, although SiO_2 inhibits the transformation to the α phase. The SiO_2 addition also reduces grain growth during soaking at 1400°C.

Many other alumina or alumina-silica type fibers are available, most of which are made by the *sol-gel* process. Sumitomo Chemical Company produces a fiber that is a mixture of alumina and silica. Starting from an organoaluminum (polyaluminoxanes or a mixture of polyaluminoxanes and one or more kinds of Si containing compounds), a precursor fiber is obtained by dry spinning. This precursor fiber is calcined to produce the final fiber. The fiber structure consists of fine crystallites of spinel. SiO_2 serves to stabilize the spinel structure and prevents it from transforming α-Al_2O_3 (Chawla, 1998). Mention has been made of the Nextel series of fiber produced by the 3M Company. These are mainly Al_2O_3 + SiO_2 and some B_2O_3. The composition and properties of these fibers are given in Table 2.4. The sol-gel manufacturing process used by 3M Co. has metal alkoxides as the starting materials. Metal alkoxides are $M(OR)_n$ type compounds where M is the metal and n is the metal valence, and R is an organic compound. Selection of an appropriate organic group is very important. It should provide sufficient stability and volatility to the alkoxide so M-OR bonds are broken and MO-R is obtained to give the desired oxide ceramics. Hydrolysis of metal alkoxides results in sols which are spun and gelled. The gelled fiber is then densified at relatively low temperatures. The high surface free energy available in the pores of the gelled fiber allows for densification at a relatively low temperature. The sol-gel process provides close control over solution composition rheology of the fiber diameter. The disadvantage is that rather large dimensional change must be accommodated and fiber integrity conserved.

Nextel 720 fiber consists of aggregates of mullite grains embedded in α-alumina grains. The grains of each phase are small and the aggregates of

Table 2.4 Properties of some oxide fibers*

Fiber Type	Composition, wt. %	Diameter, μm	Density, g cm^{-3}	Tensile Strength, GPa	Young's Modulus, GPa
Nextel 312	Al_2O_3-62.5, SiO_2-24.5, B_2O_3-13	10-12	2.70	1.7	150
Nextel 440	Al_2O_3-70, SiO_2-28, B_2O_3-2	10-12	3.05	2.0	190
Nextel 550	Al_2O_3-73, SiO_2-27	10-12	3.03	2.0	193
Nextel 610	Al_2O_3-99+	10-12	3.9	3.1	370
Nextel 650	Al_2O_3-89, ZrO_2-10, Y_2O_3-1	10-12	4.10	2.5	358
Nextel 720	Al_2O_3-85, SiO_2-15	10-12	3.40	2.1	260
Saffil	Al_2O_3-96, SiO_2-4	3	2.3	1.0	100
Saphikon	Single Crystal Al_2O_3	70-250	3.8	3.1	380
Sumitomo	Al_2O_3-85, SiO_2-15	9	3.2	2.6	250

*Manufacturer's data

similarly aligned grains act like single grains about 0.5 μm in diameter. This gives the Nextel 720 fiber very low creep rate at temperatures above 1000°C.

Sol-gel method is also used to produce silica-stabilized alumina (*Saffil*) and calcia-stabilized zirconia fibers (Birchall et al., 1985). The Saffil fiber is a δ-Al_2O_3, short staple fiber that has about 4% SiO_2 and a very fine diameter (3 μm). The aqueous phase contains an oxide sol and an organic polymer. The sol is extruded as filaments into a coagulating (or precipitating) bath in which the extruded shape gels. The gelled fiber is then dried and calcined to produce the final oxide fiber. For alumina, aluminum oxychloride [$Al_2(OH)_5Cl$] is mixed with a medium molecular weight polymer such as 2 wt. % polyvinyl alcohol. This solution is slowly evaporated in a rotary evaporator until a viscosity of about 80 Pa.s is attained. It is then extruded through a spinneret, the fibers are wound on a drum, and fired to about 800°C. The organic material is burned away and a fine-grained alumina fiber having 5-10% porosity and a diameter of 3-5 μm is obtained. The fibers produced at this stage are suitable for filter applications because of their high porosity. By heating them to 1400-1500°C, which causes a 3-4% of linear shrinkage, one obtains a refractory alumina fiber suitable for reinforcement purposes.

Yet another continuous, polycrystalline, α-alumina fiber, trade name *Almax*, is prepared by dry spinning a viscous slurry consisting of an aluminum salt, a fine powder of intermediate alumina, and an organic binder to produce the precursor fiber; this is followed by prefiring (calcining) and firing (sintering) the precursor fiber to produce an alumina fiber.

Polycrystalline oxide fibers of yttrium-aluminum garnet, $Y_3Al_5O_{12}$ (YAG), composition and a composite fiber consisting of α-alumina and YAG were prepared by sol-gel method (Towata et al., 2001). The process involved the use of α-alumina or YAG seed particles. α-alumina seed particles accelerated the phase transformation from θ-alumina to α-alumina. The YAG seed particles influenced the multi-step transformation of yttrium and aluminum oxide.

A technique called *Edge-defined Film-fed Growth* (EFG) has been used to make continuous, monocrystalline sapphire (Al_2O_3) fiber (LaBelle and Mlavsky, 1967; Gasson and Cockayne, 1970; LaBelle, 1971; Pollack, 1972; Hurley and Pollack, 1972). LaBelle and Mlavsky (1967) were the first to grow sapphire (Al_2O_3) single crystal fibers using a modified Czochralski puller and radio frequency heating. In 1971, these authors devised a growth method, called the edge-defined, film-fed growth (EFG) method. Growth rates as high as 200 mm/min have been attained. The die material must be stable at the melting point of alumina, so a molybdenum die is typically used. A sapphire seed crystal is used. A capillary supplies a constant liquid level at the crystal interface. Molten alumina wets both, molybdenum and alumina. The crystal grows from a molten film between the growing crystal and the die. The crystal shape is defined by the external shape of the die rather than the internal shape. Perhaps, the most important item about the Saphikon fiber is its single crystal nature, i.e., there are no grain boundaries. Thus, under creep conditions involving grain boundary related phenomena, such as cavitation and grain boundary sliding, will not be operative and one can expect a highly creep resistant fiber. However, single crystal alumina can undergo dislocation creep on the basal plane. Thus, it would appear that if the basal plane can be oriented such that basal slip does not occur, then one can obtain a very creep resistant fiber. Such will be the case for a single crystal fiber with its c-axis parallel to the fiber axis.

A *laser-heated floating zone method* can also be to make a variety of ceramic fibers. Gasson and Cockayne (1970) used laser heating for crystal growth of Al_2O_3, Y_2O_3, $MgAl_2O_4$, and Na_2O_3. This method has been used to grow single crystal fibers of Al_2O_3, Y_2O_3, TiC, TiB_2, mullite, Al_2O_3/YAG eutectic (Haggerty, 1972; Sayir and Farmer, 1995, Sayir et al., 1995). A CO_2 laser is focused on the molten zone and a source rod is brought into the focused laser beam. A seed crystal, dipped into the molten zone, is used to control the orientation. Crystal growth starts by moving the source and seed rods simultaneously. Mass conservation dictates that the diameter is reduced as the square root of the feed rate/pull rate ratio. Mention should be made of a containerless melting technique that has been used to grow continuous fibers of alumina and YAG-composition ($Y_3Al_5O_{12}$) directly from the melt. Containerless melting eliminates heterogeneous nucleation by container surfaces. A continuous-wave CO_2 laser beam is used for heating (Weber et al., 1998). Specimens are levitated in a gas jet and stabilized with an acoustic positioning device. The levitated samples are heated and melted with a CO_2 laser beam.

2.6 NONOXIDE FIBERS

Continuous nonoxide ceramic fibers are also available commercially. The development of Silicon carbide fiber must be regarded as a major development in the field of ceramic reinforcements during the last quarter of the twentieth century. In particular, a process developed by the late Professor Yajima in Japan, involving controlled pyrolysis of a polycarbosilane precursor to yield a flexible fine diameter fiber is considered to be the harbinger of making ceramic fibers from polymeric precursors. In this section we describe the processing, microstructure, and properties of silicon carbide and some other nonoxide fibers.

2.6.1 Silicon Carbide Fibers

Silicon carbide is by far the most important nonoxide ceramic fiber available commercially. The two main varieties of this fiber available commercially are: large diameter fiber made by chemical vapor deposition (CVD) and small diameter fiber made by controlled pyrolysis of a polymer. Another important type of SiC available for reinforcement purposes is SiC whiskers.

Silicon Carbide Fibers made by CVD

Silicon carbide fibers can be made by chemical vapor deposition on a substrate heated to about 1300°C (DeBolt et al., 1974). The substrate can be tungsten or carbon. The reactive gaseous mixture contains hydrogen and alkyl silanes. Typically, a gaseous mixture consisting of 70% hydrogen and 30% silanes is introduced at the top of the reactor, where the tungsten substrate (~13 μm diameter) also enters the reactor. Mercury seals are used at both ends as contact electrodes for the filament. The substrate is heated by combined direct current (250 mA) and very high frequency (VHF ~ 60 MHz) to obtain an optimum temperature profile. To obtain a 100 μm of SiC monofilament, it generally takes about 20s in the reactor. The filament is wound on a spool at the bottom of the reactor. The exhaust gases (95% of the original mixture + HCl) are passed around a condenser to recover the unused silanes. Efficient reclamation of the unused silanes is very important for a cost effective production process. Such CVD processes result in composite monofilaments which have built in residual stresses. The process is, of course, very expensive. Methyltrichlorosilane is an ideal raw material for this process as it contains one silicon and one carbon atom, i.e., a stoichiometric SiC will be deposited. The chemical reaction is:

$$CH_3SiCl_3 \rightarrow SiC(s) + 3HCl$$

An optimum amount of hydrogen is required. If the hydrogen is less than sufficient, chlorosilanes will not be reduced to Si and free carbon will be present in the mixture. If too much hydrogen is present, excess Si will be present in the end product. The final monofilament (100-150 μm) consists of a sheath of mainly β-SiC with some α-SiC on the tungsten core. The {111} planes in SiC are deposited parallel to the fiber axis.

A series of surface-modified silicon carbide fibers, called SCS fibers, has been used in a variety of composites. These special fibers have a complex through-thickness gradient structure. SCS-6, for example, is a thick fiber (diameter = 142 μm) and is produced by chemical vapor deposition of silicon and carbon containing compounds onto a pyrolytic graphite coated carbon core. The pyrolytic graphite coating is applied to a carbon monofilament to give a substrate of 37 μm. This is then coated with SiC by CVD to give a final monofilament of 142 μm diameter. The outer surface layer (about 1 μm thick) of SCS-6 fiber consists of C doped Si. Mann et al. (1999) verified the dramatic changes in mechanical characteristics of SCS-6 fiber, such as Young's modulus and hardness, as function of fiber radius using nanoindentation techniques. Besides SCS-6 there are two other varieties of CVD silicon carbide fibers on a carbon core that are available commercially. Typical properties of SCS-type silicon carbide fibers are given in Table 2.5. Another SiC fiber made by CVD is called *sigma* fiber; this has a tungsten filament core.

Such fibers made by CVD on a heated substrate are composite materials themselves, and can exhibit unusual behavior. Lara-Curzio and Sternstein (1993) examined the behavior of such composite fibers when subjected to thermomechanical loading. Their main conclusion was that one must take into account the strains experienced by the substrate during the CVD process of making such composite fibers. Not doing so can lead to a severe underestimate of the residual stresses in the fiber. In particular, in the case of the SCS-6 fiber, they identified the presence of large residual radial stresses at various interfaces (carbon substrate/pyrolytic graphite, pyrolytic graphite/SiC). According to these authors, the origin of rather large residual stresses was in the high temperature of CVD and the large anisotropy in the pyrolytic graphite layer.

Nonoxide Fibers via Polymer Precursors

As pointed out above, the SiC fibers obtained via CVD have very large diameter, and thus are not very flexible. Yajima et al. (1976) and Yajima (1980) developed a process of making fine, continuous, and flexible SiC-type fibers by controlled pyrolysis of a polymeric precursor. This method of using silicon based polymers to produce a family of ceramic fibers having good mechanical properties, good thermal stability, and oxidation resistance has enormous potential. Figure 2.11 shows a general flow diagram for making ceramic fibers from a polymeric precursor. Note the similarity with

Table 2.5 Properties of SCS-type fibers*

Property	SCS-6	SCS-9A	SCS-Ultra
Fiber diameter, μm	140	78	140
Density, g/cm^3	3.0	2.8	3.0
Tensile strength, GPa	3.45	3.45	5.86
Tensile modulus, GPa	380	307	415
Coefficient of thermal expansion, 10^{-6} K^{-1}	4.1	4.3	4.1

* Source: Specialty Materials Inc.

the process of making a carbon fiber from a polymeric precursor. The various steps involved in this polymer route can be listed as follows (Wax, 1985):

(a) Characterize the polymer (yield, molecular weight, purity, etc.)
(b) Melt spin polymer into a precursor fiber.
(c) Cure the precursor fiber to crosslink the molecular chains, making it infusible during the subsequent pyrolysis.
(d) Pyrolyze the precursor fiber under controlled conditions to obtain the ceramic fiber.

Specifically, the Yajima process of making SiC fiber from a polymeric precursor fiber involves the following steps and is shown schematically in Fig. 2.12. Polycarbosilane (PCS), a high molecular weight polymer,

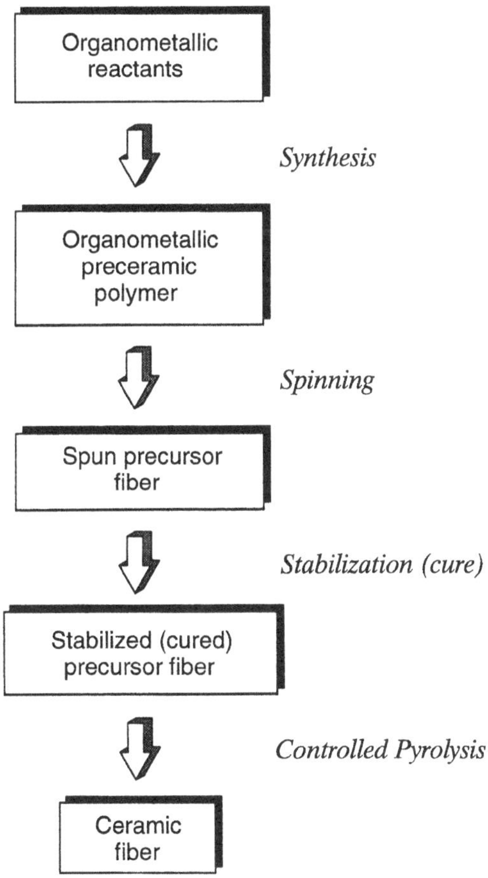

Fig. 2.11 Flow diagram for making ceramic fibers from a polymeric precursor.

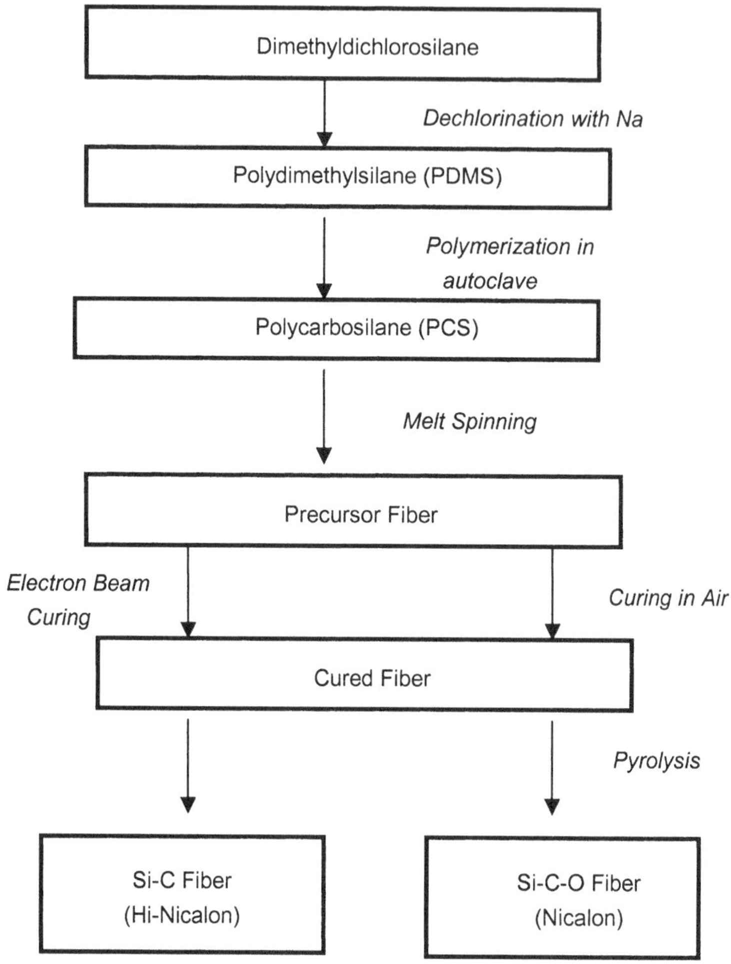

Fig. 2.12 The Yajima process of making fine diameter SiC from a polycarbosilane.

containing Si and C is synthesized. This involves the use of a commercially available material, viz., dimethylchlorosilane. Solid polydimethylsilane (PDMS) is obtained by dechlorination of dimethylchlorosilane by reacting it with sodium. Polycarbosilane is obtained by thermal decomposition and polymerization of polydimethylsilane. This is carried out under high pressure in an autoclave at 470°C in an argon atmosphere for 8-14 h. A vacuum distillation treatment at up to 280°C follows. The average molecular weight of the resulting polymer is about 1500. This is melt spun from a 500-hole nozzle at about 350°C under N_2 gas to obtain the so called pre-ceramic continuous, precursor fiber. The precursor fiber is quite weak (tensile strength ~10 MPa). This is converted to inorganic SiC by curing in air, heating to about 1000°C in N_2 gas, followed by heating to 1300°C in N_2 under stretch. During the pyrolysis, first stage of conversion occurs at around 550°C when cross-linking of polymer chains occurs. Above this temperature, the side chains containing hydrogen and methyl groups decompose. Fiber density and mechanical properties improve sharply. The conversion to SiC occurs above about 850°C.

Structure and Properties of Nicalon Fiber

The *Nicalon* multifilament fiber (single fiber diameter between 10 and 20 μm) consists of a mixture of β-SiC, free carbon, and SiO_2. The properties of Nicalon start degrading above about 600°C because of the thermodynamic instability of composition and microstructure. Ceramic grade "Nicalon" fibers, carrying the designation NLM, HI, and HI-S having low oxygen content are also available. The structure of Nicalon fiber has been studied by many researchers. The commercial variety of Nicalon has an amorphous structure while the low-oxygen variety has a microcrystalline structure (SiC grain radius of 1.7 nm) (Simon and Bunsell, 1984). Microstructural analysis shows that both the fibers contain, in addition to SiC, SiO_2 and free carbon. Laffon et al., (1989) proposed a model of Nicalon fiber consisting of β-SiC crystals mixed with some free carbon with a composition of SiC_xO_y, with x+y being equal to 4. The density of the fiber is about 2.6 g/cm^3 which is low compared to that of pure β-SiC. This is understandable in view of the fact that the composition is a mixture of SiC, SiO_2 and C. The properties of Nicalon fiber are summarized in Table 2.6. A quick comparison of Nicalon SiC fiber with CVD SiC fiber shows that the CVD fiber is superior in properties. In particular, it should be mentioned that CVD SiC fiber (stoichiometric) shows superior creep resistance *vis-à-vis* Nicalon fiber (DiCarlo 1985).

Other SiC-Type Fibers

There are many silicon carbide type fibers that are commercially available, with elemental compositions of Si-C, Si-N-C-O, Si-B-N, Si-C-O, and Si-Ti-C-O. Almost all of them are made from polymeric precursors. A multifilament silicon carbide fiber, called *Tyranno* (Yamamura et al., 1988) is made by pyrolysis of poly-titano carbosilanes and it contains between 1.54 wt.% titanium. A textile grade silicon carbide fiber, called *Sylramic*, has, according to the manufacturer, a nanocrystalline structure (crystallite size about 0.5 μm), a density of 3.0 g cm^{-3}, a tensile strength of 3.15 GPa, and a Young's modulus of 405 GPa.

Table 2.6 Typical properties of Nicalon SiC fiber

Property	Ceramic grade	HVR* grade	LVR** grade
Density, g cm^{-3}	2.55	2.32	2.45-2.55
Tensile strength, MPa	2960	2930	2960
Young's Modulus, GPa	192	186	192
Strain to failure	1.5	1.6	1.5
Coefficient of thermal expansion, 10^{-6} K^{-1}	4	--	--
Volume resistivity, Ωcm	10^3	$>10^6$	0.5-5.0

* low dielectric fiber (high volume resistivity)
** higher conductivity fiber (low volume resistivity)

2.7 WHISKERS

Whiskers are monocrystalline, short fibers with extremely high strength. This high strength, approaching the theoretical strength, comes about because of the absence of crystalline imperfections such as dislocations. Being monocrystalline, there are no grain boundaries either. Typically, whiskers have a diameter of a few μm and a length of a few mm. Thus, their aspect ratio (length/diameter) can vary between 50 and 10,000. Whiskers, however, do not have uniform dimensions or properties. This is perhaps their greatest disadvantage, i.e., the variability in properties is extremely large. Handling and alignment of whiskers in a matrix to produce a composite are other problems.

Whiskers are normally fabricated by vapor phase growth. Early in the nineteen seventies, a new process was developed, starting from rice hulls, to produce SiC particles and whiskers (Milewski et al., 1974; Lee and Cutler, 1975). The SiC particles produced by this process are very fine in size. Rice hulls are a waste by-product of rice milling. For each 100 kg of rice milled, about 20 kg of rice hull is produced. Rice hulls contain cellulose, silica, and other organic and inorganic materials. Silica from soil is dissolved and transported in the plant as monosilicic acid. This is deposited in the cellulosic structure by liquid evaporation. It turns out that most of silica ends up in hull. It is the intimate mixture of silica within the cellulose that gives the near ideal amounts of silica and carbon for silica carbide production. Raw rice hulls are heated in the absence of oxygen at about 700°C to drive out the volatile compounds. This is called coking. Coked rice hulls, containing about equal amounts of SiO_2 and free C, are heated in inert or reducing atmosphere (flowing N_2 or NH_3 gas) at a temperature between 1500 and 1600°C for about 1 h to form silicon carbide per the following reaction:

$$3C + SiO_2 \rightarrow SiC + 2CO$$

When the above reaction is over, the residue is heated to 800°C to remove any free C. Generally, both particles and whiskers are produced together with some excess free carbon. A wet process is used to separate the particles and the whiskers. Typically, the average aspect ratio of the as-produced whiskers is ~75.

The vapor-liquid-solid (VLS) process of growing whiskers has been known for sometime (Lindemanis, 1983; Milewski et al., 1985; Petrovic et al., 1985). The acronym VLS represents *vapor* feed gases, *liquid* catalyst, and *solid* crystalline whiskers. The catalyst forms a liquid solution interface with the growing crystalline phase while elements are fed from the vapor phase through the liquid-vapor interface. Whisker growth takes place by precipitation from the supersaturated liquid at the solid liquid interface. The catalyst must take in solution the atomic species of the whisker to be grown. For SiC whiskers, transition metals and iron alloys meet this requirement. Silicon and carbon are supplied in the form of SiO and CH_4 gases, respectively. The SiO gas is obtained by carbothermal reduction of SiO_2. Generally a range of whisker morphologies is obtained. The tensile strength values have been reported in range of 1.7 to 23.7 GPa (Milewski et al., 1985; Petrovic et al., 1985). Whisker lengths were about 10 mm and the equivalent circular diameter averaged 5.9 μm. Their average tensile strength and modulus were 8.4 GPa and 581 GPa, respectively. It should be pointed out that the VLS process is extremely slow.

2.8 PARTICLES

2.8.1 Particulate Silicon Carbide

Silicon carbide in particulate form has been available for a long time. It is quite cheap and commonly used for abrasive, refractory, and chemical purposes. Particulate SiC is processed by reacting silica in the form of sand and carbon in the form of coke at 2400°C in an electric furnace. The SiC produced in the form of large granules is subsequently comminuted to the desired size. Two types of SiC particulate reinforcement are shown in Fig. 2.13, with angular and rounded morphology, respectively.

2.8.2 Particulate Tungsten Carbide

Tungsten carbide is obtained by the carburization of tungsten metal, which is obtained by hydrogen reduction of tungsten oxide. In principle, tungsten carbide can be obtained directly from ore or oxide, but gaseous carburization is usually preferred. Carbon black is added to control particle size and size distribution. A mixture of tungsten powder and carbon black in the correct particle size and distribution is subjected to ball milling or attrition milling. The goal of the carburization process is to produce stoichiometric tungsten

Fig. 2.13 SEM micrograph of SiC particulate reinforcement: (a) angular morphology and (b) rounded morphology (courtesy of A. Drake and S.M.K. Emanuelsen, Saint-Gobain).

carbide with a small excess of free carbon to prevent the formation of the eta-phase. Carburization is performed in the presence of hydrogen at temperatures between 1400 and 2650°C. The hydrogen reacts with the carbon black to form gaseous hydrocarbons, which then react to form tungsten carbide:

$$W + CH_4 \rightarrow WC + 2H_2$$

After carburization, particles are normally subjected to milling for deagglomeration. Typical particle sizes range from 0.5 to 30 μm.

2.9 COMPARISON OF FIBERS

Two most important characteristics of any reinforcement material are its strength and Young's modulus. All these high performance reinforcements (fibers, whiskers, or particles) have very low density values. The reader will also recognize that the elements comprising these reinforcements pertain to the first two rows of the periodic table. Also, one should note that, irrespective of whether the reinforcements are in compound or elemental form, they are mostly covalently bonded which is the strongest type of bond. Generally, such light, strong, and stiff materials are very desirable in most applications, but particularly so in aerospace field, land transportation, energy related industry, housing and civil construction, etc. Fiber flexibility is associated with the Young's modulus and the diameter (see Sec. 2.1). In the general area of high modulus fibers, the diameter becomes the dominant parameter controlling the flexibility. For a given modulus, E, the smaller the diameter, the more flexible the fiber. Fiber flexibility is a very desirable characteristic if one wants to bend, wind, and weave a fiber in order to make a complex shaped final product.

Some of these fibers have quite anisotropic characteristics. The strength, modulus, and thermal characteristics can vary with direction, especially along the fiber axis and transverse to it. In particular, the thermal expansion coefficient of carbon is quite different in the radial and longitudinal directions. This would also be true of any single crystal fiber or whisker, e.g., alumina single crystal fiber which has a hexagonal structure. In this respect, polycrystalline fibers such as SiC and Al_2O_3 fibers are reasonably isotropic. Another important characteristic of these high performance fibers is their rather low values of strain-to-fracture, generally < 2-3 %.

2.10 STATISTICAL ANALYSIS OF FIBER STRENGTH

Fracture of brittle materials, in general, involves statistical considerations. Most reinforcements used in metal matrix composites are ceramic fibers. Such fibers fail in a brittle manner. Ceramic materials, fibrous or otherwise, will have randomly distributed defects on their surfaces or in the interior. The presence of these defects results in rather large scatter in the experimentally determined strength values of ceramic materials. Ceramic fibers also exhibit a broad distribution of tensile strength, due to the distribution of processing-induced flaws, such as micropores or notches. The large surface-to-volume ratio of fibrous materials makes them particularly susceptible to surface flaws.

We can regard a fiber of a given length to be made up of a series of chain links. When such a fiber is loaded, the link or the segment containing the longest defect will fail first and cause the fiber to fracture. The longer the fiber, the higher the probability of a link having a critical flaw size required for failure. In other words, one would expect that the mean strength of a short fiber length to be greater than the mean strength of a long fiber length. The fiber failure occurs when the weakest link fails. This is called the weakest-link assumption. It turns out that such a “weak-chain” material is well described by a statistical distribution known as the Weibull distribution, named after the person who first proposed it (Weibull, 1959). The basic assumption is that a fiber has a distribution of flaws (on the surface and/or in the interior). The Weibull distribution assumes that all segments or chain links have the same type of flaw but of different lengths. Such a "weak-link" material is well described by a statistical distribution known as the Weibull distribution. The Weibull distribution is a parametric distribution, i.e., it is an empirical distribution and does not concern itself with the origin of the defects.

According to the Weibull distribution, the *survival probability* is given by:

$$P(\sigma) = \exp\left[-\left(\frac{\sigma_0}{\sigma}\right)^m\right]$$

where σ_o and m are constants. As m increases, the distribution becomes less broad. In general, brittle materials have a lower Weibull modulus than ductile materials. Taking double logs of both sides, we have:

$$\ell n\, \ell n[P(\sigma)] = m\, \ell n\left[-\frac{\sigma_o}{\sigma}\right]$$

The Weibull distribution gives the failure probability, F(σ), of a fiber at an applied stress σ by the following expression:

$$F(\sigma) = 1 - \exp\left[-\left(\frac{\sigma_o}{\sigma}\right)^m\right]$$

where σ_o and m are statistical parameters. F(σ) is called the cumulative frequency distribution function. The parameter m is referred to as the Weibull modulus. It is a measure of the variability in the strength of the fiber. The higher the value of Weibull modulus, the higher the uniformity of strength values. The Weibull modulus can be obtained graphically by using a double log plot of such an equation that will give a straight line of slope, m. The Weibull modulus, m, is a measure of the flaw distribution in the sample. The smaller the value of m, the greater the *variability* in strength. Table 2.7 shows typical Weibull modulus for several fibers.

Table 2.7 Typical Weibull modulus (m) values for materials in fibrous form

Fiber Material	**Weibull modulus, m***
Glass	< 5
SiC, Al_2O_3, C, B	5 -10
Steel	>100

* Indicative values only.

Experimentally, to obtain the Weibull parameters, one tests a series of identical samples (each of same volume, or in the case of fiber of constant diameter, each of a constant length) to failure. From such tests we can obtain the fraction of samples that survives, P(σ),when loaded to a given stress, σ. We arrange the tensile strength values of the single filaments in an ascending order and assign a probability of failure using an estimator given by

$$F(\sigma_i) = \frac{i}{1+N}$$

where $F(\sigma_i)$ is the probability of failure corresponding to the ith strength values and N is the total number of fibers tested. Substituting this in the double log expression:

$$\ell n \ell n\left[\frac{N+1}{N+1-i}\right] = m \ell n \sigma_i + \ell n \sigma$$

A plot of $\ell n\ [(N + 1)/(N + 1 - i)]$ vs. σ_i on a log-log graph would give a straight line if the tensile strength data followed a Weibull distribution. The intercept on the y-axis will then be σ_o and the slope will be m. Figure 2.14 shows such a double log plot for some Nextel fibers of (gage length = 25 mm) (Chawla et al., 2005; Kerr et al., 2005). An example of cumulative probability of failure of Nextel 312 fiber as function of fiber strength is shown in Fig. 2.15 (Chawla et al., 2005).

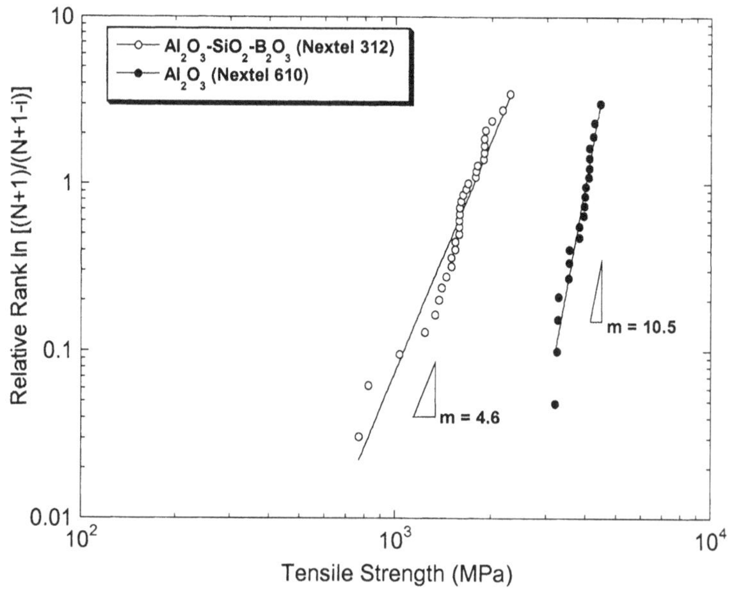

Fig. 2.14 Weibull distribution of some Nextel fibers (Chawla et al., 2005).

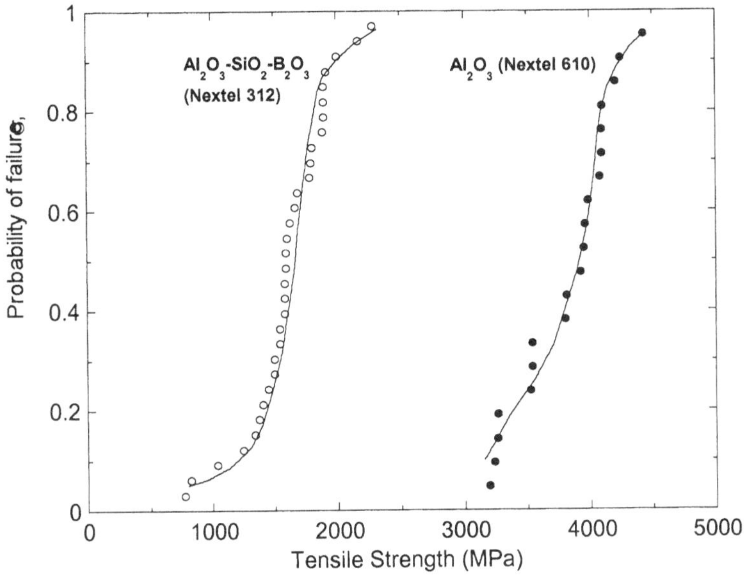

Fig. 2.15 Cumulative probability of failure of Nextel 312 fiber (Chawla et al., 2005).

References

Birchall, J.D., J.A.A. Bradbury, and J. Dinwoodie, (1985) in *Strong Fibres, Handbook of Composites*, North-Holland, Amsterdam, vol. 1, p. 115.

Chawla, K.K., (1998) *Fibrous Materials*, Cambridge University Press, Cambridge,.

Chawla, N., M. Kerr, and K.K. Chawla, (2005) *J. Am. Ceram. Soc.*, **88**, 101.

DeBolt, H.E., V.J. Krukonis, and F.E. Wawner, (1974) in *Silicon Carbide 1973*, Univ. of S. Carolina Press, Columbia, SC, p. 168.

Deurbergue, A., and A. Oberlin, (1991) *Carbon*, **29**, 691.

DiCarlo, J.A., (June 1985) *J. of Metals*, **37**, 44.

Diefendorf, R.J., and E. Tokarsky, (1975) *Polymer Eng. Sci.*, **15**, 150.

Dresher, W.H., (April 1969) *J. of Metals*, **21**, 17.

Ezekiel, H.N., and R. G. Spain, (1967) *J. Polymer Sci. C.*, **19**, 271.

Gasson, D.G., and B. Cockayne, (1970) *J. Mater. Sci.*, **5**, 100.

Haggerty, J.S., (May 1972) NASA-CR-120948.

Hurley, G.F., and J.T.A. Pollack, (1972) *Metall. Trans.*, **7**, 397.

Johnson, D.J., and C.N. Tyson, (1969) *Brit. J. App. Phys.*, **2**, 787.

Kerr, M., N. Chawla, and K.K. Chawla, (2005) *JOM*, **2**, 67.

Kumagai, M., and G.L. Messing, (1985) *J. Am. Ceram. Soc.*, **68**, 500.

Kumar, S., D.P. Anderson, and A.S. Crasto, (1993) *J. Mater. Sci.*, **28**, 423.

LaBelle, H.E., and A.I. Mlavsky, (1967) *Nature*, **216**, 574.
LaBelle, H.E., (1971) *Mater. Res. Bull.*, **6**, 581.
Laffon, C., A.M. Flank, P. Lagarde, (1989) *J. Mater. Sci.,* **24**, 1503.
Lee, J.-G., and I.B. Cutler, (1975) *Am. Ceram. Soc. Bull.,* **54,** 195.
Lara-Curzio, E., and S. Sternstein, (1993) *Composites Sci. & Tech.*, **46**, 265.
Lindemanis, A., (1983) in *Emergent Process Methods for High Technology Ceramics*, Plenum Press, New York.
Mann, A.B., M. Balooch, J.H. Kinney, and T.P. Weihs, (1999) *J. Amer. Ceram. Soc.,* **82**, 111.
Milewski, J.V., F.D. Gac, J.J. Petrovic, and S.R. Skaggs, (1985) *J. Mater. Sci.,* **20**, 1160.
Milewski, J.V., J.L. Sandstrom, and W.S. Brown, (1974) in *Silicon Carbide-1973*, University of S. Carolina Press, Columbia, SC, p. 634.
Peebles, L.H., (1995) *Carbon Fibers*, CRC Press, Boca Raton, FL.
Petrovic, J.J., J.V. Milewski. D.L. Rohr, and F.D. Gac, (1985) *J. Mater. Sci.*, **20**, 1167.
Pollack, J.T.A., (1972) *J. Mater. Sci.*, **7**, 787.
Riggs, J.P., (1985) in *Encyclopedia of Polymer Science & Engineering*, 2nd ed., , John Wiley & Sons, New York, vol. 2, p. 640.
Sayir, A., and S.C. Farmer, (1995) in *Ceramic Matrix Composites, MRS proceedings*, Mater. Res. Soc., Pittsburgh, vol. 365, p. 11.
Sayir, A., S.C. Farmer, P.O. Dickerson, and H.M. Yun, (1995) in *Ceramic Matrix Composites*, MRS proceedings, Mater. Res. Soc., Pittsburgh, vol. 365, p. 21.
Simon, G., and A. R. Bunsell, (1984) *J. Mater. Sci.*, **19**, 3649.
Singer, L., (1979) in *Ultra-High Modulus Polymers*, Applied Sci. Pub., Essex, England, p. 251.
Singer, L., (1981) *Fuel*, **60**, 839-841.
Suwa, Y., R. Roy, and S. Komarneni, (1985) *J. Am. Ceram. Soc.*, **68**, C-238.
Towata, A., H.J. Hwang, M. Yasuoka, M. Sando, and K. Niihara, (2001) *Composites A*, **32A**, 1127.
Watt, W., (1970) *Proc. Roy. Soc.*, **A319**, 5.
Watt, W., and W. Johnson, (1969) *App. Polymer Symposium*, **9**, 215.
Wax, S.G., (1985) *Amer. Cer. Soc. Bull.*, **64**, 1096.
Weber, J. K. R., J. J. Felten, B. Cho, and P. C. Nordine, (1998) *Nature*, **393**, 769.
Weibull, W., (1951) *J. App. Mech.*, **18**, 293.
Wilson, D.M., (1990) in *Proc. 14th Conf. On Metal Matrix, Carbon, and Ceramic Matrix Composites*, NASA Conference Publication, No. 3097, Part I, , pp. 105-117.
Wilson, D.M., and L.R. Visser, (2001) *Composites A*, **32A**, 1143.
Yajima, S., K. Okamura, J. Hayashi, and M. Omori, (1976) *J. Amer. Ceram. Soc.*, **59**, 324.

Yajima, S., (1980) *Phil. Trans., R. Soc., London*, **A294**, 419.
Yamamura, T., T. Ishirkawa, M. Shibuya, T. Hiasyuki, and K. Okamura, (1988) *J. Mater. Sci.*, **23**, 2589.

Chapter 3

MATRIX MATERIALS

A wide range of metals and their alloys may be used as matrix materials. In this chapter, we review some of the basic concepts and fundamentals of bonding and structure of common metals. Following this, we provide a summary of the characteristics of some of the most common metals that are used as matrix materials in metal matrix composites.

3.1 BONDING AND CRYSTALLINE STRUCTURE IN METALS

Metals are characterized by metallic bonding, i.e., valence electrons are not bound to a particular ion in the solid. A "sea of electrons" surrounds positively charged atomic nuclei. A major result of this electron cloud surrounding the atomic nuclei is that the electronic bonding in metals is non-directional. This non-directionally of bonding is very important since it contributes to isotropy in many properties.

When cooled down from their molten state, most metals assume a crystalline structure below their melting point. Some metals and alloys do not undergo crystallization and assume an amorphous structure, but only at very high cooling rates (> 10^6 Ks^{-1}). The basic difference between a crystalline and non-crystalline structure is the degree of ordering. A fully crystalline state has a high degree of order. Metal ions are quite small (diameter ~ 0.25 nm), so in a crystalline structure these ions are packed in a very regular and close-packed manner. Because of non-directional bonding, we can model the arrangement of atoms in the form of hard spheres.

There are two arrangements of packing of hard, identical spheres that result in close packed structures: Face centered cubic (FCC) and hexagonal close packed (HCP). There is a third arrangement that is observed in a number of metals, namely, body centered cubic (BCC). The BCC structure is more open than the FCC or HCP structure. Figure 3.1 shows the atomic arrangement in these three cases. Both HCP and FCC structures result in close-packed structures. The difference between the two results from the way the close packed planes are stacked. Figure 3.2(a) shows the stacking arrangement of the first layer (layer A) in a close packed structure. For the second layer of close packed atoms, one has the choice of putting the atoms at sites B or C, which are equivalent sites. The difference between FCC and HCP structures comes in the placement of the third layer of atoms. Assuming the B configuration for the second layer of atoms, one can have the third layer in A or C configuration. It turns out that the ABABAB... (or ACACAC...) results in HCP while the ABCABCABC... stacking sequence results in the FCC structure. Stacking faults or planar defects may be formed, where there is an interruption of ABCABCABC stacking (FCC) sequence such that we have an HCP region in an FCC structure.

3.2 CRYSTALLINE DEFECTS IN METALS

Most metals are crystalline in their solid state. Ideally, a metal crystal has

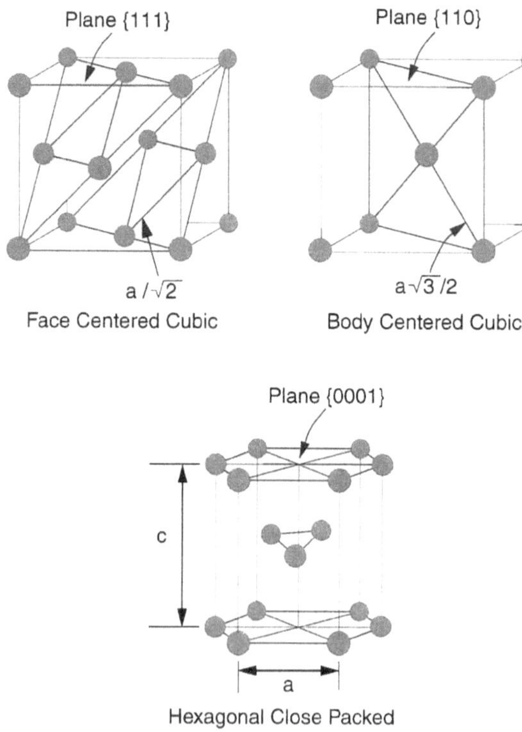

Fig. 3.1 Three common crystal structures in metals, face centered cubic (FCC), body centered cubic (BCC), and hexagonal close packed (HCP).

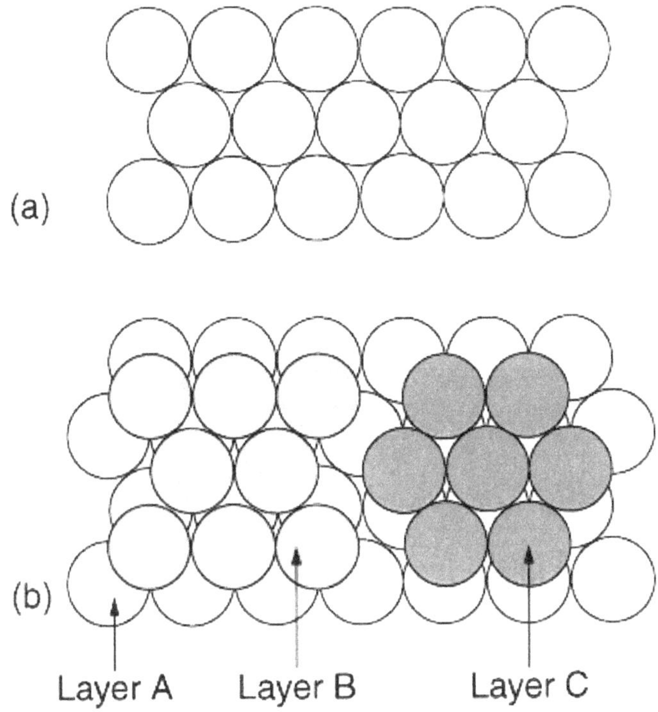

Fig. 3.2 (a) A layer of close packed atoms (b) Positions of layers B and C on top of layer A. Close packed layers in FCC has ABCABC stacking, while HCP has ABAB… stacking sequence.

atoms arranged in a very orderly fashion in a three dimensional lattice. Such perfect crystals are rarely seen in practice. Generally, various kinds of imperfections are present. It turns out that many of the important and interesting properties of crystalline materials are due to the presence of these imperfections or defects in the crystalline lattice. These defects, called lattice defects, can be zero-, uni-, bi-, or tri-dimensional. The zero-dimensional defects are an extension of the order of the lattice constant (0.2–0.4 nm) in all three spatial directions. Common examples are vacancies and interstitial atoms, etc. Unidimensional or linear imperfections have atomic extensions in one spatial direction only. The most important example of a linear defect is a dislocation. Bi-dimensional defects have atomic extension in two spatial directions, for example, grain boundaries, twin boundaries, interfaces between phases, etc. Three-dimensional defects extend into three spatial directions, for example, second phase particles, pores, etc. These defects may result under less than ideal processing conditions.

Besides geometric differences, there is a very important physical difference between zero-dimensional defects and other higher-dimensional defects in crystals. Considering the free energy involved, only point defects (i.e., zero-dimensional) can be present in a state of thermal equilibrium in a crystal, while dislocations, grain boundaries, etc., represent energy much higher than energy represented by thermal fluctuation.

A dislocation can be described in terms of two vectors:

(i) Dislocation line vector, **t**, which gives the direction of the dislocation line at any point. The vector **t** is parallel to the dislocation. In the case of a dislocation loop, the vector **t** runs along the loop length and will have opposite senses along the two opposite sides of the loop.

(ii) Burgers vector, **b**. This indicates the magnitude and direction of the displacement of the part above the slip plane of a crystal with respect to the part below the slip plane. The Burgers vector **b** of a dislocation loop is constant, while the dislocation line vector **t** can change direction continuously. The Burgers vector is always an integral atomic spacing because the lattice must maintain atomic registry through the slipped and unslipped regions.

There are two special types of dislocations:

(a) Edge dislocation: the dislocation line vector and the Burgers vector are orthogonal, i.e., at ninety degrees.

(b) Screw dislocation: the dislocation line vector and the Burgers vector are parallel.

Dislocations control many important characteristics of materials. In particular, dislocations occupy a position of fundamental importance in the mechanical behavior of crystalline materials. Their presence reduces the force necessary to cause displacement of atoms. In this sense, dislocations act like a lever. They allow a given quantity of work to be done by a small force moving through a large distance rather than a large force moving through a small distance. Thus, plastic flow or plastic deformation in crystalline solids is accomplished by means of movement of dislocations. The plastic strain depends on the displacement caused by each dislocation, the density of *mobile* dislocations, and the average distance moved by a dislocation. When these linear defects move under the action of a shear stress, they result in slip or glide between crystal planes, which in turn results in permanent (plastic) deformation.

Besides plastic deformation, there are many other physical and chemical properties that are affected by the presence of dislocations. For example, dislocations can serve as easy paths for atomic diffusion (which can affect creep behavior), precipitation reactions, and order processes. They can also be very efficient sites for nucleation of solid-state phase transformations, affect thermal and electrical conductivity, especially at very low temperatures, and affect the current carrying capacity in superconductors.

Planar or two-dimensional defects include external surfaces, grain boundaries, stacking faults, etc. The surface is treated as a planar imperfection because the surface atoms are not bonded to the maximum number of neighbors. Thus, the surface atoms are in a higher energy state than the atoms in the interior. Small angle grain boundaries consist of slight orientation mismatch (few degrees). Simple dislocation arrays form such boundaries; a wall of aligned edge dislocations forms a tilt boundary while screw dislocations form twist boundaries. When the degree of misorientation between grains is too large to be accommodated by dislocations, high angle grain boundaries are formed. These high angle grain boundaries have a higher energy at the grain boundaries than small angle grain boundaries. A twin boundary is a special type of grain boundary across which there is mirror symmetry of the lattice. The region of material between these boundaries is termed a twin. Twin boundaries can result from mechanical shear (mechanical twins) or annealing following deformation (annealing twins). Annealing twins are commonly observed in FCC crystals while mechanical twins are seen in BCC and HCP metals.

3.3 STRENGTHENING MECHANISMS IN METALS

Metals can be strengthened in a number of ways. We provide a summary of these inasmuch as many of these mechanisms influence the behavior of the metal matrix, and thus, of the composite. For a more thorough treatment of strengthening mechanisms, the reader is referred to Meyers and Chawla (2005).

3.3.1 Dislocation Strengthening

The presence of dislocations generates internal stress fields. Extra work needs to be done to move a dislocation through such internal stress fields of other dislocations. Various theories have been proposed to explain the strengthening due to increased dislocation density that results, for example, from subjecting a metal to cold working. Although the dislocation structure of a deformed metal, polycrystalline or single crystal, is inhomogeneous, the strengthening due to dislocation density results in the flow stress being proportional to the square root of the dislocation density, ρ:

$$\tau = \tau_o + \alpha G b \sqrt{\rho}$$

where τ is the shear flow stress, τ_o is a friction stress, G is the shear modulus, b is the Burgers vector, ρ is the dislocation density, and α is a constant on the order of 0.5. As an example, copper, in an annealed condition, has a dislocation density, $\rho = 10^7\ cm^{-2}$ while after some cold working, its dislocation density, ρ increases to $10^{12}\ cm^{-2}$.

Plastic deformation in metals commonly occurs by the movement of dislocations. In HCP metals, plastic deformation can also occur by twinning. The presence of dislocations results in what is called strain hardening or work hardening in metals. Sometimes we also refer to this as dislocation strengthening. Obstacles to dislocation motion (other dislocations, solute atoms, precipitates, grain boundaries, etc.) produce strengthening at room temperature and intermediate temperatures. At high temperatures, thermal energy can help overcome these obstacles (i.e., dislocation climb).

3.3.2 Grain Boundary Strengthening

Grain boundaries can be a very effective source of strengthening at moderate temperatures. Commonly, this is referred to as the Hall-Petch equation and is written as:

$$\sigma_y = \sigma_o + k d^{-1/2}$$

where σ_y is the yield strength, σ_o is the friction stress, k is the Hall-Petch coefficient, and d is the average grain size. The dependence of strength on inverse square root of the grain size is strictly true only over a certain range of grain sizes. At very small grain size, especially in the nanometer grain scale, this equation does not hold. The exponent of d may also change with the grain size.

3.3.3 Solute and Precipitation Strengthening

Solute atoms (solid solution strengthening), precipitates (precipitation hardening) and dispersoids (dispersion strengthening) can also impede dislocation motion and strengthen the metal.

Precipitation hardening is a versatile strengthening technique for some metallic alloys, especially aluminum, steels, and nickel-based alloys. Precipitation hardening comes about an aging treatment, which involves precipitating, out of a homogeneous supersaturated solid solution, a series of metastable and stable precipitates. Various structures offer different levels of resistance to dislocation motion. An example of precipitates formed in an aluminum alloy, 2080 alloy matrix in a composite after T6 heat treatment, is shown in the transmission electron micrograph in Fig. 3.3. This material was also subjected to cyclic fatigue, and dislocations, constrained by the particles, are also shown. Figure 3.4 shows, schematically, the variation of hardness with aging time for 2080 aluminum alloy for three different temperatures. Also shown are the different precipitate types that occur during the aging treatment. Peak hardness or strength corresponds to a critical dispersion of semi-coherent precipitates.

The shape of the aging curve can be explained as follows. Immediately after quenching, only solid solution hardening is predominant. At the onset of aging, coherent zones appear in the initial stages of aging, that are nothing but clusters of solute atoms on certain crystallographic plains of the matrix

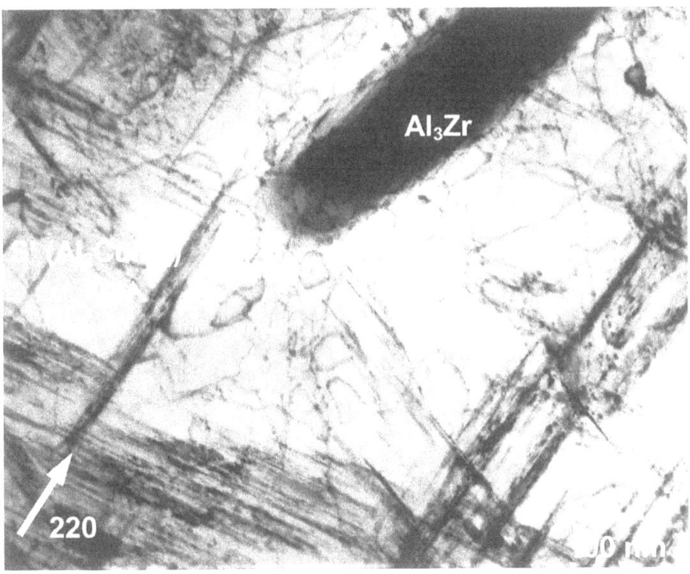

Fig. 3.3 Precipitates (S') and dispersoids (Al_3Zr) formed in aluminum alloy 2080 (Ganesh and Chawla, 2004). The transmission electron micrograph shows the precipitate structure in the matrix of a fatigued 2080/SiC/10$_p$ composite (<220> zone axis). Dislocation motion is constrained by the precipitates and dispersoid particles.

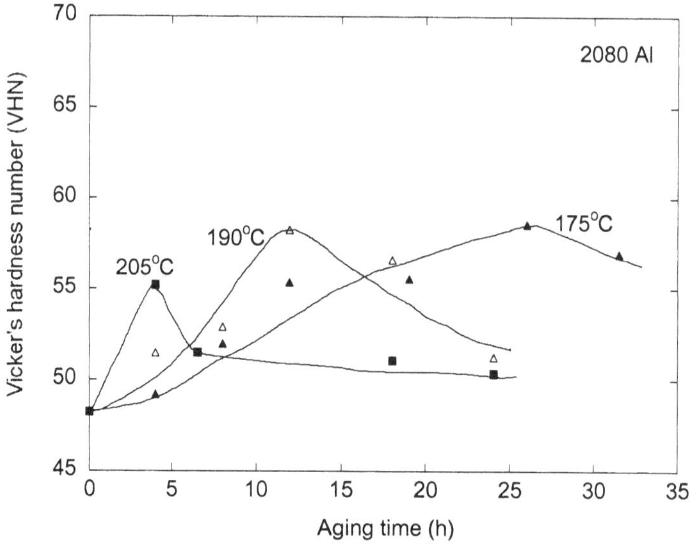

Fig. 3.4 Schematic of the variation of hardness with aging time in a 2080 (Al-Cu-Mg) aluminum alloy. Peak hardness or strength corresponds to a critical dispersion of coherent or semi-coherent precipitates. With increasing temperature, the time to peak aging decreases, because of the acceleration in aging.

(e.g., clustering of copper atoms on {100} planes of aluminum). These zones are transition structures and are referred to as Guinier-Preston zones. We call them *zones* rather than precipitates in order to emphasize the fact that zones represent a small cluster of solute atoms which have not yet taken the form of precipitate particles. The very small GP zones are coherent with the matrix, i.e., the lattice planes cross the interface in a continuous manner and small elastic, coherency strains in the matrix are present. With increasing aging time, these zones grow or thicken by diffusion. As they grow, the misfit strains between the precipitate and matrix increase. Along some planes, coherency is lost, and dislocations form at the interface to take up the strain. Thus, the elastic energy associated with coherency strains along specific planes is reduced. The overall strain field around the particle, however, increases because of the newly-formed dislocations, as well as coherency strains on other planes. This increases the resistance to dislocation motion and increases the strength of the alloy. Further growth of these semi-coherent zones or precipitates results in a complete loss of coherency; an incoherent interface forms between the precipitate and the matrix.

The hardness increases as the GP zone size increases with time, making it more difficult for the dislocations to shear them. With a further increase in aging time, equilibrium precipitates with incoherent interface start appearing and the mechanism of dislocation bowing (called Orowan bowing) around the particles becomes operative. The peak hardness or strength is associated with a critical dispersion of coherent or semi-coherent precipitates. At this point, the contributions from shearing and by-pass of precipitates is equal. Further aging results in an increase of the interparticle distance and a lower strength results as dislocation bowing becomes easier.

Dislocations move on specific slip planes in a metal under the action of a shear stress. If the slip plane has obstacles penetrating it, e.g., precipitates or dispersoids, a dislocation moving on this slip plane must interact with these obstacles. The dislocation, under the action of an applied shear stress, τ, bows between the obstacles in the slip plane. If we ignore the changes in the orientation of the dislocation because of bowing, the shear stress required to bow the dislocation to a radius r is given by:

$$\tau = \frac{Gb}{2r}$$

where G is he shear modulus of the metal and b is its Burgers vector. This mechanism of dislocation bowing around particles is called Orowan bowing.

3.3.4 Fracture in Metals

Generally, metals are quite ductile and fracture by a process known as microvoid coalescence. In this process, also referred to as dimple facture, microvoids nucleate at precipitate particles, inclusions, etc. Under applied stress, these microvoids grow and coalesce leading to final fracture with characteristic dimples on the fracture surface. Figure 3.5(a) shows schematically the process while 3.5(b) shows an SEM micrograph of an actual fracture surface obtained after tensile testing in an aluminum alloy. Note the presence of precipitates, in the center of dimples, where microvoids nucleated.

3.4 COMMON MATRIX MATERIALS

3.4.1 Aluminum and Aluminum Alloys

Aluminum alloys, because of their low density and excellent strength, toughness, and corrosion resistance, have been used extensively in the automotive and aerospace fields. Of special mention are Al-Cu-Mg and Al-Zn-Mg-Cu alloys, which are very important precipitation hardenable alloys.

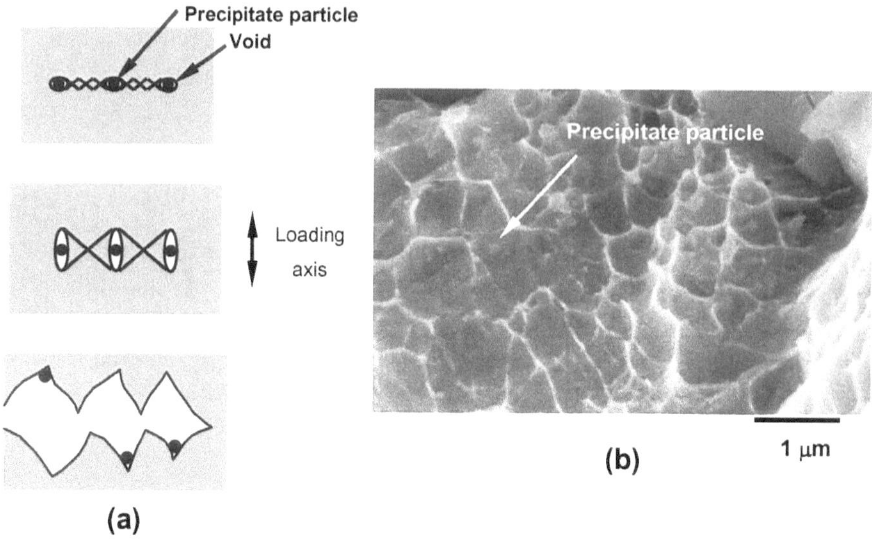

Fig. 3.5 (a) Schematic of nucleation, growth, and calescence of microvoids at precipitate particles in a ductile metallic alloy, (b) Characteristic dimples on the fracture surface in an aluminum alloy. Note the presence of precipitate particles that served as the nucleating sites for microvoids. (Chawla et al., 2002).

Aluminum alloys can be classified as cast, wrought, or age-hardenable alloys. Some of the common age-hardening or precipitation-hardenable treatments for Al alloys are the following:

1. T4: Solutionizing and quenching, followed by aging at room temperature, or "natural aging."

2. T6: Solutionizing and quenching, followed by aging at a temperature above room temperature (120-190°C), or "peak-aging."

3. T7x: Solutionizing, quenching, and overaging.

4. T8xx: Solutionizing, quenching, cold working, and peak-aging.

The solutionizing temperature is typically between 440 and 540°C, while the quenching medium can be water or a synthetic quenchant.

Liquid aluminum, like most metals, has a very low viscosity, which makes for easy casting practice. The temperature dependence of viscosity of liquid aluminum is given by (Smithells, 1976):

$$\eta = 0.1492\exp(1984.5/T)$$

where η is the viscosity in MPa.s, R is the gas constant (= 8.3144 J/K mol), and T is the temperature in Kelvin. At its melting point, pure aluminum has a viscosity of $\sim\eta$ = 12 MPa.s. The addition of ceramic particles or inclusions raises the viscosity very quickly. This has important implications in the processing of particle reinforced metal matrix composites (see chapter 4).

3.4.2 Titanium Alloys

Titanium is one of the most important aerospace materials. Pure titanium has a density of 4.5 gcm^{-3} and a Young's modulus of 115 GPa. For titanium alloys, the density can vary between 4.3 and 5.1 gcm^{-3} while the Young's modulus can range between 80 and 130 GPa. Thus, titanium and its alloys have relatively high strength/weight and modulus/weight ratios. Titanium has a relatively high melting point (1672°C) and retains strength to high temperatures with good oxidation and corrosion resistance. All these factors make it an ideal material for aerospace applications. Titanium alloys are used in jet engine (turbine and compressor blades), fuselage parts, etc. It is, however, an expensive material.

At extremely high speeds, such as in a supersonic military aircraft, the skin of an airplane heats up so much that aluminum alloys are no longer an option. Titanium alloys must be used at such high temperatures. In a supersonic plane, flying at speeds over Mach 2, the temperatures will be even higher than what titanium alloys can withstand. Titanium aluminides are one of the candidate materials in this case.

Titanium has two polymorphs; alpha (α) titanium has an HCP structure and is stable below 885°C and beta (β) titanium which has a BCC structure and is stable above 885°C. Aluminum raises the $\alpha \rightarrow \beta$ transformation temperature, i.e., aluminum is an alpha stabilizer. Most other alloying elements (Fe, Mn, Cr, Mo, V, No, Ta) lower the $\alpha \rightarrow \beta$ transformation temperature, i.e., they stabilize the phase. Thus, three general alloy types can be produced, viz., α, $\alpha+\beta$, and β titanium alloys. The Ti-6%Al-4% V, called the *work horse* Ti alloy of the aerospace industry, belongs to the ($\alpha+\beta$) group. Most titanium alloys are used after hot working in the ($\alpha+\beta$) region, which breaks the structure and distributes the β phase in an extremely fine form.

Titanium has a great affinity for oxygen, nitrogen, and hydrogen. Parts per million of such interstitials in titanium can change mechanical properties drastically. In particular, these microstructural changes may result in severe embrittlement. This is why welding of titanium by any technique requires protection from the atmosphere. Electron beam techniques, in vacuum, are also frequently used.

3.4.3 Magnesium and its Alloys

Magnesium and its alloys form another group of light metals. Magnesium is one of the lightest metals, its density being 1.74 gcm^{-3}. Magnesium alloys, especially castings, are used in automotive and aircraft gearbox housings, chain saw housing, laptop casings, electronic equipment, etc. Magnesium has a hexagonal close-packed structure, which gives it limited ability to deform plastically by slip at room temperature.

3.4.4 Cobalt

Cobalt is a very common metal matrix that is used in WC/Co composites, also known as cemented carbides, which are used as inserts for cutting tool and in oil drilling. During processing of cemented carbides, cobalt is used in a powder form. Cobalt powder can be produced chemically by hydrogen reduction or atomization of cobalt liquid. The chemistry and crystal structure of Co powder are different from that of Co matrix in a WC/Co composite. Milling and processing by liquid-phase sintering or high-temperature, high-pressure compaction can affect the chemistry of the finished Co matrix.

Pure Co is stable below 417°C in the HCP crystal structure. Above this temperature, the high-temperature face centered cubic (FCC) structure is stable. It turns out that the FCC cobalt becomes stable at room temperature because of dissolution of carbon from WC during processing of WC/Co composites. The FCC structure has more slip systems available, which results in higher ductility. In WC/Co composites, the small amount of Co matrix holds the WC particles in place and provides toughness, which stems from its ability to deform plastically.

3.4.5 Copper

Copper has a face centered cubic structure. Because of its high electrical conductivity (only silver and gold are better), it is used extensively as an electrical conductor. It also has good thermal conductivity, which makes it suitable for thermal management applications. It can be cast and worked easily, because it is quite ductile. One of the major applications of copper in a composite is as a matrix material in niobium-based superconductors. Copper-zinc alloys (brass) and copper-tin alloys (bronze) are solid solution strengthened and among the earliest alloy metals used.

3.4.6 Silver

Silver is another face centered cubic metal. It is a very good electrical and thermal conductor, highly ductile, and has good corrosion resistance. It has found a new use as a matrix material in high temperature oxide superconductors.

3.4.7 Nickel

Nickel is also an FCC metal, which gives it good ductility. More importantly, nickel-based alloys show an excellent combination of properties. Nickel-based superalloys, primarily nickel-iron-cobalt alloys, have superior high-temperature creep resistance, making them suitable for turbine blades.

3.4.8 Niobium

Although niobium alloys are not commonly used as a matrix material in MMCs, they are used as filaments in superconducting composites. We give a brief description of these alloys in Sec. 3.4 below.

3.4.9 Intermetallics

Intermetallics (sometimes referred to as intermetallic compounds) are formed when two dissimilar metals are combined following rules of chemical valence. Generally, the bonding in intermetallics is not metallic, but ionic and covalent in nature. Such an alloy is called an intermetallic compound (Villers and Calvert, 1985; Sauthoff, 1995).

Intermetallics generally have a stoichiometric composition and appear as a line compound in the phase diagram. Some intermetallics have a range of composition. In general, intermetallics have a complex crystal structure and are brittle due to their ionic and covalent bonding. Some intermetallics have been identified to have rather unique attributes of high-temperature strength and stiffness, low density, and excellent oxidation resistance. Used as a matrix material for making composites, they have the potential to increase the operating temperatures over conventional materials.

Intermetallics can have a disordered or ordered structure. Ordered intermetallic alloys possess structures characterized by long-range ordering, i.e., different atoms occupy specific positions in the lattice. Because of their ordered structure, dislocations in intermetallics are much more restricted than in disordered alloys. This results in retention (and, in some cases, even an increase) in strength at elevated temperatures, a very desirable feature. For example, nickel aluminide shows a marked increase in strength up to 800°C. An undesirable feature of intermetallics is their extremely low ambient temperature ductility.

An important disordered intermetallics is molybdenum disilicide ($MoSi_2$). It has a high melting point and shows good stability at temperatures greater than 1200°C, in an oxidizing atmosphere. It is commonly used as a heating element in furnaces. The high oxidation resistance comes from a protective SiO_2 film that is formed at high temperatures.

3.5 SUPERCONDUCTIVITY

Since one of the major applications of metal matrix composites is as filamentary superconducting composites, we briefly describe the materials related aspects of the phenomenon of superconductivity.

Certain metals and alloys lose all resistance to the flow of electricity when cooled to within a few degrees of absolute zero. This phenomenon is called superconductivity and the materials exhibiting this phenomenon are called superconductors. Unlike normal metals, these superconductors can carry a high-density current without any electrical resistance. Superconductors can potentially carry as much as 100 times the amount of electricity of ordinary copper or aluminum wires of the same size. This, in turn, can be exploited in devices such as motors and generators, and to transmit electricity in power lines. Superconductors can generate very high magnetic fields that are common in high energy physics and fusion energy programs. Other fields of application include magnetic resonance imaging, magnetohydrodynamic generators, rotating machines, magnetic levitation vehicles, and magnets in general.

Conventional superconductors require cooling in liquid helium (4.2 K), a very costly proposition. In 1986, however, a new class of ceramic superconductors was discovered which exhibited a critical temperature above that of liquid nitrogen (77 K). These are called the high-temperature superconductors and the phenomenon is referred to as high temperature superconductivity (HTS). Unlike the low-temperature superconductors, which are metallic or semimetallic, these new compounds are ceramics. The first HTS discovered had a critical temperature of 35 K (–238°C). Then in 1987, a compound that became superconducting at 94 K (–179°C) was discovered. This discovery was particularly significant because this compound could be cooled with cheap and readily available liquid nitrogen. This was followed by the discovery of bismuth-based compounds; thallium- and yttrium-based compounds with even higher critical temperature were discovered. The high-temperature superconductors (HTS) can function at temperatures as high as 140 K (–133°C). This allows cooling to be done more economically and efficiently by liquid nitrogen, compared with conventional or low-temperature superconductors which need liquid helium as a cooling medium. In 2001, another low temperature superconductor was discovered with composition MgB_2 (Nagamatsu et al., 2001). Its T_c of 39 K is slightly higher than that of Nb-based metallic superconductors.

Niobium/titanium alloys are one of the important low-temperature superconductors. Low-temperature superconductors must be cooled to below 20 K in order to become superconducting. They are widely used in magnetic resonance imaging (MRI) machines, and in the fields of high-energy physics and nuclear fusion. Additional commercial use has been limited largely by the high refrigeration costs associated with liquid helium, which is needed to cool the materials to such low temperatures.

3.5.1 Types of Superconductors

The phenomenon of superconductivity was discovered in mercury by Onnes in 1911. Since then, many elements and hundreds of solid solutions or compounds have been discovered that show this phenomenon of total disappearance of electrical resistance below a critical temperature T_c. Figure 3.6 shows the variation of electrical resistivity with temperature of a normal metal and that of a superconducting material. T_c is the critical temperature, a characteristic constant of each material, above which the material becomes a normal conductor. Kunzler et al. (1961) discovered the high critical field capability of Nb_3Sn and thus opened up the field of practical, high-field superconducting magnets. Superconductors came into the realm of economic viability when techniques were developed to put the superconducting species in the form of ultrathin filaments in a copper matrix.

The technologically most important property of superconductors is their capacity to carry an electric current without normal I^2R (I is the current in amp and R is the resistance in ohm) losses up to a critical current density, J_c. This critical current is a function of the applied field and temperature. The

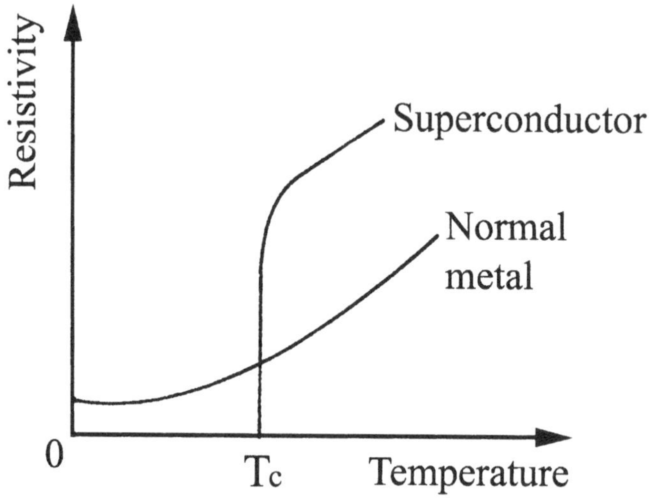

Fig.3.6 Variation of electrical resistivity with temperature for a normal metal and that of a superconducting material.

commercially available superconductors in the 1980s could demonstrate critical current densities of $J_c > 10^6$ A cm^{-2} at 4.2 K and an applied field of H = 5 T. There are three parameters that limit the properties of a superconductor, namely, the critical temperature (T_c), the critical electrical current density (J_c), and the critical magnetic field (H_c); see Fig. 3.7. As long as the material does not cross into the shaded area indicated in Fig. 3.7, it will behave as a superconductor.

There are two types of superconductors:

Type I: These are characterized by low T_c values and they lose their superconductivity abruptly at H_c.

Type II: These behave as diamagnetic materials up to a field H_{c2}. Above this field, the magnetic field penetrates gradually into the material and concomitantly the superconductivity is gradually lost, until at the critical magnetic field H_{c2} the material reverts to its normal state. All major applications of superconductivity involve the use of type II superconductors.

3.5.2 Flux and Flux-Pinning

When magnetic lines of force from a strong magnetic field, called flux lines,

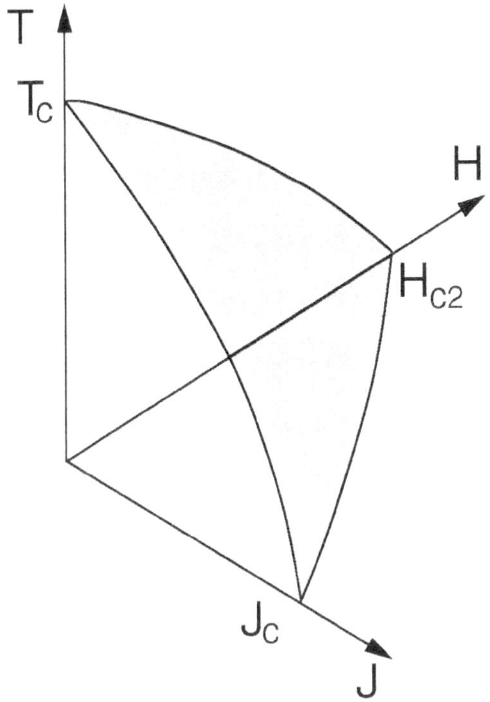

Fig. 3.7 The state of superconductivity is described by three critical parameters: magnetic field (H), temperature (T), and current density (J). As long as the material does not cross into the shaded area, it will behave as a superconductor.

try to penetrate the surface of a Type 2 superconductor, we get what is called a flux-lattice. This periodic lattice, consisting of tubular regions of flux lines, stems from the tiny magnetic moments in each flux line, which repel each other and array themselves in an orderly fashion, usually in the close packed hexagonal arrangement. Flux pinning is required in superconductors because movement of flux, sometimes called “flux-creep,” is an energy-dissipative process; that is, heat is produced. Any disturbance of a superconductor, say, by motion or a change in the applied field, leads to a rearrangement of magnetic flux lines in the superconductor. Flux movement in a superconductor, irrespective of the source of flux movement, results in a temperature increase. This leads to a reduced critical current and more flux movement. The net result is that the superconductor is heated above T_c and reverts to its normal state. A practical solution to this problem is to make the superconductor in the form of ultrathin filaments so that the amount of energy (heat) dissipated by flux motion is too small to cause this runaway behavior; see Fig. 3.8. Flux-pinning occurs via defects in the crystalline structure of the superconductor such as grain boundaries or impurities. The

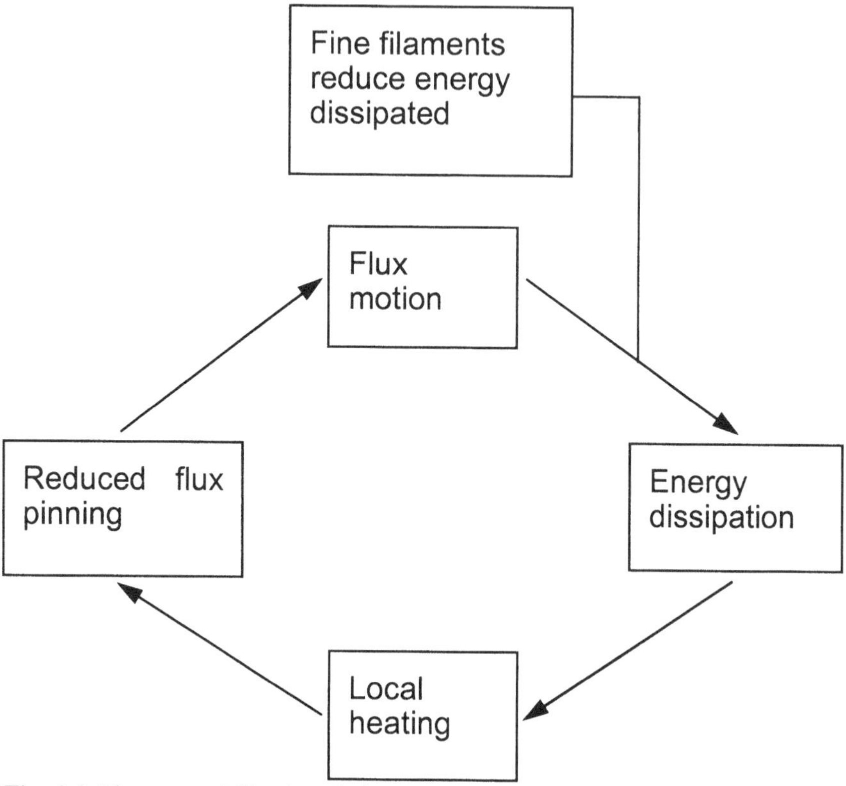

Fig. 3.8 The cryostabilization design concept: fine filaments reduce the energy dissipated due to magnetic flux movement.

high-purity copper or silver matrix provides a high-conductivity alternate path for the current. In the case of a quench, that is, superconductor reverting to the normal state, the metal matrix carries the current without getting excessively hot. The superconductor is cooled again below its T_c and carries the electric current again. This is the so-called cryogenic stability or cryostabilization design concept; namely, the superconductor is embedded in a large volume of low-resistivity metal matrix and a coolant is in contact with all windings.

3.5.3 Ductile Alloys: Niobium-Titanium Alloys

Niobium-titanium alloys provide a good combination of superconducting and mechanical properties. A range of compositions is available commercially: Nb-44% Ti in the United Kingdom, Nb-46.5% Ti in the United States, and Nb 50% Ti in Germany. Figure 3.9 shows the Nb-Ti phase diagram as well as the critical temperature and upper critical field as a

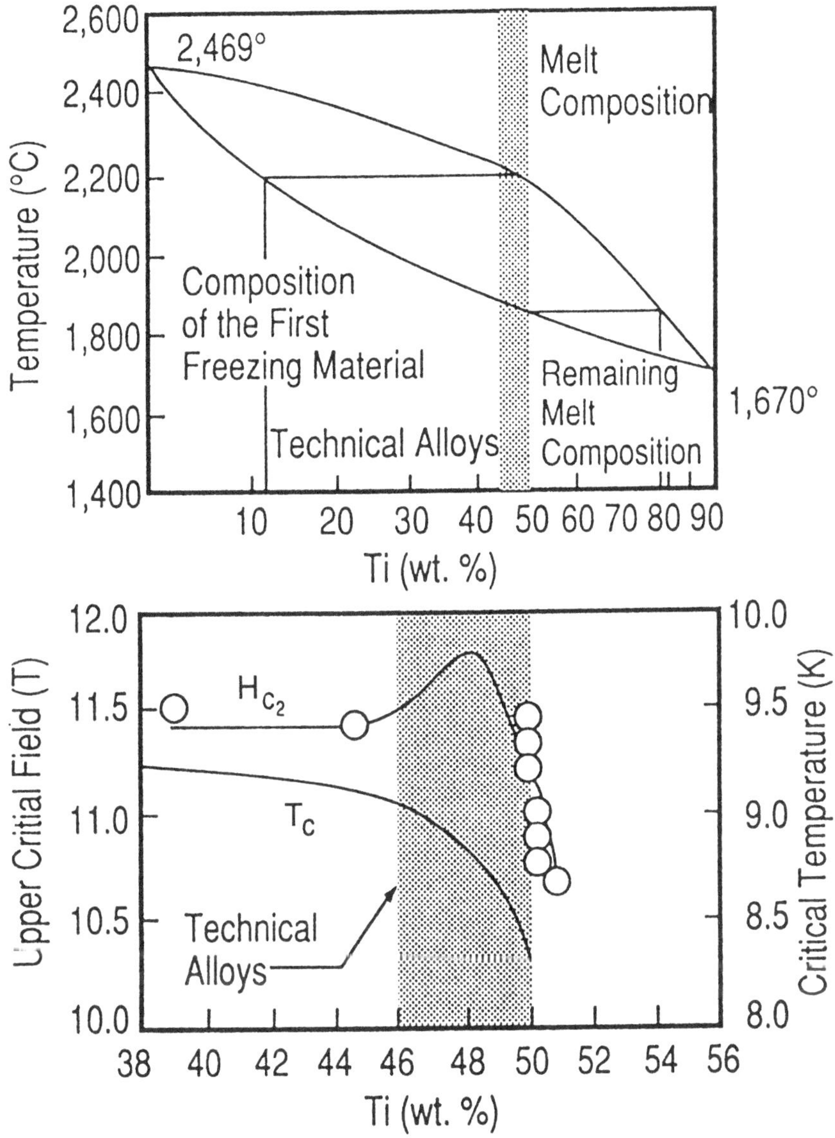

Fig. 3.9 The top figure shows the Nb-Ti phase diagram while the bottom figure shows the critical temperature and upper critical field as a function of alloy composition. Note the range (46-50 wt.% Ti) for the superconducting alloys. (after Hillmann, 1981).

function of alloy composition. Note the range (46-50 wt.% Ti) for the superconducting alloys.

In all these alloys, a J_c > 1000 A mm^2 at 4.2 K and an applied field of 7 T can be obtained by a suitable combination of mechanical working and annealing treatments. Strong flux pinning and, therefore, high J_c are obtained in these alloys by means of dislocation cell walls and precipitates. The flux pinning by precipitates becomes important in high-Ti alloys because the Nb-Ti phase diagram indicates precipitation of α-Ti in these alloys. As pointed out above, the condition of stability against flux motion requires that the superconductor be manufactured in the form of a composite system. Extremely fine superconducting filaments embedded in a copper matrix provide flux stability and reduced losses caused by varying magnetic fields becomes important in high-Ti alloys because the Nb-Ti phase diagram indicates precipitation of α-Ti in these alloys. As pointed out above, the condition of stability against flux motion requires that the superconductor be manufactured in the form of a composite system. Extremely fine superconducting filaments embedded in a copper matrix provide flux stability and reduced losses caused by varying magnetic fields.

3.5.4 A-15 Superconductors: Nb_3Sn

For applications involving fields greater than 12 T, the ordered intermetallic compounds having an A-15 crystal structure are better suited than the Nb-Ti type. Nb_3Sn has a T_c of 18 K. It is easy to appreciate that the higher the T_c, the lower the refrigeration costs. Nb_3Sn is the most widely used superconductor for high fields and high temperatures. A characteristic feature of these intermetallic compounds is their extreme brittleness (typically, a strain-to-failure of 0.2% with very little plasticity). Compare this to Nb-Ti which can be cold worked to a reduction in area over 90%. Initially, the compound Nb_3Sn was made in the form of wires or ribbons either by diffusion of tin into niobium substrate in the form of a ribbon or by chemical vapor deposition. V_3Ga on a vanadium ribbon was also produced in this manner. The main disadvantages of these ribbon type superconductors were: (a) flux instabilities due to one wide dimension in the ribbon geometry and (b) limited flexibility in the ribbon width direction. Later on, with the realization that flux stability could be obtained by the superconductors in the form of extremely fine filaments, the filamentary composite approach to A-15 superconductor fabrication was adopted. This breakthrough involving the composite route came through early in the 1970s. Tachikawa (1970) showed that V_3Ga could be produced on vanadium filaments in a Cu-Ga matrix while Kaufmann and Pickett (1970) demonstrated that Nb_3Sn could be obtained on niobium filaments from a bronze (Cu-Sn) matrix. The processing of some important superconducting composite systems is described in chapter 4, while their applications are given in chapter 11.

References

Chawla, N., J.J. Williams, and R. Saha, (2002) *Metall. Mater. Trans.*, **33A**, 3861.

Ganesh, V.V., and N. Chawla, (2004) *Metall. Mater. Trans.*, **35A**, 53-62.

Hillmann, H., (1981) in *Superconductor Materials Science*, Plenum, New York, p. 275.

Kaufmann, A.R., and J.J. Pickett, (1970) *Bull. Am. Phys. Soc.*, **15**, 833.

Kunzler, J.E., E. Bachler, F.S.L. Hsu, and J.E. Wernick, (1961) *Phys. Rev. Lett.*, **6**, 89.

Meyers, M.A., and K.K. Chawla, (2005) *Mechanical Behavior of Materials*, 2nd edition, Prentice Hall, Upper Saddle River, NJ.

Nagamatsu, J., N. Nakagawa, T. Muranaka, Y. Zenitani, and J. Akimitsu, (2001) *Nature*, **410**, 63.

Sauthoff, G., (1995) *Intermetallics*, VCH Publishers, New York, NY.

Smithells Metals Reference Book, (1976) Butterworths, Boston, p. 944.

Tachikawa, K., (1970) in *Proceedings of the 3rd ICEC*, Illife Science and Technology Publishing, Surrey, U.K.

Pearson's Handbook of Crystallographic Data for Intermetallic Phases, (1985), vol. 1,2,3, P. Villers and L.D. Calvert (eds.), ASM, Metals Park, OH.

Chapter 4

PROCESSING

Metal matrix composites can be made by liquid, solid, or gaseous state processes. In this chapter we describe some important processing techniques for fabricating MMCs.

4.1 LIQUID STATE PROCESSING

Metal matrix composites can be processed by incorporating or combining a liquid metal matrix with the reinforcement. There are several advantages to using a liquid phase route in processing. These include near net-shape (when compared to solid state processes like extrusion or diffusion bonding)), faster rate of processing, and the relatively low temperatures associated with melting most light metals, such as Al and Mg. The most common liquid phase processing techniques can be subdivided into four major categories:

- *Casting or liquid infiltration*: This involves infiltration of a fibrous or particulate preform by a liquid metal. In the case of direct introduction of short fibers or particles into a liquid mixture, consisting of liquid metal and ceramic particles or short fibers, is often stirred to obtain a homogeneous distribution of particles. In centrifugal casting, a gradient in reinforcement particle loading is obtained. This can be quite advantageous from a machining or performance perspective.

- *Squeeze casting or pressure infiltration*: This method encompasses pressure-assisted liquid infiltration of a fibrous or particulate preform. This process is particularly suited for complex shaped components, selective or localized reinforcement, and where production speed is critical.

- *Spray co-deposition*: In this process the liquid metal is atomized or sprayed while a particle injector introduces ceramic particles in the spray stream to produce a granulated mixture of composite particles. The composite particles are then consolidated using another suitable technique, such as hot-pressing, extrusion, forging, etc.

- *In situ processes*: In this case, the reinforcement phase is formed *in situ* either by reaction during synthesis or by controlled solidification of a eutectic alloy.

We now discuss each of these processing classes in detail.

4.1.1 Casting

Conventional Casting

Casting of MMCs can typically be accomplished with conventional equipment used to cast metallic alloys (Surappa and Rohatgi, 1981). It is typically used with particulate reinforcement because of the difficulty in casting fibrous performs without pressure. The particles and matrix mixture are cast into ingots and a secondary mechanical process, such as extrusion or rolling, is applied to the composite, see Fig. 4.1.

Casting of MMCs requires some modifications to existing conventional casting processes, which can be summarized as follows:

- Alloys that minimize reactivity with the reinforcement must be used. Al-Si alloys (with Si levels of up to 9%) are typically used with SiC reinforcement. Figure 4.2 shows that with increasing temperature, an increase in the amount of silicon is required to prevent Al_4C_3 formation

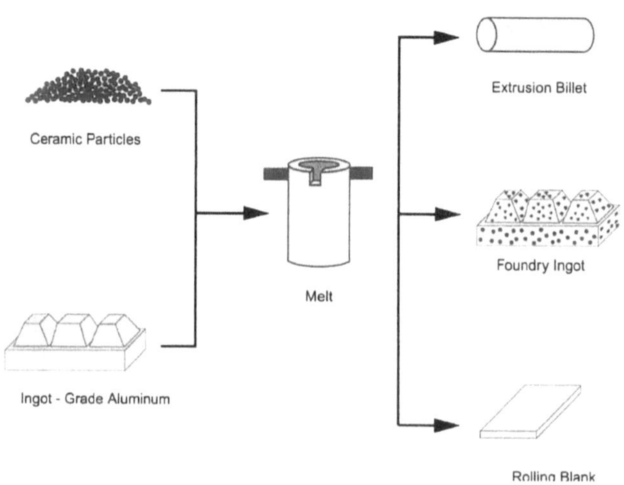

Fig. 4.1 Conventional casting route for processing particle reinforced MMCs.

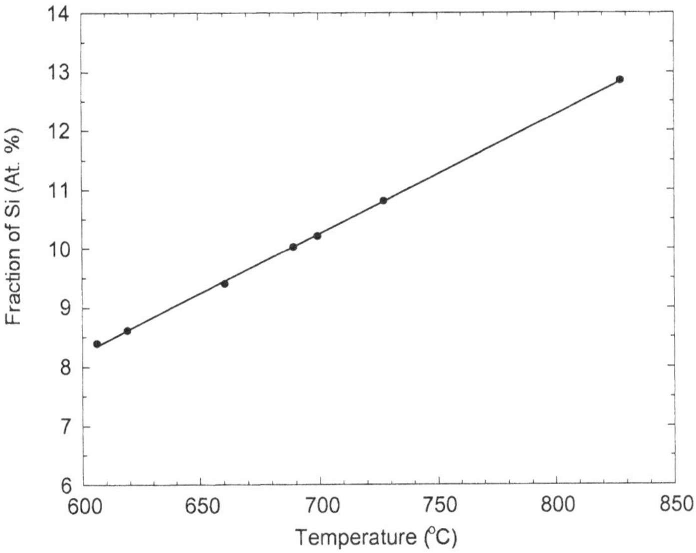

Fig. 4.2 Fraction of Si, at a given temperature, required to prevent formation of Al_4C_3 in an Al-Si/SiC composite (after Lloyd, 1997).

(see chapter 5).

- Addition of reinforcement particles increases the viscosity of the melt. For a viscous non-metallic liquid with nonmetallic particles, the viscosity of the composite, η_c, is given by (Thomas, 1965):

$$\eta_c = \eta_m \left(1 + 2.5V_p + 10.05V_p\right)$$

where η_m is the viscosity of the unreinforced metal and V_p is the volume fraction of particles. Thus, the temperature of the composite melt should be above a certain limit (~745°C for Al-Si/SiC) to keep the melt from becoming too viscous. Covering the melt with an inert gas atmosphere reduces oxidation of the melt.

- Stirring of the composite melt is often required, Fig. 4.3 (Mehrabian et al., 1974). The density of SiC (3.2 g/cm^3), for example, is higher than that of Al (2.2 g/cm^3), so the particles will sink unless the melt is agitated. Alternating currents in a magnetic field (Katsura, 1982) and mechanical vibration (Pennander and Anderson, 1991) have also been used to improve wetting and permeability of the reinforcement in the liquid matrix.

Centrifugal Casting

One of the disadvantages of MMCs with ceramic reinforcement is that they are typically more difficult to machine than the unreinforced alloy. In centrifugal casting, optimal placement of the reinforcement can be achieved by inducing a centrifugal force immediately during casting, Fig. 4.4, to intentionally obtain a gradient in reinforcement volume fraction (Divecha et al., 1981; Tsunekawa et al., 1988). In brake rotors, for example, wear resistance is needed on the rotor face, but not in the hub area. Thus, in areas where reinforcement is not as crucial, such as in the hub area, machining would be easier without the reinforcement. Figure 4.5 shows the microstructure at different points in a centrifugally-cast brake rotor, showing

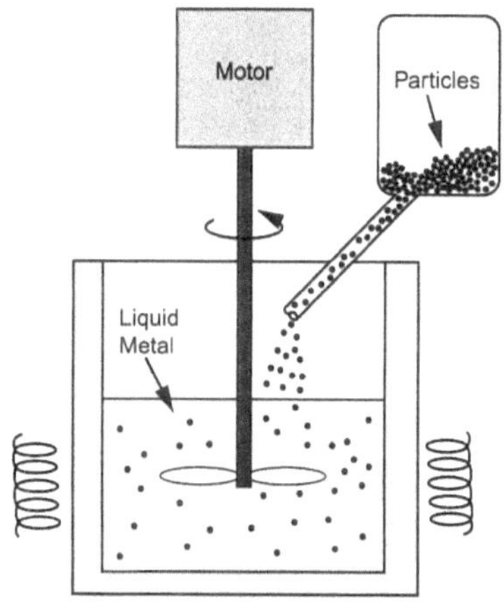

Fig. 4.3 Stirring of composite melt with ceramic particles to minimize settling of the particles during processing.

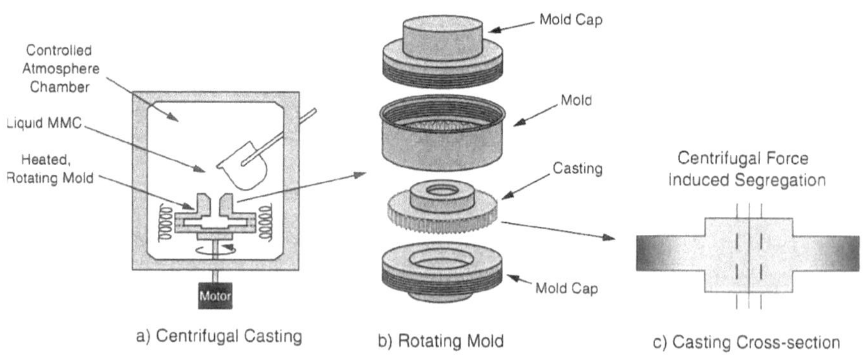

Fig. 4.4 (a) Schematic of centrifugal casting process, (b) rotating mold, and (c) cross-section of finished casting with intentionally-segregated reinforcement.

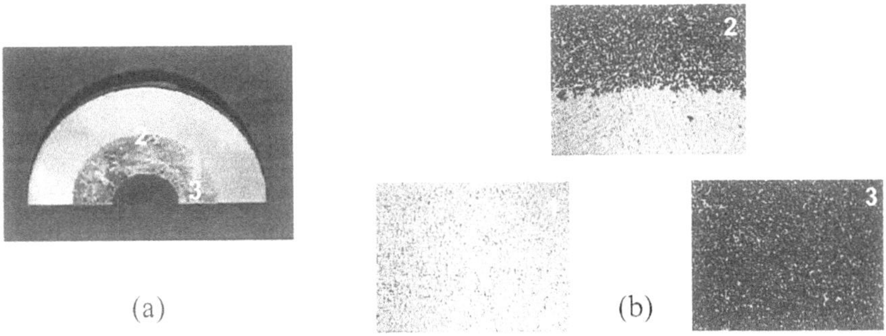

Fig. 4.5 (a) Centrifugally cast brake rotor with selective placement of reinforcement, and (b) regions of the microstructure in the rotor: (1) matrix-rich, (2) interface, (3) reinforcement-rich (courtesy of D. Herling).

a pure aluminum alloy matrix region, interface region, and reinforced region. The process is relatively inexpensive, with a potential for composite materials at as low as $ 2/kg.

Microstructure Evolution during Casting

During casting the reinforcement (particle or short fiber) is typically mobile, so that motion or pushing of the particle by the liquid needs to be considered. If the particles are pushed by the solidifying liquid, the particles are segregated to the remaining liquid that solidifies. In eutectic systems, for example, the remaining liquid is highly rich in solute. The resulting microstructure, then, depends on the relative size of the primary dendrites relative to the particle size. For composites cooled at relatively slow cooling rates, the dendrite size is much larger than the particle size and the solidified microstructure consists of several areas of high particle clustering, Fig. 4.6 (Lloyd, 1994). For faster cooling rates, the matrix cell size is much smaller (on the order of the particle size) so the degree of particle pushing is diminished significantly, resulting in a much more homogeneous distribution of particles, Fig. 4.7.

Experimental studies on particle pushing in particle reinforced MMCs show the following trends: (i) in eutectic and hypereutectic matrix alloys, such as Al-Si, particle capture takes place for all growth conditions, (ii) in hypoeutectic Al alloy systems, where a non-planar solid/liquid interface is typically present, particle pushing takes place, and (iii) in several systems, there is a critical velocity of the solid/liquid interface, V_c, below which particle pushing takes place and above which particle capture occurs. For a planar solid/liquid interface, V_c can be related to the particle diameter, d, by the following relationship (Mortensen and Jin, 1992):

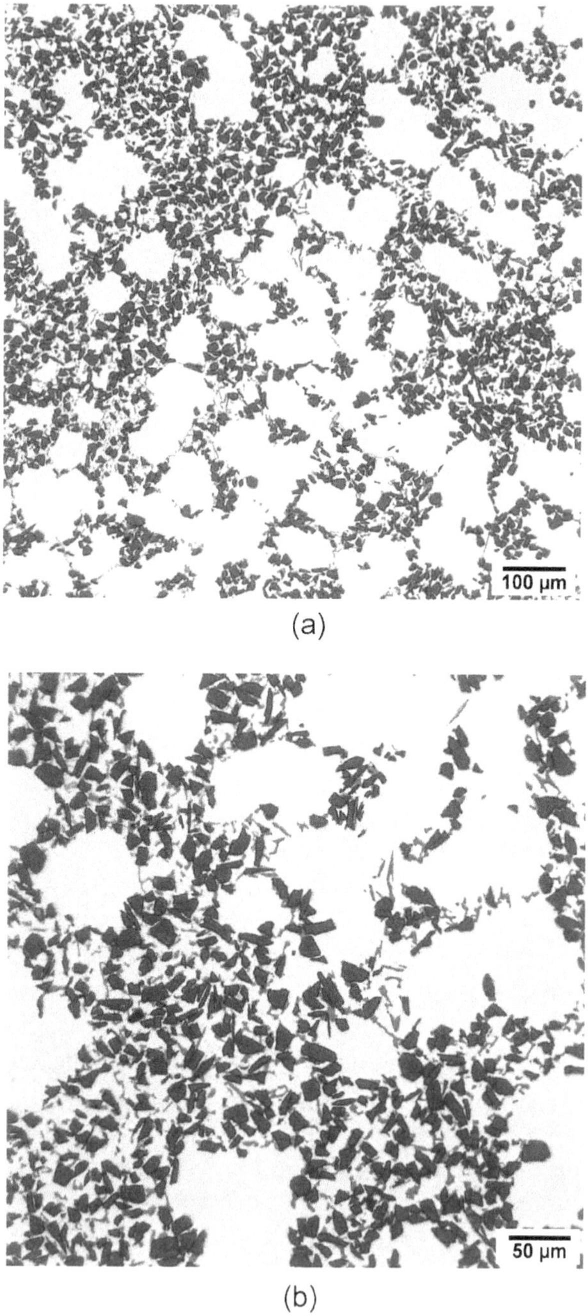

Fig. 4.6 Highly clustered SiC particle distribution in Al resulting from slow cooling, due to particle pushing: (a) lower magnification and (b) higher magnification (courtesy of D.J. Lloyd).

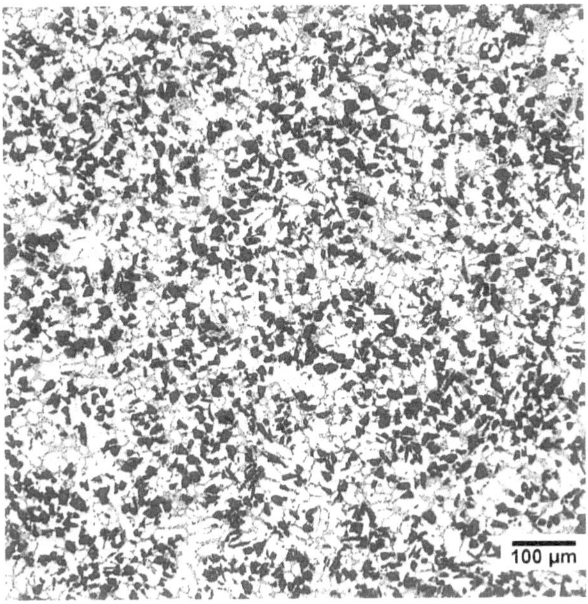

(a)

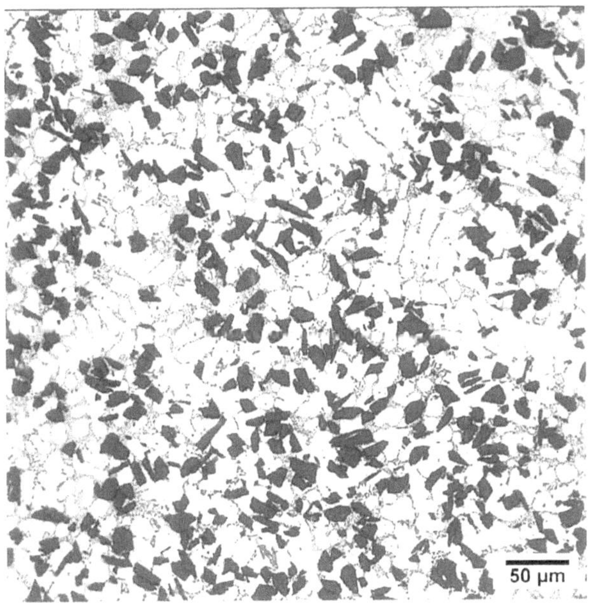

(b)

Fig. 4.7 Uniform SiC particle distribution in Al resulting from faster cooling: (a) lower magnification and (b) higher magnification (courtesy of D.J. Lloyd).

$$V_c d^n = C$$

where C is a system-dependent constant and the exponent n varies between 0.5-3.

Two criteria have generally been proposed for particle pushing. The first is related to the free energy change as the solid grows and comes into contact with the particle in the liquid. Uhlmann et al. (1964) and Potschke and Rogge (1989) proposed the following criterion for particle pushing, based on interfacial free energies:

$$\gamma_{ps} > \gamma_{pl} + \gamma_{sl}$$

where γ is the interfacial energy, and the subscripts ps, pl, and sl refer to particle/solid, particle/liquid, and solid/liquid interfaces, respectively. This model is generally only valid when the interface velocities are small and it is not of great practical interest because the interfacial free energies are largely unknown.

Several investigators have taken a different, more practical approach, of trying to predict the critical interface velocity for particle pushing (Uhlmann et al., 1964; Cisse and Bolling, 1971; Stefanescu et al., 1988; Shangguan et al., 1992; and Kim and Rohatgi, 1999). This approach is based on experimental observations that a critical interface velocity exists, below which the particles are pushed, and above which the particles are engulfed. Cisse and Bolling (1971) modeled the balance between the viscous drag on the particle and the counteracting repulsing force exerted by the particle on the solid/liquid interface front, and proposed a critical velocity for particle pushing/engulfment:

$$V_c^2 = \frac{4kT\gamma_{sl}a_o}{9\pi\eta^2 R^3}\frac{\alpha(1-\alpha)^3}{(1-3\alpha)}$$

where η is the viscosity of the liquid, T is temperature, γ is the free energy of the solid/liquid interface, R is the radius of the particle, a_o is the distance between the particle and the solid/liquid interface, k is the curvature of the solid/liquid interface, and α is the ratio of particle radius to interface radius. Kim and Rohatgi (1999) modeled the shape of the solid/liquid front in terms of the ratio of thermal conductivity of the particle to that of the liquid, the temperature gradient and surface tension of the solid/liquid interface, and the heat of fusion of the liquid. They derived the following expression:

$$V_c = \frac{\Delta\gamma a_o (kR+1)}{18\eta R}$$

where $\Delta\gamma$ is the difference in interface energy, $\Delta\gamma = \gamma_{sp} - \gamma_{lp} - \gamma_{sl}$. Kim and Rohatgi (1999) showed that their model was closest to experimental measurements of the critical velocity, compared to other models, yet, their prediction was still a factor of 2 lower than that of the experimental values. This was attributed to the fact that (a) a constant particle temperature was used in the calculations, when in reality, the particle temperature may be changing with time, and (b) factors such as particle shape, particle roughness, and heat convection between the particle and the interface, were not considered in the model.

4.1.2 Liquid Infiltration

Liquid Infiltration (Lanxide™ Process)

Another pressureless infiltration process consists of reactive or non-reactive infiltration of a reinforcement preform. In this process a particulate or fribrous filler is infiltrated with pure Al or Al-Mg alloy, Fig. 4.8. Alloy matrices may also be formed in this process. For example, when pure Al is

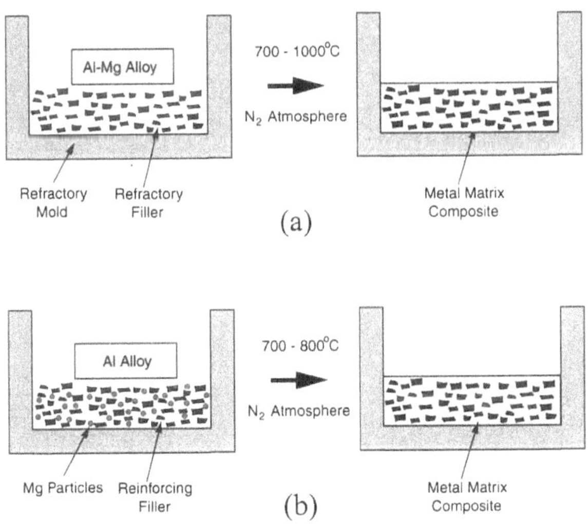

Fig. 4.8 Pressureless infiltration of MMCs: (a) Alloy matrix infiltration of particulate preform and (b) pure matrix infiltration of metallic alloy particle and ceramic particulate preform.

intermingled with Mg particles (in addition to the ceramic reinforcement particles) an Al-Mg alloy matrix is formed. The process is conducted in N_2 atmosphere to minimize interfacial reactions, particularly since the pressureless process involves long infiltration times at high temperatures (Aghajanian et al., 1989). When pure Al is infiltrated, the infiltration temperatures are between 700-800°C, while infiltration of Al-Mg alloys is conducted between 700-1000°C (Lloyd, 1997). Typical infiltration rates are less than 25 cm/h.

Squeeze Casting

Squeeze casting or pressure infiltration involves forcing the liquid metal matrix into a short fiber or particulate preform (Mortensen et al., 1988; Masur et al., 1989; Cook and Werner, 1991). The main advantages of this method over conventional casting are the shorter processing times (which is of particular interest for production of materials in high volumes), ability to fabricate relatively complex shapes, minimal residual porosity or shrinkage cavities due to the applied pressure, and minimization of interfacial reaction products between reinforcement and matrix (due to the shorter processing times). Before infiltration takes place, the reinforcement preform must be prepared. Figure 4.9 shows two processes for making the preform: (a) press forming and (b) suction forming. In the press forming process, an aqueous slurry of fibers is agitated and poured into a mold, pressure is applied to squeeze out the water, and the preform is dried. In the other process, suction

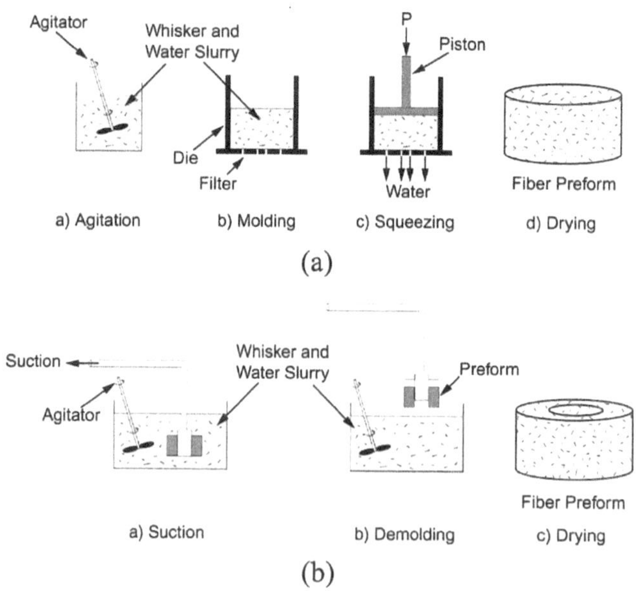

Fig. 4.9 Processes for making particulate performs: (a) press forming and (b) suction forming.

is applied to the mixture that consists of the reinforcement, fiber, and water. The mixture is then demolded and dried.

In order to obtain infiltration of the preform, the molten metal must have a relatively low viscosity and good wettability of the reinforcement. A schematic of the liquid infiltration process is shown in Figure 4.10. The reinforcement preform is placed in a mold, and the liquid is poured into a preheated die located on the bed of a hydraulic press. Infiltration takes place by mechanical force or by using a pressurized inert gas. Applied pressures on the order of 70-100 MPa are typically used. Having the preform temperature lower than that of the matrix liquidus temperature is highly desirable in order to minimize interfacial reaction and to obtain a fine matrix grain size.

Squeeze casting can also be used to obtain composites with relatively high reinforcement volume fractions (> 40%) (Mortensen et al., 1988; Saha et al., 2002). In conventional casting, achieving homogeneous particle packing and distribution at high volume fractions of reinforcement is problematic. Figure 4.11 shows the microstructure of a 6060 Al matrix composite reinforced with 50% SiC particles. Note the fine interparticle spacing and absence of any residual porosity. This technique has also been used to fabricate continuous fiber reinforced MMCs (Deve′ and McCullough, 1995). Figure 4.12 shows a microstructure of Al_2O_3 fibers (Nextel 610) in a pure Al matrix infiltrated by 3M. The distribution of fibers is relatively uniform, although some "channels" of pure matrix were formed during infiltration.

Although the squeeze casting technique has been primarily used for infiltrating low melting point metals, such as Al, continuous fiber reinforced intermetallic matrix composites such as Al_2O_3 reinforced TiAl, Ni_3Al, and Fe_3Al have also been processed (Nourbakhsh et al., 1990). The process is similar to that for discontinuously reinforced MMCs. A mixture of the matrix alloy with wetting-enhancing additives is melted in a crucible, while

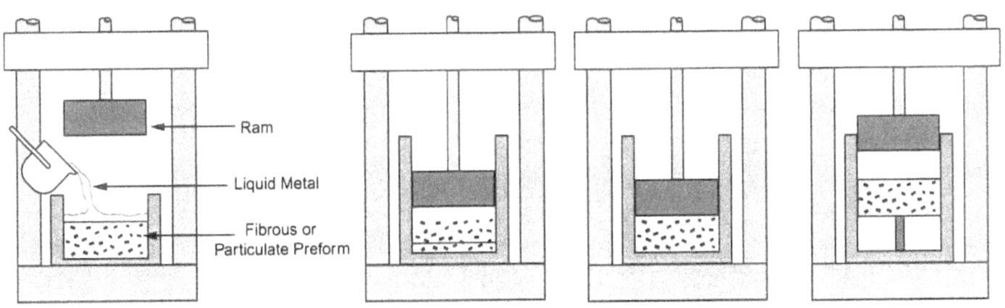

Fig. 4.10 Schematic of squeeze casting process.

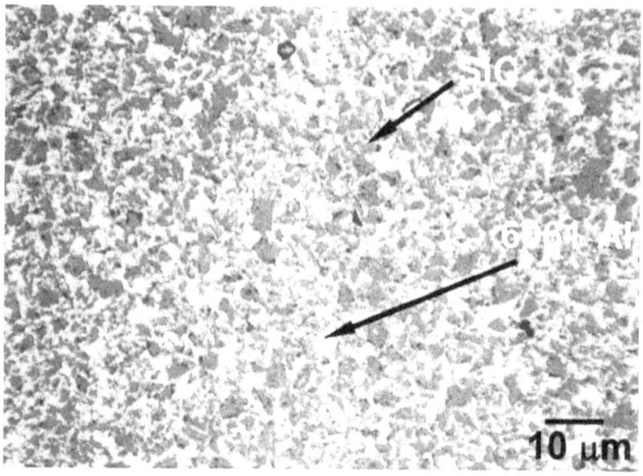

Fig. 4.11 Microstructure of 6061/SiC/50_p composite. Notice the large volume fraction of particles and relatively small interparticle spacing (Saha et al., 2000).

the fibrous preform is heated separately. The molten metal is then poured onto the fibers, and pressure is applied simultaneously, via argon gas, to infiltrate the preform.

Infiltration Mechanics

Infiltration of the liquid matrix through a particulate or fibrous preform can be thought of as liquid infiltration of a porous medium. As such, the capillarity of the liquid and its permeability in the preform will affect the infiltration characteristics of the metal through the porous reinforcement. The reader is referred to an excellent review by Michaud (1993) for details. Here we provide some salient points on the mechanics of infiltration in MMCs. Consider infiltration of a liquid phase with velocity v_L through a small volume ΔV, Fig. 4.13 (Michaud, 1993). The volume consists of fiber, matrix, and porosity, so the sum of the volume fractions, V_f, V_m, and V_p, is equal to 1. The velocity of the solid preform, v_s, is also taken into consideration, although we can neglect deformation of the preform during infiltration, so $v_s = 0$. A fraction of the liquid metal, g_s, will gradually solidify during infiltration so the volume fraction of solid matter in the composite (fibers plus solidified metal) is given by $V_{sf} = V_f + g_s V_m$.

We can now discuss the liquid infiltration processes, in terms of the parameters described above, using the Forscheimer equation, used to describe flow in a porous medium:

$$f - \nabla P = \left[\frac{\eta V_m (1 - g_s)}{K} + B \rho_m \sqrt{(v_l - v_s)^2} \right] (v_l - v_s)$$

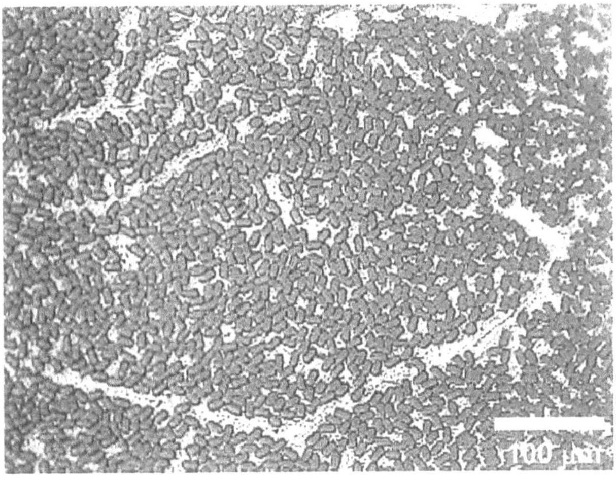

(a)

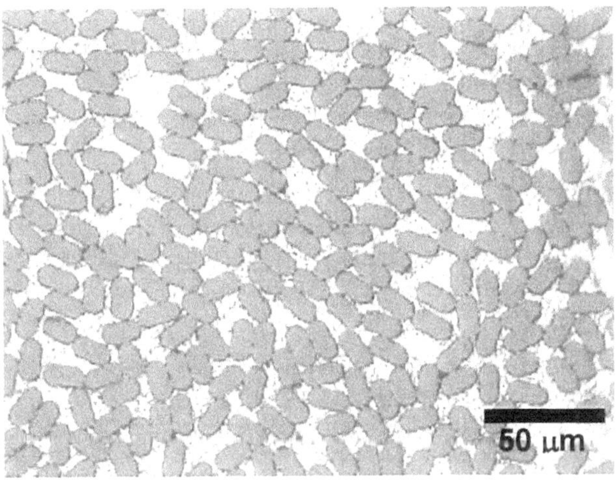

(b)

Fig. 4.12 Microstructure of liquid infiltrated Al_2O_3 fiber (Nextel 610) reinforced pure Al matrix composite. (a) low magnification and (b) high magnification. The distribution of fibers is relatively uniform, although some "channels" of pure matrix were formed during infiltration (indicated by arrows).

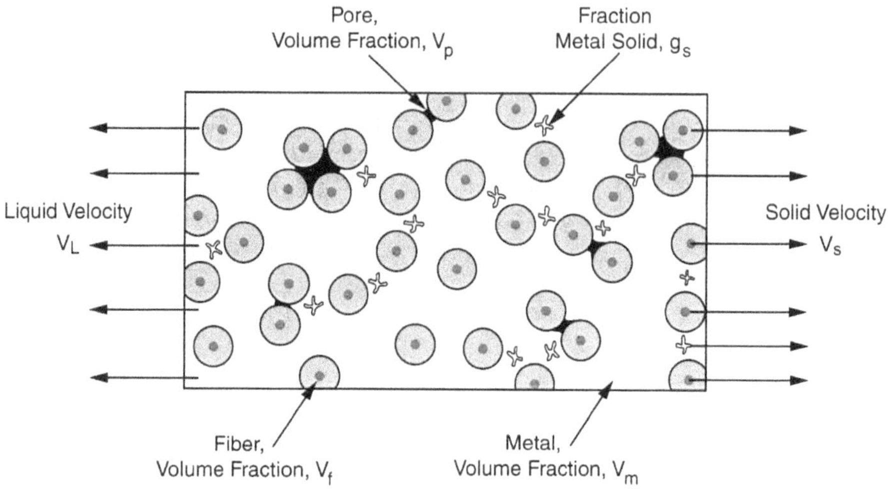

Fig. 4.13 Schematic of parameters associated with liquid infiltration of a metal matrix through a porous reinforcement preform (after Michaud, 1993).

where f is the volumetric gravitational, centrifugal, or electromagnetic force, P is the average pressure applied in the volume ΔV, η is the viscosity of the liquid metal, ρ_m is the density of the liquid metal, and v_l and v_s are the velocity of the liquid and solid, respectively. K is termed the permeability of the preform and B is a constant. K is a function of the reinforcement preform microstructure, wetting characteristics, reactivity, etc., as well as V_m, the matrix volume fraction, and characteristics of the solidified metal. The relevant Reynolds number for a preform with characteristic length d (for example, fiber diameter), is given by:

$$R_e = \frac{d\rho_m\sqrt{(v_l - v_s)^2}}{\eta V_f}$$

where V_f is the fiber volume fraction and the other symbols have the significance given earlier. When R_e is below a critical value of the Reynolds number, $R_{e,critical}$, (approximately equal to 1 for the preform considered here), the second term in the Forscheimer equation can be neglected, and we obtain Darcy's law:

$$v_l - v_s = \frac{K}{\eta V_m (1 - g_s)}(f - \nabla P)$$

The left-hand side of the equation can be thought of as a volumetric density of the fluid (i.e., $\frac{\text{volume}}{\text{area} \cdot \text{time}}$). It can be recognized that Darcy's law is an analog of Ohm's law for electrical conduction, i.e., permeability is analogous to electrical conductivity. Thus, in general, higher permeability, lower viscosity, and higher applied pressure will contribute to faster infiltration of the liquid matrix into the reinforcement preform.

Analysis of heat transfer during the matrix infiltration process is important because as the liquid metal infiltrates and solidifies it will release heat into its surroundings. The heat flow will cause an increase in temperature of the system, allowing incoming liquid to remain in the liquid state while infiltrating the preform. In practice, liquid flow ceases when enough cooling from the preform and/or partially solidified composite takes place, or when the preform is completely filled. In general, heat transfer within the volume element ΔV is dictated by conduction, convention, exothermic characteristics of interfacial reactions, and the fraction of total solidified metal, g_s. The governing equation for this heat transfer is (Michaud, 1993):

$$\nabla \cdot (k_c \nabla T) = \rho_c c_c \frac{\partial T}{\partial t} + \rho_m c_m V_m (1 - g_s) v_l \cdot \nabla T$$
$$+ (\rho_f c_f V_f + \rho_m c_m V_m g_s) v_s \cdot \nabla T - \rho_m \Delta H \frac{\partial (g_s V_m)}{\partial t} - \dot{Q}$$

where k_c is the thermal conductivity of the composite, ΔH is the latent heat of solidification of the metal, and $\dot{Q}$ is the rate of heat released from any chemical reaction. ρ and c refer to the density and heat capacity, respectively, and the subscripts f, m, and, c denote the reinforcement, metal, and composite, respectively.

Mass transfer also needs to be considered when analyzing the liquid infiltration process. Because of the relatively short time scales involved with most infiltration processes, the contribution from diffusion to mass flow within ΔV can be neglected. Rather, convection plays a more prominent role, due to solution or rejection of a solute in the metal (for an alloy) and consumption during an interfacial chemical reaction. The governing relation for mass transport is:

$$\frac{\partial \overline{C}}{\partial t} = -\nabla \cdot [(1 - g_s) C_l v_l + g_s C_s v_s] + r_A$$

where $\overline{C}$ is the average matrix composition, r_A is the rate of change in solute due to the chemical reaction, and C_l and C_s are the composition of the liquid and solid phases, respectively, as determined from the phase diagram of the alloy at a given temperature.

Microstructure Evolution during Liquid Phase Infiltration

The microstructure of the composite during infiltration depends on the local solidification and cooling processes within the preform. The liquid metal will solidify when it comes in contact with the preform, which is at a lower temperature. A schematic of the infiltration front, temperature distribution in the preform, and grain size distribution in the matrix is shown in Figure 4.14 (Mortensen and Jin, 1992). At the infiltration front, the metal will solidify first (Region 1), while in the middle of the preform (Region 2), a relatively constant temperature is obtained. The semi-solid metal close to the infiltration gate (Region 3), however, will continuously be in contact with the liquid metal. The situation is more complicated for a binary eutectic alloy. The infiltration front, temperature distribution and liquid metal composition, for a binary alloy matrix composite, are shown in Figure 4.14(b). The temperature gradient is similar to that of the pure metal, but a significant amount of solute segregation takes place at the infiltration front. Thus, the remaining incoming liquid is depleted of solute.

Grain size distribution will also be affected by the local temperature distribution and cooling rate. For infiltration of the pure alloy, Figure 4.14(b), in Region 1 a fine matrix grain size is obtained because the molten metal comes into contact with the cooler reinforcement preform, which results in rapid solidification of the matrix. Heterogeneous nucleation at the reinforcement does not occur in most MMCs processed by liquid phase (Mortensen and Jin, 1992). In most Al matrix composites, for example, the grain size is usually much larger than the reinforcement size. The exception to this is in Al-Si matrix composites, where the primary Si phases nucleates preferentially at the reinforcement. Thus, in these composites, the composite has a finer distribution of Si phase, compared to the unreinforced alloy. Figure 4.15 shows the cell size in 6061 matrix composites, reinforced with 10 and 20% SiC particles (Lloyd, 1989). Note that the composite has a finer overall microstructure than the unreinforced alloy, at all cooling rates.

In some cases, particularly where fiber reinforcement is involved, the reinforcement actually impedes heat convection from the liquid metal, resulting in a larger degree of columnar dendritic growth than in the unreinforced alloy (Cole and Bolling, 1965). In Region 2, the grain size is slightly larger because of the higher temperatures and somewhat lower cooling rate. Finally, at the gate a coarser microstructure will result, as this

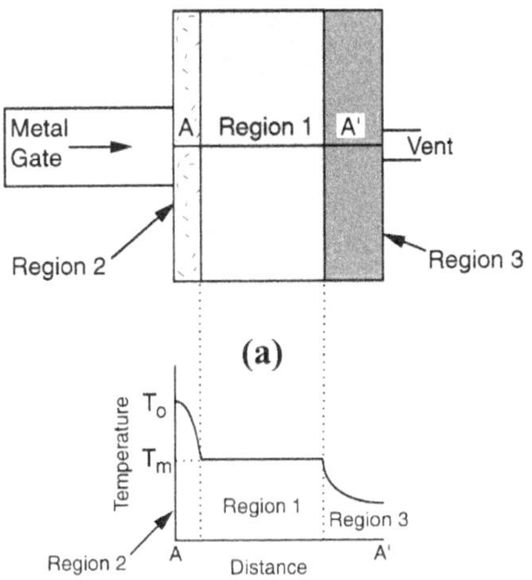

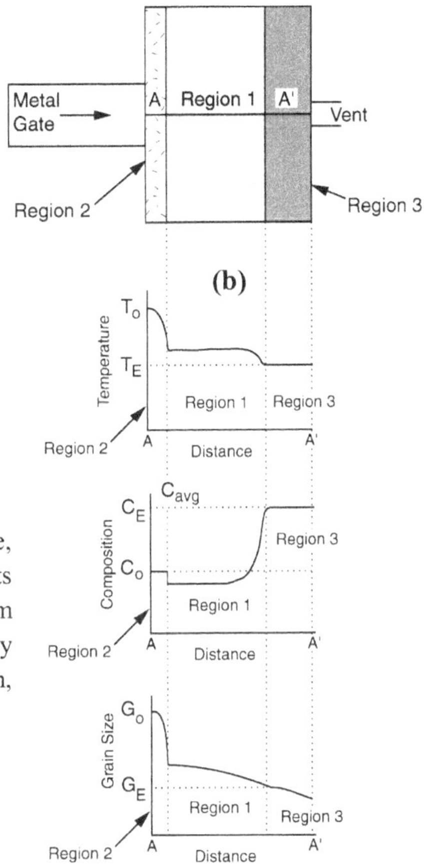

Fig. 4.14 Schematic of temperature, compositional, and grain size gradients during infiltration of reinforcement preform with: (a) pure metal matrix and (b) alloy metal matrix (after Mortensen and Jin, 1992).

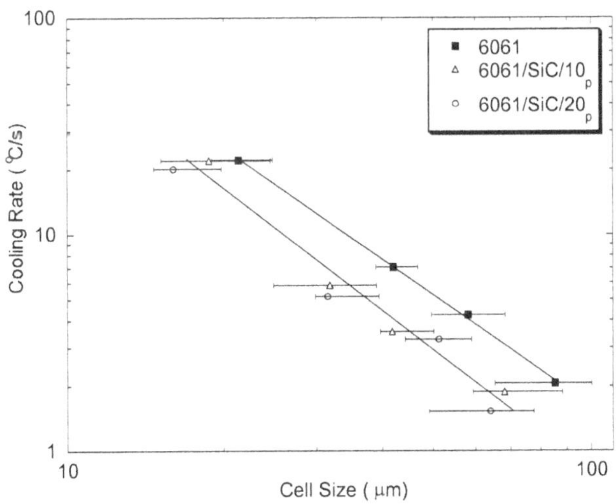

Fig. 4.15 Cooling rate vs. matrix cell size in Al 6061 matrix composites, reinforced with 10 and 20% SiC particles (Lloyd, 1989). Note that the composite has a finer overall microstructure than the unreinforced alloy, at all cooling rates, because of pinning of the reinforcement particles.

region will be the last to solidify and will cool at a relatively slow rate. Figure 4.16 shows the microstructure of a Saffil (Al_2O_3) fiber reinforced Al-4.5Cu matrix, where the matrix microstructure well within the preform is much finer than that close to the infiltration gate.

The microstructure of the matrix is not only a function of the temperature and composition gradients, but also on the velocity of the liquid and the interfiber spacing. Shekhar and Trivedi (1989, 1990) studied the solidification processes in model fiber and particle-containing organic transparent systems. In fiber reinforced systems, as the interfiber spacing approaches the primary dendrite or cell spacing, perturbations in the advancing liquid cause morphological changes upon solidification. The effect of velocity and interfiber spacing (indicated by the ratio of interfiber spacing, d, to the primary dendrite spacing λ) can be represented by a

microstructure map, Figure 4.17 (Trivedi et al., 1990). For very large interfiber spacing, the morphology is dependent on the liquid velocity, going from planar to cellular to dendritic with increasing velocity. When the interfiber spacing is very small, the cellular regime is larger since both planar and dendritic structures will transform to cellular structures. At intermediate spacing, dendritic microstructures will be observed at lower velocities, when the ratio of d/λ is slightly greater than 1.

4.1.3 Spray Co-deposition

Spray deposition has been used for some time to fabricate metallic alloys in powder form (Lavernia et al., 1992). The metal or alloy is melted and the liquid stream is atomized with water or an inert gas. Rapid solidification of the liquid takes place, resulting in a fine solid powder. This technique has been modified, by injecting reinforcement particles or co-depositing the particles with the matrix alloy, Figure 4.18, (Lloyd, 1997). The advantage of this technique is the high rate of production, which can approach 6-10 kg/min, and the very fast solidification rate, which minimizes any reaction between particle and matrix. The as-processed billet is not fully dense, so secondary processing is required to fully densify and homogenize the composite. The distribution of particles in the liquid droplet is very much dependent on the size of the reinforcement and at what point in the process the reinforcement is injected into the matrix (Lloyd, 1997). Whiskers, for example, are too fine to be optimally injected and distributed in the liquid droplets. When the particles are injected into the matrix immediately after atomization, the matrix is still liquid, so the particles are able to envelope themselves into the liquid droplet, and a relatively homogenous distribution

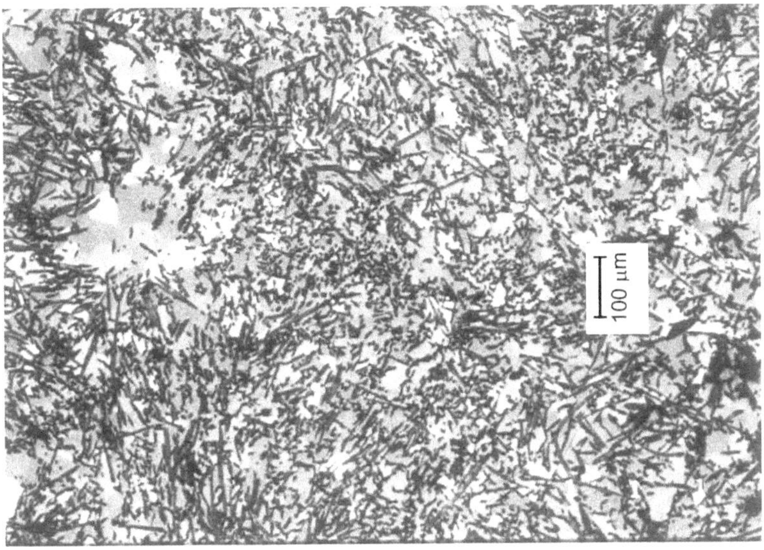

(a)

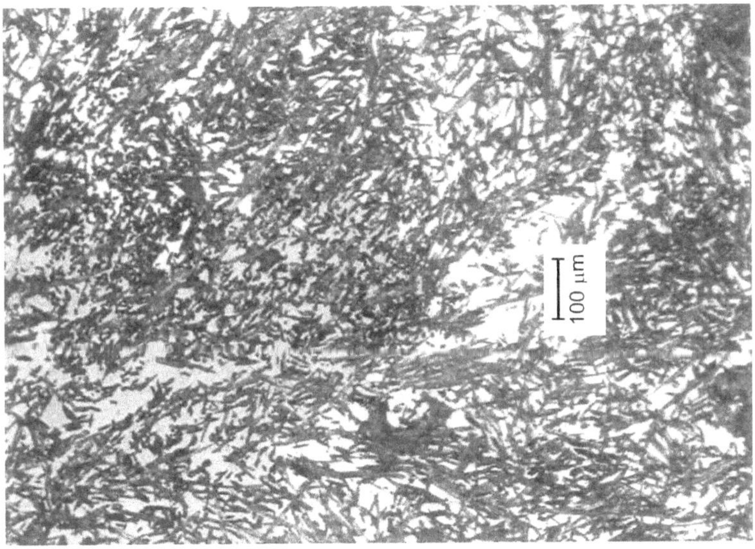

(b)

Fig. 4.16 Microstructure of a Saffil (Al_2O_3) fiber reinforced Al-4.5Cu matrix. The matrix grain size well within the preform (a), is much finer than that close to the infiltration gate (b), since the latter goes through remelting and much slower cooling (courtesy of V. Michaud).

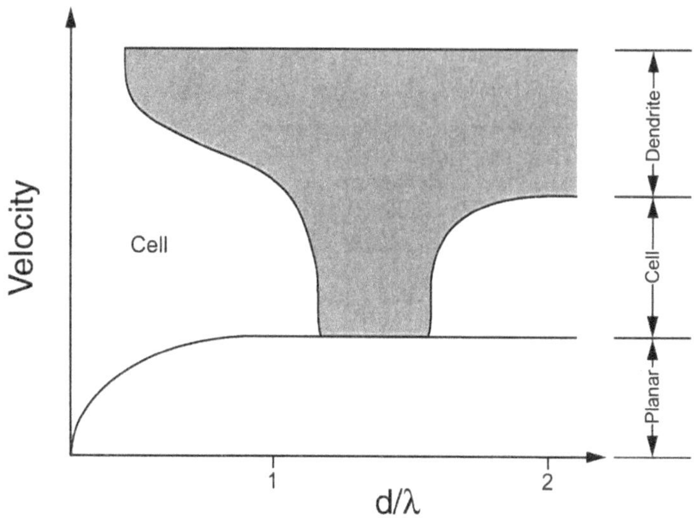

Fig. 4.17 Effect of solidification front velocity and interfiber spacing (represented by the ratio of interfiber spacing, d, to dendrite spacing, λ) on matrix microstructure. For very large d/λ, the morphology goes from planar to cellular to dendritic with increasing velocity. When the interfiber spacing is very small, the cellular regime is larger since both planar and dendritic structures will transform to cellular structures (after Trivedi et al., 1990).

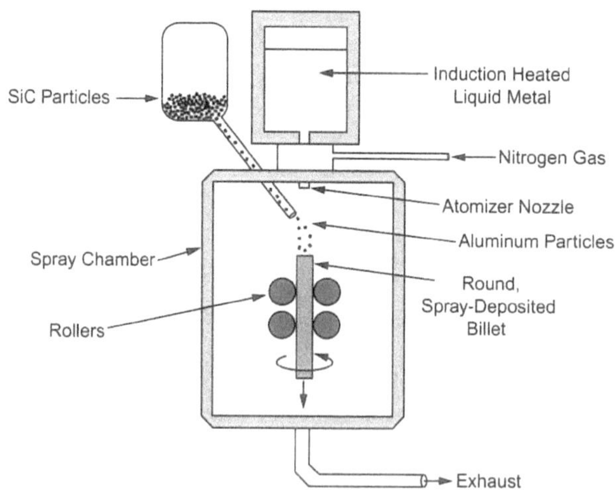

Fig. 4.18 Spray co-deposition of SiC particles and Al liquid droplets, to form composite particles.

of particles in the matrix is obtained, Figure 4.19(a). Injection of the particles later in the process, i.e., when the matrix droplet is semisolid, does not allow easy ingression of the particles into the matrix, causing the particles to reside along the periphery of the matrix particle, Figure 4.19(b). A consolidated composite with the latter particle size distribution will have inferior properties as a result of a high degree of particle clustering. This process is quite flexible in that the metal and reinforcement sprayers can be tailored to obtain in situ laminates or even functionally graded materials. The process can be quite expensive, however, owing to the high cost of the capital equipment.

4.1.4 *In Situ* Processes

In situ processes fall into two major categories: Reactive and non-reactive processes. In the reactive processes two components are allowed to react exothermically to form the reinforcement phase. The XD process is an example of this process (Martin Marietta, 1987). Typically, a rather high volume fraction of ceramic particles is formed in the matrix alloy, and this master alloy is diluted with the matrix alloy to obtain a composite of desired reinforcement volume fraction. Typical examples involve TiB and TiC as reinforcement particles, which are formed by the following reactions:

$$2B + Ti + Al \rightarrow TiB_2 + Al$$

and

$$C + Ti + Al \rightarrow TiC + Al$$

Processing variables such as reaction temperature can be used to tailor the desired reinforcement particle size, which is usually in the 0.25-1.5 μm

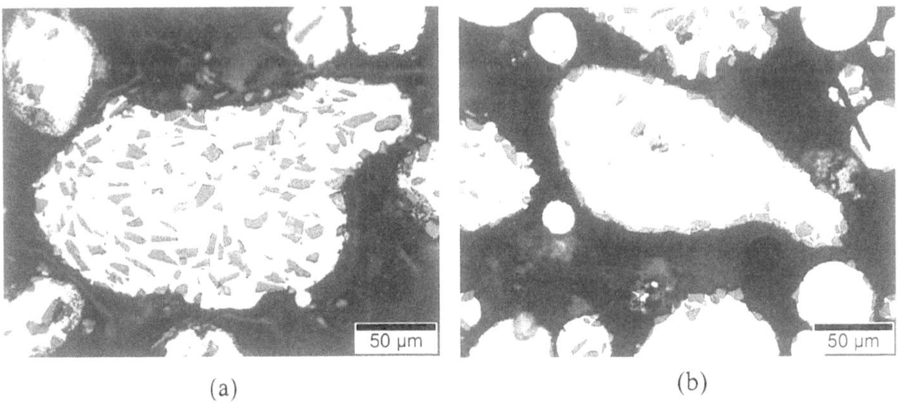

Fig. 4.19 Microstructure of composite particles obtained from spray co-deposition: (a) homogeneous particle distribution for particle injection while matrix is in liquid state and (b) particles at periphery of matrix droplet, due to particle injection when the matrix is in a semi-solid state (courtesy of D. Lloyd).

range. The matrix commonly is Al, Ni, or an intermetallic matrix. An alternative process that combines spray co-deposition and reaction is to atomize an Al-Ti alloy (at sufficiently high temperature) with a C-containing gas to form TiC particles (Chawla, 1997). The advantage of *in situ* reaction processes, in general, is that the reaction eliminates problems typically associated with wetting of the particle, so a relatively clean and strong interface is typically formed (Christodolou et al., 1988). The number of composite systems where reaction processing is beneficial, however, is limited, and the relatively fine size distribution of particles produced can significantly increase the viscosity of the melt.

Non-reactive *in situ* processes take advantage of two-phase systems, such as eutectic or monotectic alloys, to form the fiber and matrix in situ (McLean, 1983). Controlled directional solidification is conducted to separate the two phases, as shown in Fig. 4.20. A precast and homogenized material is melted in a graphite crucible, and contained in a quartz tube in vacuum or inert gas atmosphere. Heating is typically conducted by induction and the thermal gradient is obtained by chilling the crucible. Electron beam heating may also be used, particular when using reactive metals such as titanium. One can control the fineness of the microstructure, i.e., the size and spacing of the reinforcement (at constant volume fraction) by controlling the solidification rate. These rates typically lie in the range of 1-5 cm/h due to the need to maintain a stable solidification growth front. Figure 4.21 shows a cross-section of the microstructure of an *in situ*-processed composite.

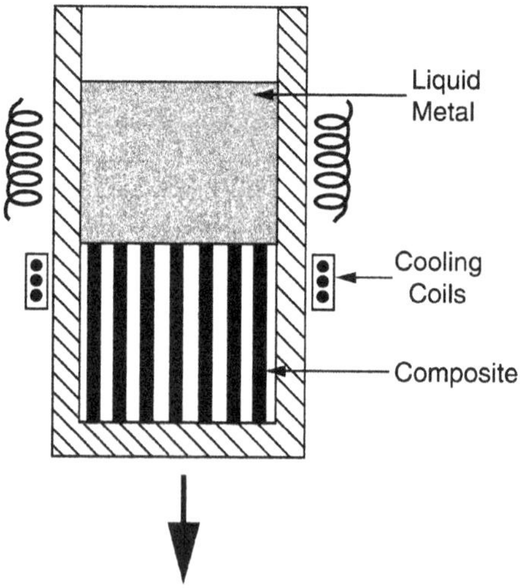

Fig. 4.20 Schematic of directional solidification process to obtain *in situ* composites.

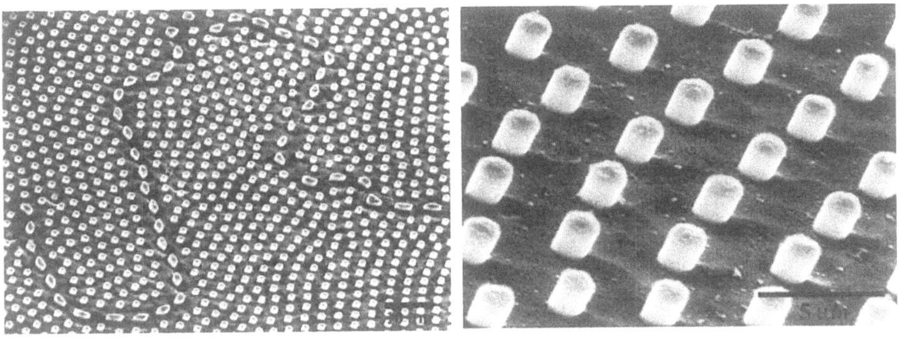

(a) (b)

Fig. 4.21 Cross-section of the microstructure of an *in situ* processed composite: (a) lower magnification and (b) higher magnification. The matrix has been etched out locally to show the fibers. Note that the fibers at grain boundaries are somewhat coarser due to earlier nucleation and growth, relative to the fibers within the grain (courtesy of J.B. Andrews).

4.2 SOLID STATE PROCESSING

The main drawback associated with liquid phase techniques is the difficulty in controlling reinforcement distribution and obtaining a uniform matrix microstructure (Michaud, 1993). Furthermore, adverse interfacial reactions between the matrix and the reinforcement are likely to occur at the high temperatures involved in liquid processing. These reactions can have an adverse effect on the mechanical properties of the composite (Sahoo and Koczak, 1991; Chawla, 1997). The most common solid phase processes are based on powder metallurgy techniques (Ghosh, 1993). These typically involve discontinuous reinforcements, due to the ease of mixing and blending, and the effectiveness of densification. The ceramic and metal powders are mixed, isostatically cold compacted, and hot-pressed to full density. The fully-dense compact then typically undergoes a secondary operation such as extrusion or forging (Lloyd, 1997). Novel low-cost approaches, such as sinter-forging, have aimed at eliminating the hot pressing step, with promising results (Chawla et al., 2003).

4.2.1 Powder Metallurgy Processing

Powder processing involves cold pressing and sintering, or hot pressing to fabricate primarily particle- or whisker-reinforced MMCs (Hunt, 1994). The matrix and the reinforcement powders are blended to produce a homogeneous distribution. Figure 4.22 shows a blended powder mixture of Al, 50Al-50Cu (wt.%), and SiC particles. The 50Al-50Cu articles are added to provide a uniform distribution of the alloying additions, while the pure Al

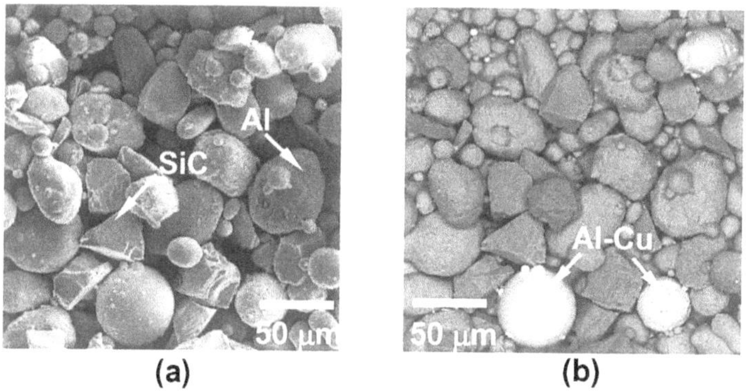

Fig. 4.22 Powder mixture of Al, SiC, and 50wt.%Al-50wt.%Cu particles (Chawla et al., 2003): (a) secondary electron image, and (b) backscattered electron image, showing the Al-Cu particles.

particles enable easier cold pressing. The blending stage is followed by cold pressing to produce what is called a green body, which is about 80% dense and can be easily handled, Fig. 4.23. The cold pressed green body is canned in a container, sealed, and degassed to remove any absorbed moisture from the particle surfaces. One of the problems with bonding metallic powder particles, such as Al particles, to ceramic particles, such as SiC, or to other Al particles is the oxide "skin" that is invariably present on the Al particle surface (Kim et al., 1985; Anderson and Foley, 2001). Degassing and hot pressing in an inert atmosphere contributes to the removal of Al hydrides present on the particle surface, making the oxide skin more brittle and, thus, more easily sheared (Estrada et al., 1991; Kowalski et al., 1992). The material is hot pressed, uniaxially or isostatically, to produce a fully dense composite and extruded. The rigid particles or fibers do not deform, causing the matrix to be deformed significantly.

The ratio of the reinforcement particle size to matrix particle size is very important in achieving a homogenous distribution of particles in the matrix. A comparison of the effect of Al-to-SiC particle size ratio on the composite microstructure is shown in Fig. 4.24. With the larger Al particle size, the SiC particles are "pushed" and packed in the interstices between the larger Al particles, yielding a more clustered microstructure, Fig. 4.24(a). A particle size ratio closer to one yields a more homogeneous microstructure, Fig. 4.24(b). The degree of homogeneity in particle distribution can be quantified by a clustering parameter. Several techniques have been used to quantify SiC particle clustering in metal matrix composites (Dirichlet, 1850; Lewandowski et al., 1989; Spowart et al., 2001; Yang et al., 2001). One important technique is called tesselation and is described in detail in Chapter

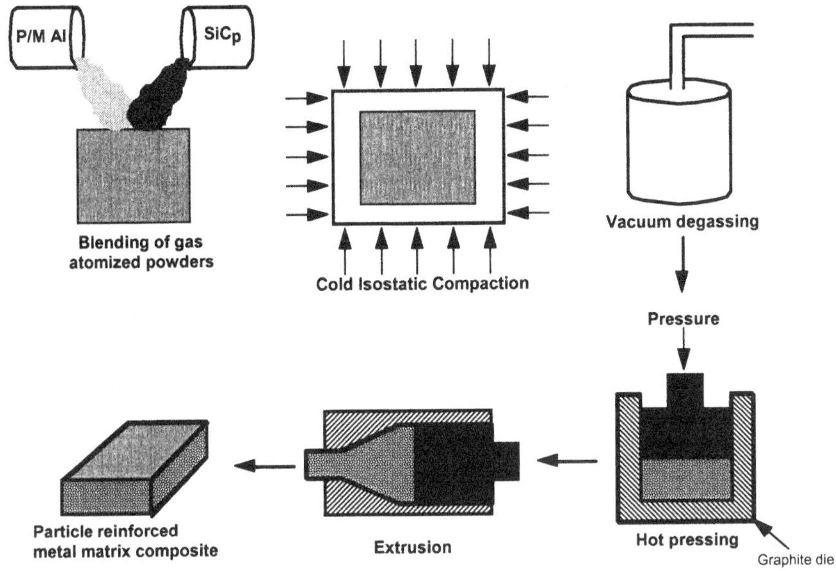

Fig. 4.23 Powder processing, hot pressing, and extrusion process for fabricating particulate or short fiber reinforced MMCs.

7. Measurement of the nearest-neighbor distance of each SiC particle is the simplest means of assessing particle clustering but it can be deceptive, since the mean nearest neighbor spacing ignores the distribution in nearest neighbor spacings in the microstructure. Yang et al. (2001) showed that the coefficient-of-variance of the mean near-neighbor distance (COV_d) is particularly sensitive and effective in characterizing particle clustering. This parameter is also relatively insensitive to particle volume fraction, size, and morphology. COV_d can be described by the following equation (Yang et al., 2001):

$$COV_d = \frac{\sigma_d}{d}$$

where σ_d^2 is the variance in the mean nearest-neighbor distance, and d is the average of the mean near-neighbor distance of the particles sampled.

The degree of clustering, as measured by COV_d, as a function of Al:SiC particle size ratio is shown in Fig. 4.25. For large particle size ratios, i.e., Al particle size much greater than SiC, or vice-versa, the degree of clustering is relatively high. It is interesting to note that the minimum degree of clustering (COV_d) tends to a particle size ratio of about one.

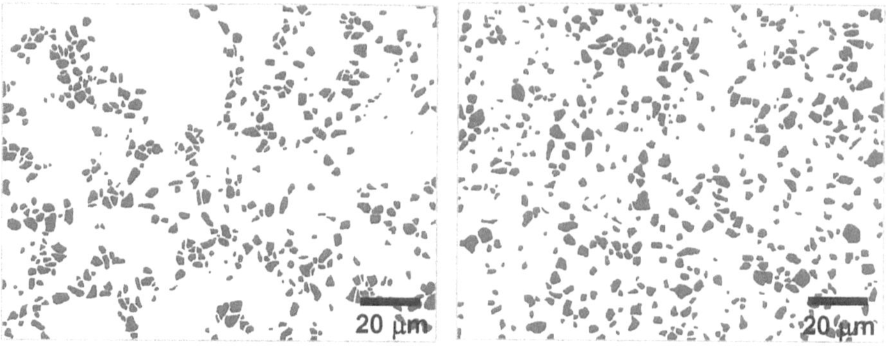

Fig. 4.24 Particle clustering in Al/SiC/15_p with different Al-SiC particle size ratios: (a) Al-SiC ratio of 6.6 (d_{Al} = 33 µm, d_{SiC} = 5 µm) and (b) Al-SiC ratio of 1.4 (d_{Al} = 7 µm, d_{SiC} = 5 µm). Increasing the Al:SiC ratio results in a greater degree of SiC clustering.

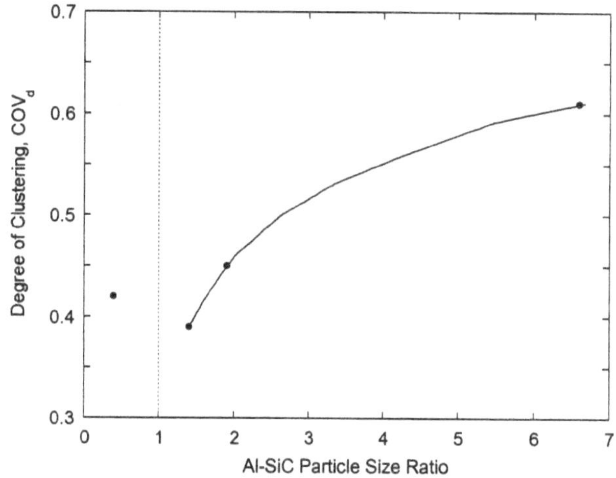

Fig. 4.25 Degree of clustering versus Al-to-SiC particle size ratio. The degree of clustering tends to a minimum at a particle size ratio close to one.

4.2.2 Extrusion

Extrusion processing has been used extensively as a means of secondary deformation processing of MMCs (Ghosh, 1993; Hunt, 1994; Lloyd, 1997). It is particularly advantageous because the combination of pressure and temperature results in shear between Al/Al particles and Al/SiC particles, which contributes to fracture of the oxide skin on the Al particles, and the bonding between particle and matrix is enhanced. Because of the large strains associated with this process, however, extrusion has been used primarily to consolidate composites with discontinuous reinforcement, in order to minimize reinforcement fracture. Even in discontinuously reinforced materials, fracture of short fibers or particles often takes place, which can be detrimental to the properties of the composite.

Three types of extrusion are shown in Fig. 4.26, direct, conventional, and hydrostatic. In direct extrusion, the extruded material is deformed against a flat plate with a small orifice. This causes a "dead-metal zone" which consists of a region where the metal cannot flow. In conventional extrusion, the die is tapered to minimize the dead-metal zone. Die friction is present, however, which will result in higher deformation stresses at the metal/die interface. In MMCs, this effect is particularly exacerbated due to the presence of the particles and the large shear stresses at the metal/container interface result in material discontinuities. This can result in ragged edges, called the "christmas tree" effect, Fig. 4.27. This is more predominant with an increase in reinforcement particle content, due to the increase in die friction and, thus, shear stresses. Finally, the problems of die friction can be minimized by conducting the extrusion inside a high pressure fluid. This results in a close to hydrostatic stress state, and minimal effect of friction in the billet because the contact area between material and die is minimized.

Several microstructural changes take place during extrusion. These include alignment of particles along the extrusion axis, particle fracture (depending on the reinforcement particle size and the strains involved), and refinement and recrystallization of matrix grains (Chawla et al., 1998; Tham et al., 2002; Ganesh and Chawla, 2004; Ganesh and Chawla, 2005). Figure 4.28 shows the microstructure of SiC particle reinforced Al alloy at three different reinforcement volume fractions: 10, 20, and 30%. Note the preferred orientation of reinforcement particles along the extrusion axis.

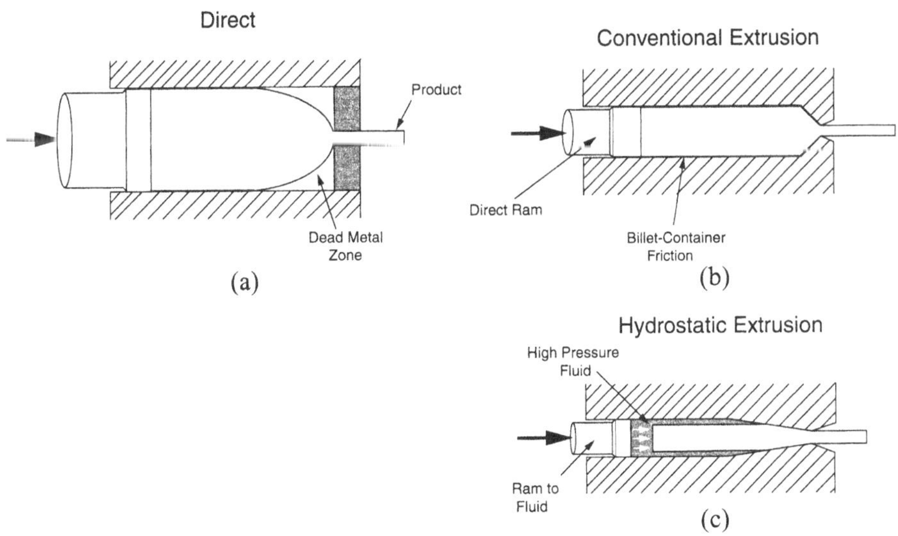

Fig. 4.26 Three types of extrusion processing: (a) direct, (b) conventional, and (c) hydrostatic. The latter minimizes die-friction and "dead-metal" zones.

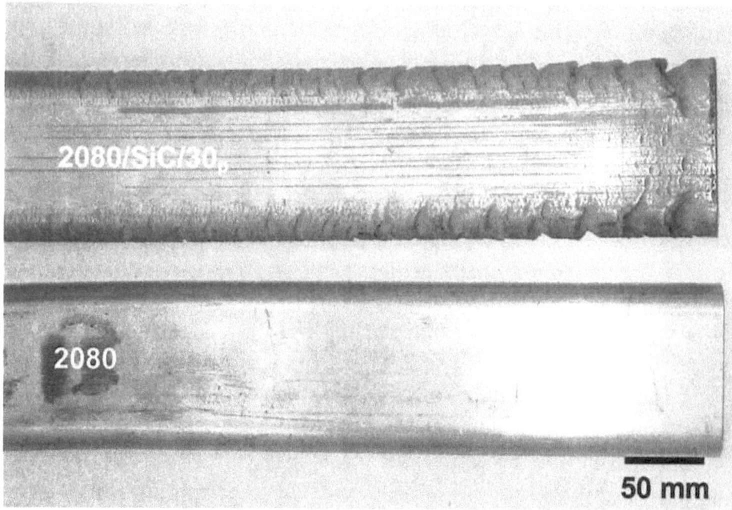

Fig. 4.27 Comparison of extruded Al 2080 alloy and 2080/SiC/30_p. Note the ragged edges or "christmas tree" effect due to enhanced die friction in the composite.

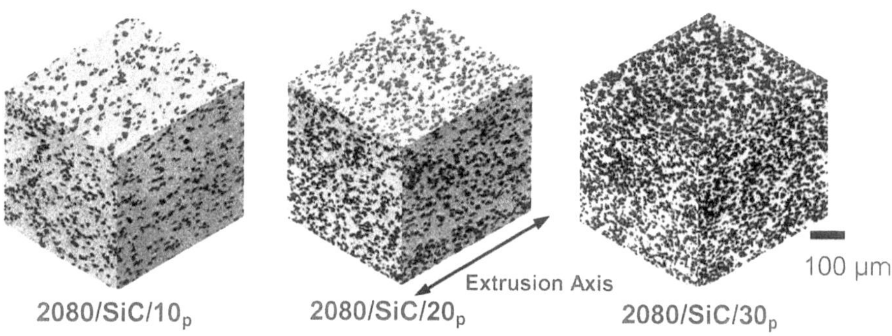

Fig. 4.28 Microstructure of SiC particle reinforced Al alloy for three reinforcement volume fractions: 10, 20, and 30 %. Note the preferred orientation of reinforcement particles along the extrusion axis (Ganesh and Chawla, 2004).

Quantitative analysis of the degree of orientation of the particles, defined as the angle of a given particle to the longitudinal or transverse axis, shows that the degree of alignment of particles in the longitudinal plane, at a given volume fraction of particles, was much higher than that in the transverse plane, Fig. 4.29. However, the degree of orientation in the longitudinal plane decreased with an increase in volume fraction of particles. This can be explained by noting that the larger the fraction of particles, the lower the mean free path available for particle rotation and alignment along the extrusion axis. The anisotropy in particle alignment has a profound impact on Young's modulus and tensile strength (Logsdon and Liaw, 1986; Ganesh and Chawla, 2005), and these are described in Chapter 7. Extrusion-induced particle fracture can reduce the strength of the composite significantly, often to levels below that of the unreinforced alloy (see Chapter 7).

The incorporation of reinforcement particles also has an "indirect" effect on the metal matrix grain size. As described above, with the incorporation of the hard reinforcement particles, a larger degree of plastic flow is required to deform the matrix around the particles (Chawla et al., 1998). This also results in an overall refinement in matrix grain size, shown in Fig. 4.30. The

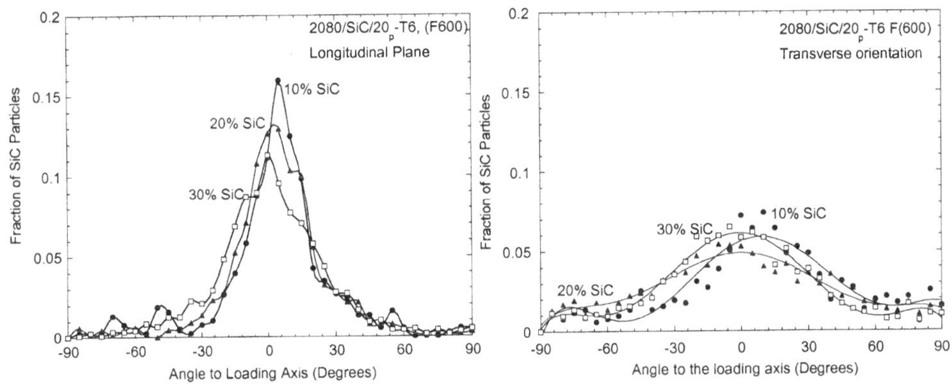

(a) (b)

Fig. 4.29 Quantitative analysis of the degree of orientation of the particles (orientation is defined by the angle of a given particle to loading axis): (a) longitudinal and (b) transverse. The degree of alignment of particles in the longitudinal plane, at a given volume fraction of particles, is much higher than that in the transverse plane (Ganesh and Chawla, 2004).

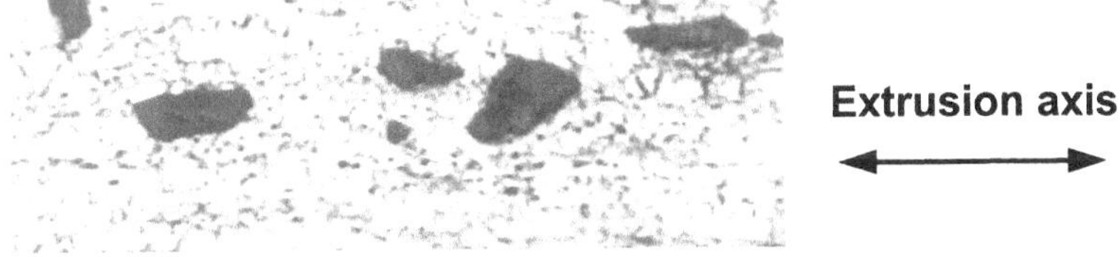

Fig. 4.30 Microstructure of extruded 2080/SiC/10_p composite (Ganesh and Chawla, 2004). The matrix grain size is refined due to the constraint on deformation from the SiC particles. Dynamic recrystallization during hot extrusion results in a finger grain structure at the particle/matrix interface.

grain size in the composite is also inhomogeneous, being smaller near the particle/matrix interface and increasing in size with increasing distance from the particle/matrix interface. The large degree of shear produced by extrusion between particles and matrix is conducive to a very strong mechanical bond between the particle and matrix. A strong mechanical bond between particle and matrix is highly desirable in strengthening of MMCs, because it maximizes the degree of load transfer from the matrix to the particle, and thus, increases the chance that a given particle will be loaded to its fracture stress (Williams et al., 2002).

During hot extrusion the particles also act as nucleating sites for recrystallization of new matrix grains, which results in a much finer grain size along the particle/matrix interface (Liu et al., 1989; Humphreys et al., 1990). This is shown in the orientation imaging map of the matrix grains in the composite, Fig. 4.31. Macroscopically, the matrix exhibits a [100]<111> texture, typical of deformation processed FCC materials. Perpendicular to the extrusion direction, the texture is random, indicating an overall fibrous texture in the material. A closer look at the grain orientation at the particle/matrix interface shows that the grains at the interface are randomly oriented, due to dynamic recrystallization, Fig. 4.32. The grains further away from the interface, however, exhibit the typical [100]<111> texture.

A disadvantage of the conventional powder metallurgy process followed by extrusion is the high cost of the process. The high cost is driven primarily by the difficulty in machining these wear-resistant composites and the material waste generated to produce a component. Thus, while the extrusion process has been used to fabricate composites for very demanding applications, particularly in the aerospace sector, it is not conducive for applications where low cost and high volume are as important as performance.

4.2.3 Forging

Forging is another common secondary deformation processing technique used to manufacture metal matrix composites. Once again, this technique is largely restricted to composites with discontinuous reinforcement. In conventional forging, a hot-pressed or extruded product is forged to near-net shape (Helinski et al., 1994).

A new, low cost sinter-forging technique has been developed whereby a powder mixture of the reinforcement and matrix powder is cold compacted, sintered, and forged to nearly full density, see Fig. 4.33 (Chawla et al., 2003). The main advantage of this technique is that forging is conducted to produce a near-net shape material, and machining operations and material

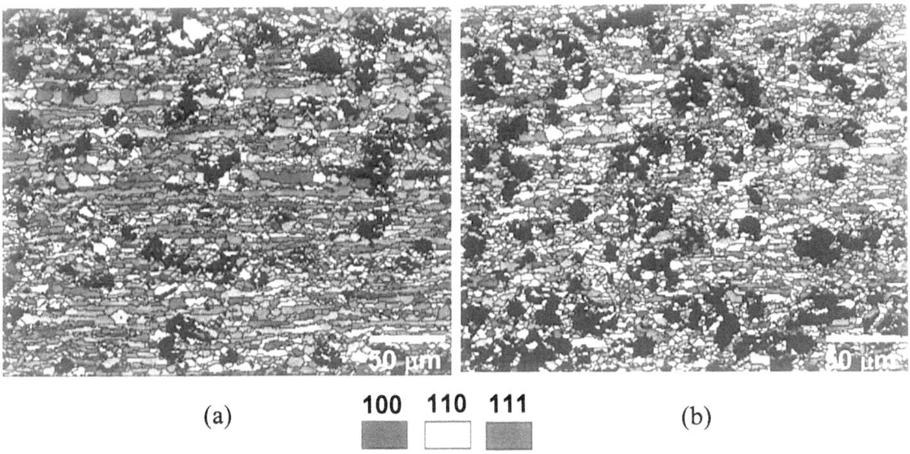

Fig. 4.31 Orientation imaging map of matrix grains in a 2080/SiC$_p$ composite: (a) parallel to extrusion axis and (b) perpendicular to extrusion axis. The matrix exhibits a [100[<111> texture, typical of deformation processed FCC materials. Perpendicular to the extrusion direction, the texture is random, indicating an overall fibrous texture.

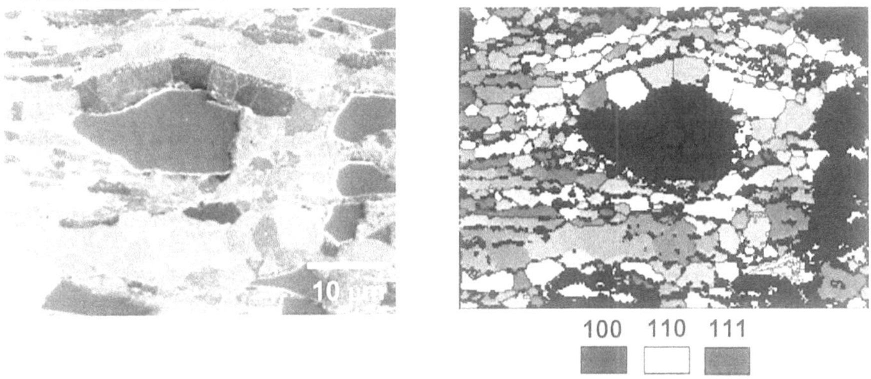

Fig. 4.32 Grain orientation at the particle/matrix interface showing that the grains at the interface are randomly oriented, due to dynamic recrystallization.

waste are minimized. A similar process is used to manufacture ferrous connecting rods for large volume passenger car applications (James, 1985).

The microstructure of forged composites exhibits some preferential alignment of SiC particles perpendicular to the forging direction. A comparison with extruded composites, of similar particle size, indicated that the alignment in the sinter-forged composites was not as significant as that produced by extrusion (Chawla et al., 2003). This is to be expected since a much larger amount of plastic deformation is induced during extrusion. The low cost, sinter-forged composites have tensile and fatigue properties that are comparable to those of materials produced by extrusion (Chawla et al., 2003).

The morphology of the Al grains is also affected by forging. The grains are "pancake-shaped," Fig. 4.34, with the major axis of the grains being perpendicular to the forging axis. The observed anisotropy appears to be a direct result of the constraint on lateral deformation provided by the closed die forging, so the grains are unable to deform equally in all directions.

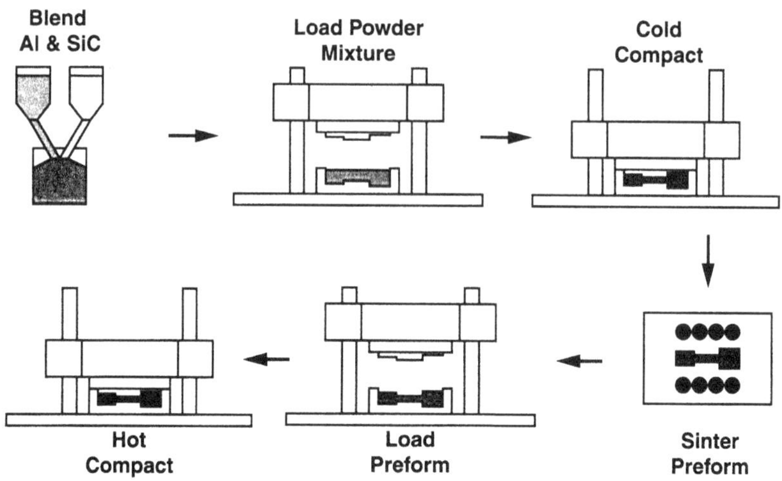

Fig. 4.33 Schematic of near net-shape sinter-forging technique. A mixture of reinforcement powder and matrix powder is cold compacted, sintered, and forged to nearly full density (Chawla et al., 2003).

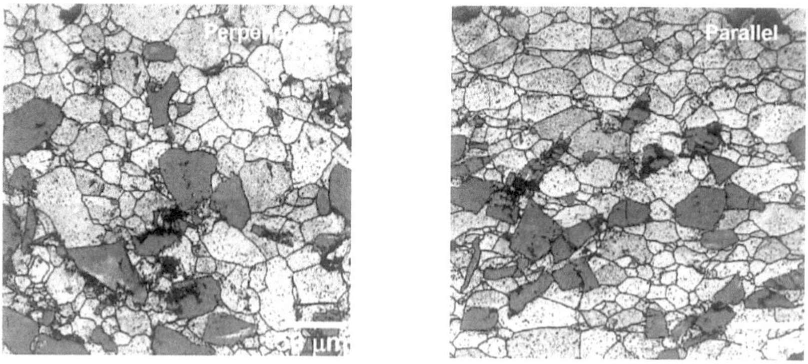

Fig. 4.34 Morphology of Al matrix grains after forging (Chawla et al., 2003): (a) perpendicular to forging axis and (b) parallel to forging axis. The latter shows that the grains are "pancake-shaped," with the major axis of the grain being perpendicular to the forging axis.

4.2.4 Pressing and Sintering

A relatively inexpensive and simple technique involves pressing and sintering of powders. These composite systems are typically sintered in a temperature range to obtain some degree of liquid phase. The liquid phase flows through the pores in the compact resulting in densification of the composite (unless interfacial reaction takes place). Special mention should be made of WC/Co composites, commonly known as cemented carbides. They are really nothing but very high volume fraction of WC particles distributed in a soft cobalt matrix. These composites are used extensively in machining and rock and oil drilling operations. They are made by liquid phase sintering because liquid cobalt wets WC very well, the contact angle being 0°, with little or no interfacial reaction. Figure 4.35 shows the microstructure of such a composite, 90% WC and 10% Co. Note the angular nature of the WC particles (Deng et al., 2001).

4.2.5 Roll Bonding and Co-extrusion

Roll bonding is a common technique used to produce a laminated composite consisting of different metals in layered form (Chawla and Godefroid, 1984). Such composites are called sheet laminated metal-matrix composites. Roll bonding and hot pressing have also been used to make laminates of Al sheets and discontinuously reinforced MMCs (Hunt et al., 1991; Manoharan et al., 1990). Figure 4.36 shows the roll bonding process of making a laminated MMC. Figure 4.37 shows a transmission electron micrograph (TEM) of a roll-bonded Al 1100 and Al 2024 laminated composite, which exhibits the interface with very good integrity (Chawla, 1991). A larger dislocation density is observed in the Al 1100 layer, which underwent a larger

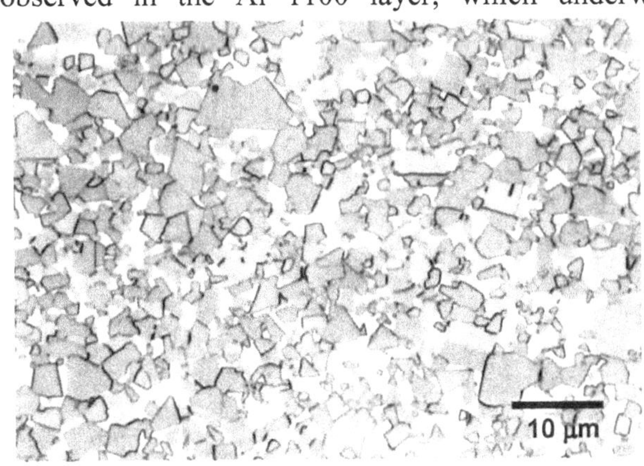

Fig. 4.35 Microstructure of liquid phase sintered 90% WC reinforced Co matrix composite. Note the angular nature of the WC particles (Deng et al., 2001).

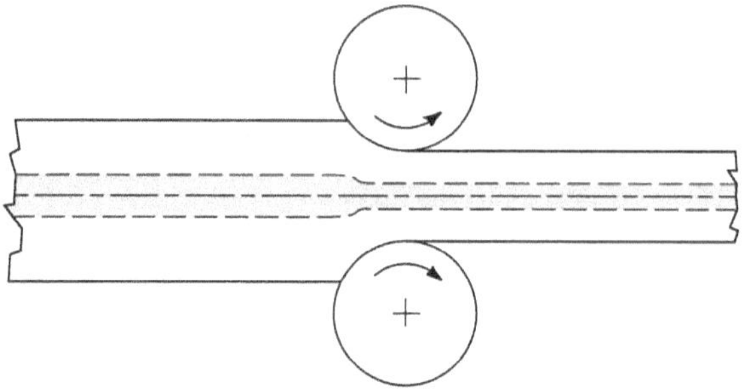

Fig. 4.36 Roll bonding process of making a laminated MMC where a metallurgical bond is produced between the layers.

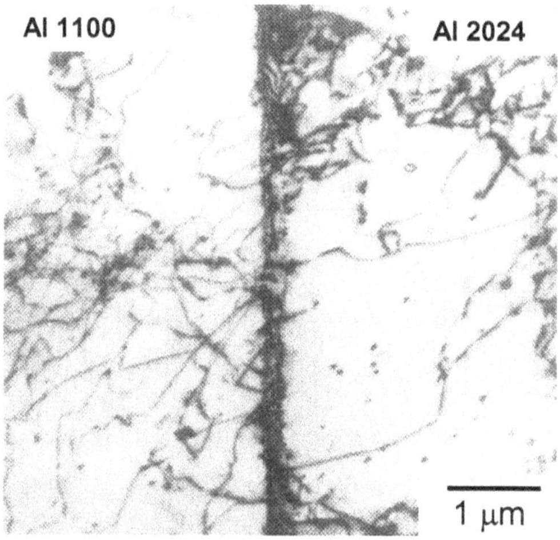

Fig. 4.37 Transmission electron micrograph (TEM) of a roll-bonded Al 1100 and Al 2024 laminated composite, which exhibits an interface with very good integrity (Chawla, 1991). A larger dislocation density is observed in the Al 1100 layer, which is softer than Al 2024.

deformation during processing than Al 2024.

Other examples of deformation processed metal-matrix composites are niobium-based conventional filamentary superconductors with a copper matrix and high-T_C superconductors with a silver matrix. There are two main types of the conventional niobium-based superconductors: Nb-Ti/Cu and Nb_3Sn/Cu. Niobium-titanium (~50–50) form a ductile system. Rods of Nb-Ti are inserted in holes drilled in a block of copper, evacuated, sealed, and subjected to a series of drawing operations interspersed with appropriate annealing treatments to obtain the final diameter of the composite superconductor, Fig. 4.38. In the case of Nb_3Sn/Cu, a process called the bronze route is used to make this composite. Nb_3Sn, an A-15–type intermetallic, cannot be processed like Nb-Ti because of its extreme brittleness. Instead, the process starts with a bronze (Cu–13% Sn) matrix; pure niobium rods are inserted in holes drilled in bronze, evacuated, sealed, and subjected to wire drawing operations as in the case of Nb-Ti/Cu, Fig. 4.39. The critical step is the final heat treatment (~700°C) that drives out the tin from the bronze matrix to combine with niobium to form stoichiometric,

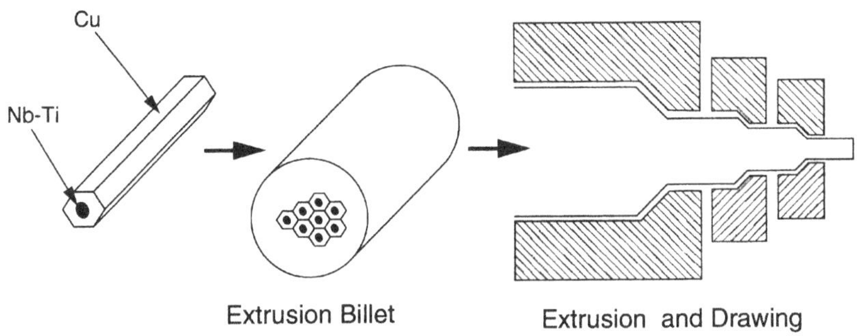

Fig. 4.38 Extrusion-drawing route for fabricating niobium-titanium superconductors. Rods of Nb-Ti are inserted in holes in a block of copper, evacuated, sealed, and subjected to a series of drawing operations interspersed with appropriate annealing treatments to obtain the final composite superconducting wire.

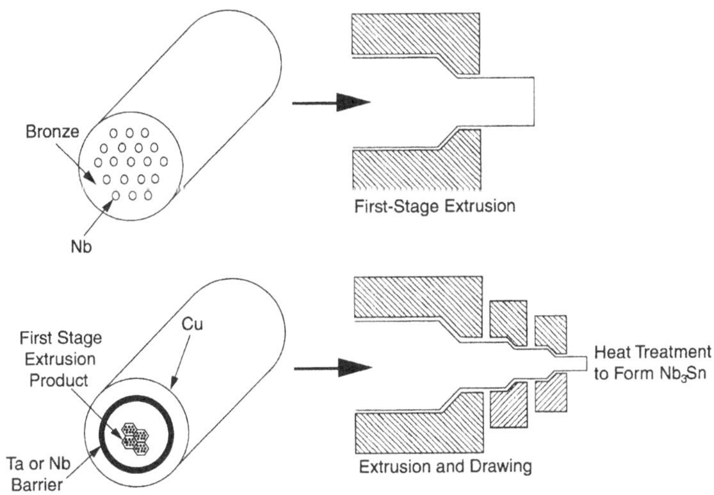

Fig. 4.39 Process for fabricating Nb_3Sn/Cu superconductors. Pure niobium rods are inserted in a bronze (Cu-13% Sn) matrix, evacuated, sealed, and subjected to wire drawing operations as in the case of Nb–Ti/Cu. The critical step is the final heat treatment (~700°C) that drives out the tin from the bronze matrix to form stoichiometric, superconducting Nb_3Sn, leaving behind copper matrix.

superconducting Nb_3Sn, leaving behind copper matrix. Fig. 4.40 shows the microstructure of a Nb_3Sn/Cu superconductor.

In ceramic oxide (high-T_c) superconductors, the superconducting filaments are also brittle, so it is desirable to synthesize the brittle ceramic at the very end of this process. The metallic precursor (MP) method, involves melting the metallic elements (i.e., Y, Ca, Ba, Cu, and Ag), melt spinning the molten alloy into ribbon form, and pulverizing to obtain a homogeneous alloy powder. The precursor is packed into a silver can, sealed, and extruded into a hexagonal rod. Figure 4.41(a) shows a low magnification cross-section of a cable made by the MP process. The higher magnification view, Fig. 4.41(b), shows a transverse section of $Y_{0.9}Ca_{0.1}Ba_2Cu_4Ag_{0.65}$ (Y-124) filament/silver matrix composite containing 962,407 filaments. Another important process for ceramic superconductor metal matrix composites is the oxide-powder-in-tube (OPIT) method (Sandhage et al., 1991). In this process, the oxide powder of appropriate composition (stoichiometry, phase content, purity, etc) is packed inside a metal tube (generally silver), sealed, and degassed, Fig. 4.42. Commonly, swaging and drawing are used for making wires and rolling is used for tapes. Heat treatments, intermediate and/or subsequent to deformation, are given to form the correct phase, promote grain interconnectivity and crystallographic alignment of the oxide, and obtain proper oxygenation (Sandhage et al., 1991).

We now provide a summary of the work on so-called second generation (2G) HTS that followed the work on first generation HTS involving silver matrix containing superconducting filaments. The driving force for the 2G

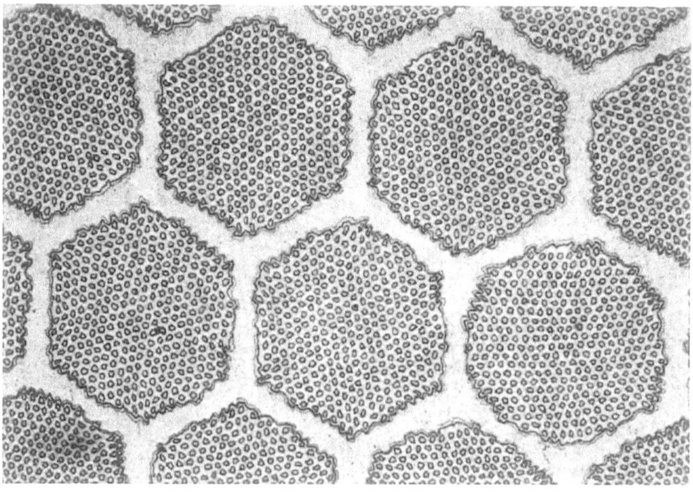

Fig. 4.40 Microstructure of Nb_3Sn filament/Cu matrix composite superconductor used in high-field magnets (courtesy of Hitachi Cable Co.).

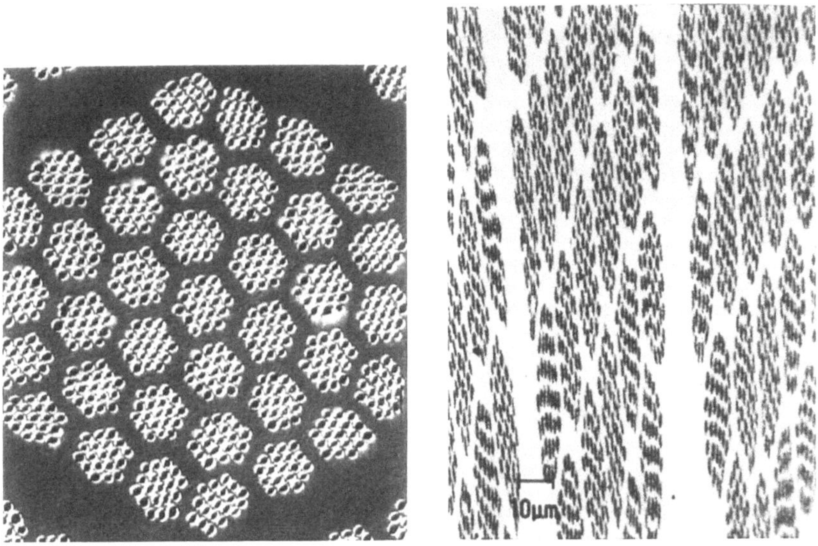

(a) (b)

Fig. 4.41 Microstructure of $Y_{0.9}Ca_{0.1}Ba_2Cu_4Ag_{0.65}$ (Y-124) filament/silver matrix composite cable made by melt spinning: (a) lower magnification and (b) higher magnification (courtesy of American Superconductor Co.). The cable consists of 962,407 filaments.

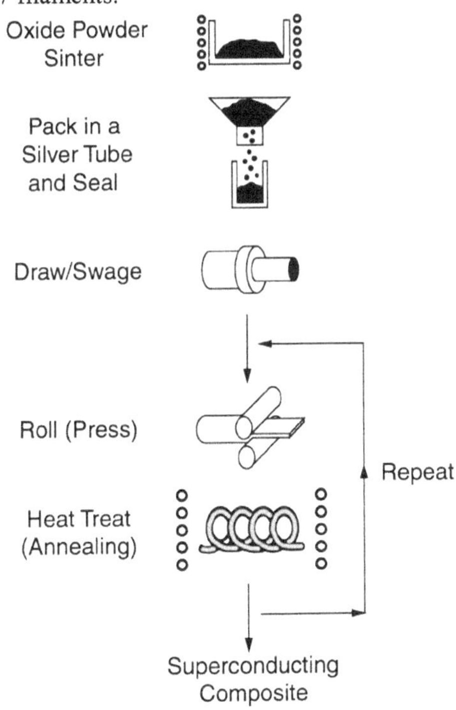

Fig. 4.42 Oxide-powder-in-tube (OPIT) method used to fabricate silver matrix high-T_C superconducting composites. The oxide powder is packed inside a metal tube (generally silver), sealed, and degassed.

work was the realization that a high temperature superconductor with randomly oriented grains will have randomly oriented grain boundaries; such misoriented grain boundaries form obstacles to the flow of superconducting current. If the grains can be aligned, i.e., produce a textured HTS, the flow of current will improve. The approach arrived at, after a considerable amount of work, consisted of epitaxial growth of YBCO on a textured template of the substrate.

The so-called second generation HTS wire consists of a long substrate in the form of a tape that has a highly textured surface. A technique developed at Oak Ridge National Laboratory, called RABiTS™ (rolling-assisted biaxial textured substrates) enables the superconducting phase to have a high degree of grain alignment (Goyal et al., 1996; Goyal et al., 1999). This textured grain structure is required for an efficient flow of current through the superconductor. The RABiTS™ process results in substrates for the HTS wire that are chemically compatible with high-temperature superconductors and exhibit sharp biaxial texture. The process is quite complex. Tape of a metal such as nickel is prepared by special rolling and heat treatment. Next, a buffer layer technology, developed specifically for these textured metals, was used to provide a chemical barrier between the nickel and the superconductor while maintaining the texture. For this, a thin layer of palladium is deposited using electron beam evaporation or sputtering. Metal oxide buffer layers, for example ceria and yttria-stabilized zirconia (YSZ), are placed on the tapes by pulsed-laser deposition. These oxide layers maintain the biaxial texture and provide a chemical barrier between the superconductor and the metal tape. This then is a RABiTS™ substrate, ready for application of the superconductor. The high-temperature superconductor yttrium-barium-copper-oxide (YBCO) is then deposited on the conditioned surface by pulsed-laser deposition. On this an epitaxial YBCO coating is deposited, which is highly textured and thus achieves the desired high superconducting current density. The term epitaxy means that the orientation of the YBCO coating matches that of the underlying substrate. Thus, the second generation (2G) high temperature superconductor (HTS) wires consists a tape-shaped substrate, on which a thin coating of $YBa_2Cu_3O_7$ (YBCO) is deposited epitaxially so that the grains of YBCO are highly aligned. This then is the basic principle for making of a second generation HTS wire: Use a long-length substrate tape with a highly textured interface and an epitaxial YBCO coating replicating the interface texture. This results in a superconductor with high current density.

4.2.6 Diffusion Bonding

Diffusion bonding is a common solid-state processing technique for joining similar or dissimilar metals. Interdiffusion of atoms, at an elevated temperature, from clean metal surfaces in contact with each other leads to bonding (Partridge and Ward-Close, 1993; Guo and Derby, 1995). The principal advantages of this technique are the ability to process a wide variety of matrix metals and control of fiber orientation and volume fraction. Among the disadvantages are long processing times, high processing temperatures and pressures (which makes the process expensive), and limitation on complexity of shapes that can be produced. There are many variants of the basic diffusion bonding process, although all of them involve simultaneous application of pressure and high temperature. These include:

- Foil-fiber-foil process – a matrix alloy foil or powder cloth (a mixture of matrix powder and a fugitive organic binder) and fiber arrays (composite wire) are stacked in a predetermined order, as shown in Fig. 4.43. The stacked layers are vacuum hot pressed so that diffusion bonding may take place. Hot isostatic pressing (HIP), instead of uniaxial pressing, can also be used. In HIP, gas pressure against a can consolidates the composite piece inside the can. With HIP it is relatively easy to apply high pressures at elevated temperatures with variable geometries. This process is clearly suited for flat products, although corrugated structures have also been produced.

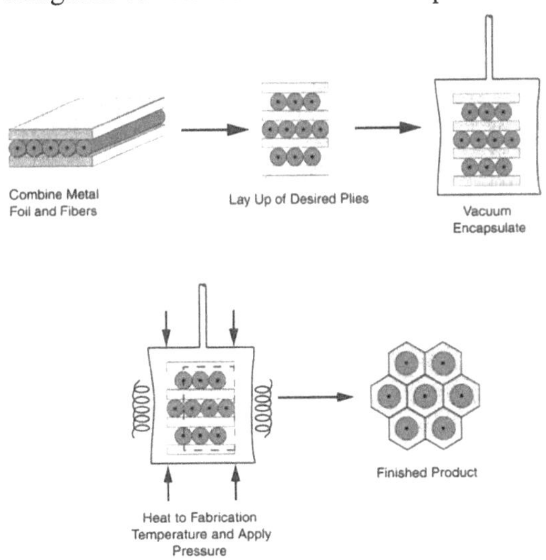

Fig. 4.43 Foil-fiber-foil diffusion bonding process: (**a**) apply metal foil and cut to shape, (**b**) lay up desired plies, (**c**) vacuum encapsulate and heat to fabrication temperature, (**d**) apply pressure and hold for consolidation cycle, and (**e**) cool, remove, and clean part.

- Matrix coated fiber process – here the fibers are coated with the matrix material by plasma spraying or some type of physical vapor deposition (PVD) process, Fig. 4.44. The uniformity of fiber spacing is easier to control and the fibers can be handled more easily, without inducing defects deleterious to fiber strength. Depending on the flexibility of the fibers, the coated fibers may also be wound on a drum. Since the matrix volume fraction is dependent on the thickness of matrix on each fiber, very high fiber volume fraction (~80 vol.%) may be obtained. It should be noted that the fiber or fiber coatings may react or be damaged by the high impact velocity of the liquid droplets.

Fiber distribution is extremely important in controlling mechanical properties. In particular, fiber-to-fiber contact or very close spacing between fibers can result in very high, localized stress concentrations which will result in fiber cracking and/or matrix damage during processing, and premature damage, cracking and failure of the composite, under an applied load.

4.2.7 Explosive Shock Consolidation

A fairly novel, high strain rate, rapid solidification technique is explosive shock consolidation (Thadhani, 1988; and Thadhani et al., 1991). In this technique, dynamic compaction and/or synthesis of powders can be achieved by means of shock waves generated by either explosives in contact with the powder or high velocity impact from projectiles, Fig. 4.45. This process is particularly attractive for consolidating hard materials such as ceramics, or composites. Figure 4.46 shows an optical micrograph of a Ti_3Al particle

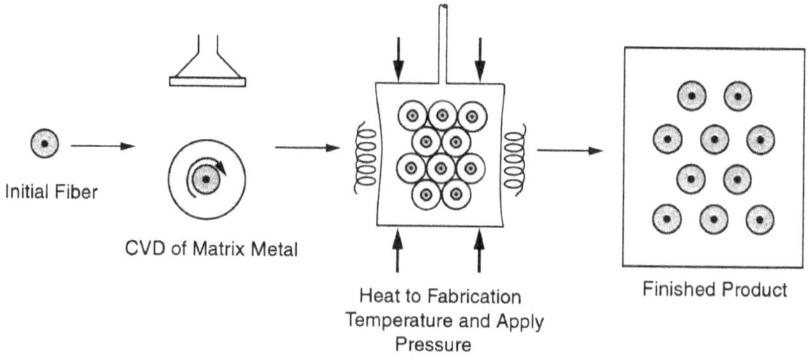

Fig. 4.44 Matrix coated fiber process. The fibers are coated with the matrix material by plasma spraying or some type of physical vapor deposition (PVD) process.

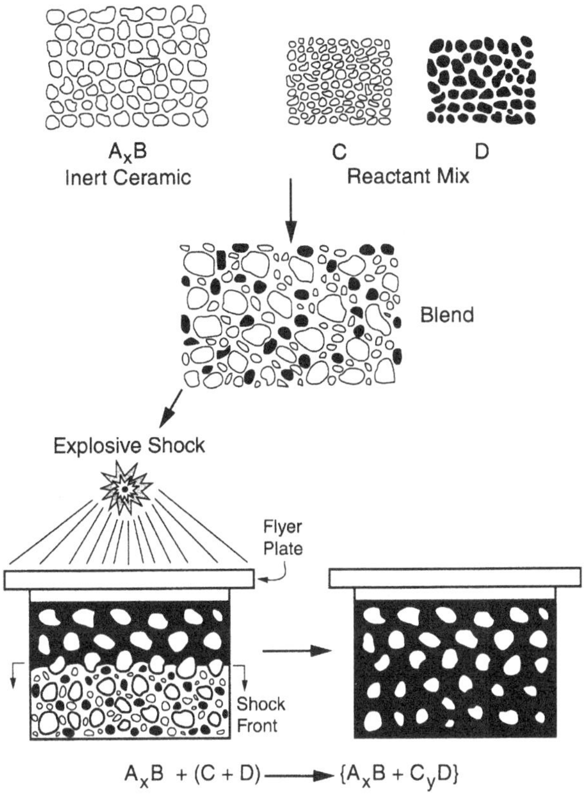

Fig. 4.45 Dynamic compaction and/or synthesis of powders achieved by means of large shock waves generated by either explosives in contact with the powder or high velocity impact from projectiles.

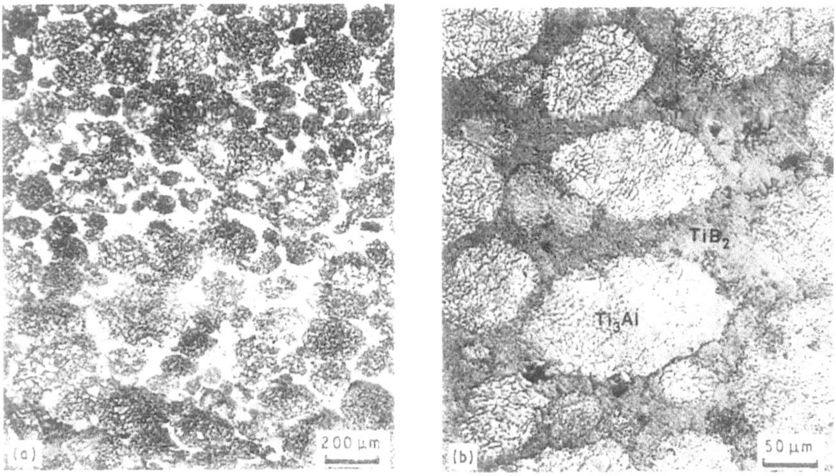

Fig. 4.46 Optical micrographs of Ti_3Al particle reinforced TiB_2 composite (Thadhani et al., 1991). Ti_3Al particles were mixed with TiB_2 powder and formed a continuous matrix phase around the Ti_3Al particles.

reinforced TiB_2 composite. In this composite, the Ti_3Al particles were mixed with TiB_2 powder. The TiB_2 then formed a continuous matrix phase around the Ti_3Al particles.

4.3 GASEOUS STATE PROCESSING

4.3.1 Physical Vapor Deposition

Plasma spraying is the primary form of gaseous state processing. The main application of plasma spraying was described above, to form matrix-coated fibers, which are subsequently hot-pressed to form the final product. In addition, laminated composites, particularly on the nanometer scale have been processed by PVD. PVD processes (specifically, sputter deposition-based processes) offer an extremely wide range of possibilities for fabricating nanolaminate microstructures with tailored chemistry, structure, and thickness of the individual layers and interfaces. Additional important PVD processing parameters include reactive deposition (Ji et al., 2001), plasma-assisted deposition (O'Keefe and Rigsbee, 1994), and substrate heating (Misra et al., 1998).

Some metal/metal layered systems, such as Ni/Cu and Ni/Ti, have been processed at the nanoscale by sputter deposition with great success (Misra et al., 1998; Misra and Nastasi, 1999). An example of a nanoscale Cu-Ni multilayer with a bilayer period of 5 nm is shown in Fig. 4.47 (Misra et al., 1998). Note the well-defined layered structure. The corresponding selected area diffraction pattern (inset) shows a <001> growth direction and cube-on-cube orientation relationship between FCC Cu and FCC Ni. A challenge in the synthesis of nanolaminates via the PVD approach is the control of intrinsic residual stresses. Some control over residual stresses has been achieved by energetic particle bombardment, either *in situ* or post-deposition using an ion source. In the case of magnetron sputtering, a negative substrate bias may be sufficient to change the residual stress from tensile to compressive (Misra and Nastasi, 1999). A film deposited with low bombardment energy yields a tensile residual stress and a microstructure with nanoscale columnar porosity/cracking for a sputtered 150 nm thick Cr film. The same material sputtered with a negative bias, on the other hand, yields a nanocrystalline film with an equiaxed grain structure, near-zero residual stress, and, thus, no intergranular porosity.

While metal/metal systems exhibit high strengths, metal/ceramic systems have the advantage of combining the superior properties of both ceramics and metals (Chou et al., 1992; He et al., 1998; Mearini and Hoffman, 1993;

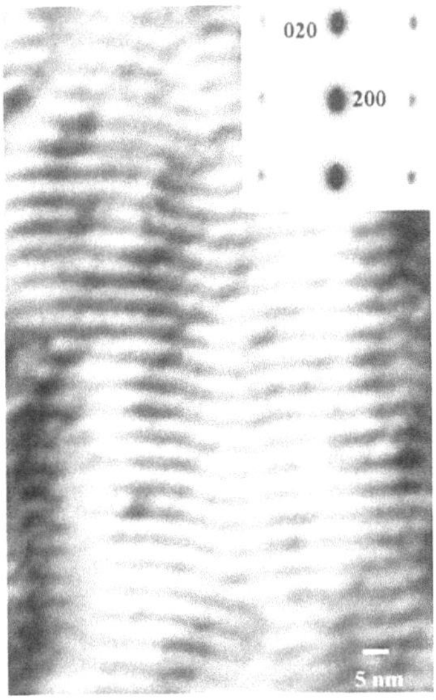

Fig. 4.47 TEM micrograph of Cu-Ni multilayer with a bilayer period of 5 nm, by PVD (courtesy of A. Misra). Note the well-defined layered structure.

Ding et al., 1995; Deng et al., 2005). Systems such as Al/SiC and Al/Al_2O_3 have been investigated. Deng et al. (2005) fabricated nanolayered Al/SiC composites by RF magnetron sputter deposition. The deposition rates of Al and SiC are shown in Fig. 4.48. Note that because of the higher hardness of the SiC, removal of the SiC atoms from the substrate is more difficult, for a given beam energy, resulting in a lower deposition rate than for Al. The microstructure of an Al/SiC nanolayered composite is shown in Fig. 4.49. The sample surface was cut using a focused ion beam (FIB). The SiC layers are about 33 nm in thickness, while the Al layers are 100 nm. The layers appear to be well bonded at the interface.

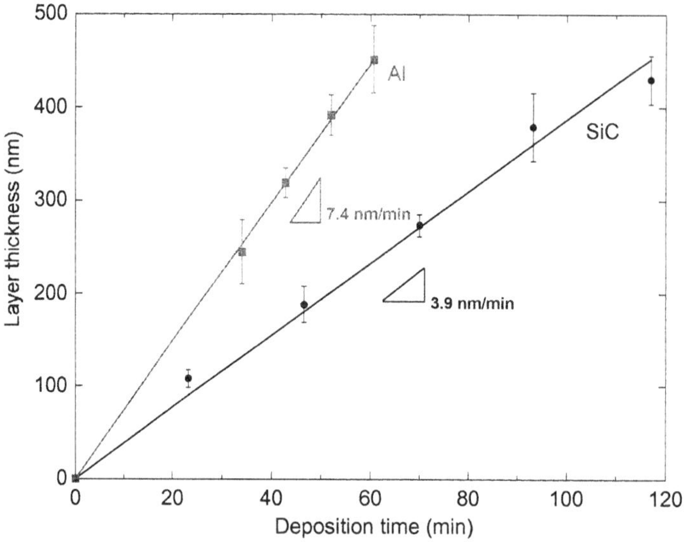

Fig. 4.48 Layer thickness versus deposition time from physical vapor deposition (PVD) of Al and SiC (Deng et al., 2005). Since the SiC is harder, the deposition rate is lower.

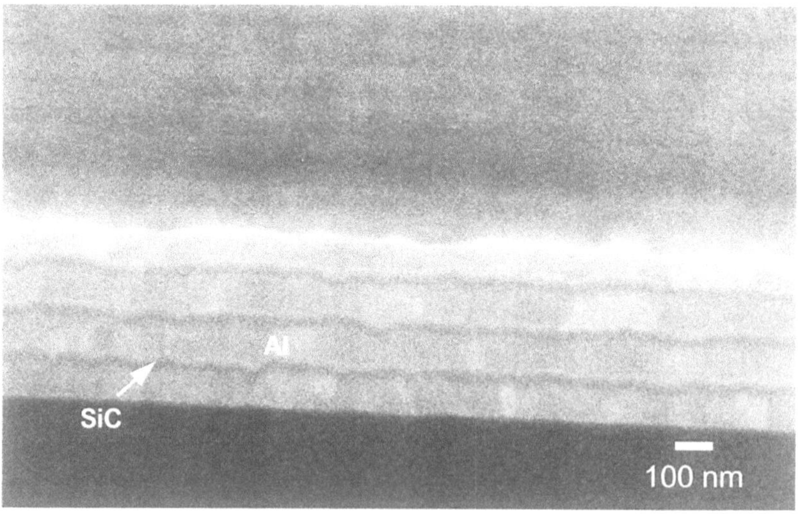

Fig. 4.49 SEM micrograph showing the microstructure of an Al/SiC nanolayered composite, cut using a focused ion beam (FIB) (Deng et al., 2005). The SiC layers are about 33 nm in thickness, while the Al layers are 100 nm.

References

Aghajanian, M.K., J.T. Burke, D.R. White, and A.S. Nagelberg, (1989) *SAMPE Quarterly*, **34**, 817-823.
Anderson, I.E., and J.C. Foley, (2001) *Surf. Int. Anal.*, **31**, 599-608.
Chawla, K.K., (1991) in *Metal Matrix Composites: Mechanisms and Properties*, (R.K. Everett and R.J. Arsenault, eds.) Academic Press, New York, pp. 235-253.
Chawla, K.K., and L.B. Godefroid, (1984) *Proceedings of the 6th International Conference on Fracture,* Pergamon Press, Oxford, U.K., p. 2873.
Chawla, K.K., (1998) *Composite Materials: Science and Engineering*, Springer-Verlag.
Chawla, N., C. Andres, J.W. Jones, and J.E. Allison, (1998) *Metall. Mater. Trans.*, **29A,** 2843.
Chawla, N., J.J. Williams, and R. Saha, (2003) *J. Light Metals*, **2**, 215-227.
Chou, T.C., T.G. Nieh, T.Y. Tsui, G.M. Pharr, and W.C. Oliver, (1992) *J. Mater. Res.*, **7**, 2765-2773.
Christodolou, L., P.A. Parrish, and C.R. Crowe, (1988) *Mat. Res. Soc. Symp. Proc.*, vol. 120, 29.
Cisse, J., and G.F. Bolling, (1971) *J. Cryst. Growth*, **11**, 25-28.
Cole, G.S., and G.F. Bolling, (1965) *Trans. Metall. Soc. AIME*, **233**, 1568-1572.
Cook, A.J., and P.S. Werner, (1991) *Mater. Sci. Eng.*, **A144**, 189.
Deng, X., C. Cleveland, T. Karcher, M. Koopman, N. Chawla, and K.K. Chawla, (2005) *J. Mater. Eng. Perf.*, **14**, 1-7.
Deng, X., B.R. Patterson, K.K. Chawla, M.C. Koopman, Z. Fang, G. Lockwood, A. Griffo, (2001) *Int. J. Refrac. Metals & Hard Mater.*, **19**, 547-552.
Deve, H.E. and C. McCullough, (1995) *JOM*, **47**, 33-37.
Ding, Y., D.O. Northwood, and A.T. Alpas, (1995) *Mater. Sci. Forum*, **189**, 309-314.
Dirichlet, G.L., (1850) *J. Reine Angew. Math.*, **40**, 209-227.
Divecha, A.P., S.G. Fishman, and S.D. Karmarkar, (Sept. 1981) *J. Metals*, **9**, 12.
Estrada, J.L., J. Duszczyk, and B.M. Korevaar, (1991) *J. Mater. Sci.*, **26**, 1631-1634.
Ganesh, V.V., and N. Chawla, (2004) *Metall. Mater. Trans.*, **35A**, 53-62.
Ganesh, V.V., and N. Chawla, (2005) *Mater. Sci. Eng.*, **A391**, 342-353.
Ghosh, A.K., (1993) in *Fundamentals of Metal Matrix Composites*, Butterworth-Hinemann, Stoneham, MA, pp. 3-22.
Goyal, A., D.P. Norton, D.K. Christen, E.D. Specht, M. Paranthaman, D.M. Kroeger, J.D. Budai, Q. He, F.A. List, R. Feenstra, H.R. Kerchner, D.F.

Lee, E. Hatfield, P.M. Martin, J. Mathis, and C. Park, (1996) *Appl. Supercon.*, **4**, 403-427.
Goyal, A., R. Feenstra, F.A. List, D.P. Norton, M. Paranthaman, D.F. Lee, D.M. Kroeger, D.S. Beach, J.S. Morell, T.G. Chirayil, D.T. Verebelyi, X. Cui, E.D. Specht, D.K. Christen, and P.M. Martin, (July 1999) *JOM*, 19-23.
Guo, Z.X., and B. Derby, (1995) *Prog. Mater., Sci.*, **39**, 411-495.
He, J.L., W.Z. Li, H.D. Li, and C.H. Liu, (1998) *Surf. Coatings Tech.*, **103**, 276-280.
Helinski, E.J., J.J. Lewandowski, T.J. Rodjom, and P.T. Wang, (1994) in *World P/M Congress, vol. 7*, (C. Lall and A. Neupaver, eds.), Metal Powder Industries Federation, Princeton, NJ, pp. 119-131.
Humphreys, F.J., W.S. Miller, and M.R. Djazeb, (1990) *Mater. Sci. Tech.*, **6**, 1157-66.
Hunt, W.H., T.M. Osman, and J.J. Lewandowski, (Mar. 1991) *JOM*, 30–35.
Hunt, W.H., (1994) in *Processing and Fabrication of Advanced Materials*, The Minerals and Metal Materials Society, Warrendale, PA., pp. 663–683.
James, W.B., (1985) *Int. J. Powder Metall.*, **21**, 163.
Ji, Z., J.A. Haynes, M.K. Ferber and J.M. Rigsbee, (2001) *Surf. Coatings Tech.*, **135**, 109-117.
Katsura, M., (1982) Japan Pat. 57-25275.
Kim, J.K., and P.K. Rohatgi, (1999) *Metall. Mater. Trans.*, **29A**, 351-358.
Kim, Y.W., W.M. Griffith, and F.H. Froes, (1985) *J. Metals*, **37**, 27-33.
Kowalski, L., B.M. Korevaar, and J. Duszczyk, (1992) *J. Mater. Sci.*, **27**, 2770-2780.
Lavernia, E.J., J.D. Ayers, and T.S. Srivatsan, (1992) *Int. Mater. Rev.*, **37**, 1-44.
Lewandowski, J.J., C. Liu, and W.H. Hunt, (1989) *Mater. Sci. Eng.*, **A107**, 241-255.
Liu, Y.L., N. Hansen, and D.J. Jensen, (1989) *Metall. Trans.*, **20A**, 1743-1753.
Lloyd, D.J., (1989) *Comp. Sci. Tech.*, **35**, 159-179.
Lloyd, D.J., (1994) *Int. Mater. Rev.*, **39**, 1.
Lloyd, D.J., (1997) in *Composites Engineering Handbook* (P.K. Mallick, ed.), Marcel Dekker, New York, pp. 631-669.
Logsdon, W.A., and P.K. Liaw, (1986) *Eng. Frac. Mech.*, **24**, 737-751.
McLean, M., (1983) *Directionally Solidified Materials for High Temperature Service*, The Metals Soc., London.
Martin Marietta Corp., (1987) U.S. Patent 4,710,348.
Manoharan, M., L. Ellis, and J.J. Lewandowski, (1990) *Scripta Metall. Mater.*, **24**, 1515–1521.

Masur, L.J., A. Mortensen, J.A. Cornie, and M.C. Flemings, (1989) *Metall. Trans.* **20**, 2549-2557.
Mearini, G.T., and R.W. Hoffman, (1993) *J. Elec. Mater.*, **22** 623-629.
Mehrabian, R., R.G. Riek, and M.C. Flemings, (1974) *Metall. Trans.*, **5**, 1899-1905.
Michaud, V.C., (1993) in *Fundamentals of Metal Matrix Composites*, Butterworth-Heinneman, Stoneham, MA, pp. 3-22.
Misra, A., M. Verdier, Y.C. Lu, H. Kung, T.E. Mitchell, M. Nastasi, and J.D. Embury, (1998) *Scripta Mater.*, **39**, 555.
Misra, A., and M. Nastasi , (1999) *J. Mater. Res.*, **14**, 4466.
Mortensen, A., J.A. Cornie, and M.C. Flemings, (1988) *J. Metals*, **40**, 12.
Mortensen, A., and I. Jin, (1992) *Inter. Mater. Rev.*, **37**, 101-128.
Nourbakhsh, S., F.L. Loang, and H. Margolin, (1990) *Metall. Trans.*, **21A,** 213.
O'Keefe, M.J., and J.M. Rigsbee, (1994) *J. App. Polymer Sci.*, **53**, 1631-1638.
Partridge, P.G., and C.M. Ward-Close, (1993) *Int. Mater. Rev.*, **38**, 1-24.
Pennander, L., and C.-H. Anderson, (1991) in *Metal Matrix Composites – Processing, Microstructure and Properties*, 12th Risø Int. Symp. on Materials Science, (N. Hansen et al., eds.), Risø, Denmark, 575.
Potschke, J., and V. Rogge , *J. Cryst. Growth*, (1989) **94**, 726-738.
Saha, R., E. Morris, N. Chawla, and S.M. Pickard, (2002) *J. Mater. Sci. Lett.*, **21,** 337-339.
Sahoo, P., and M.J. Koczak, (1991) *Mater. Sci. Eng.*, **A144**, 37-44.
Sandhage, K.H., G.N. Riley, Jr., and W.L. Carter, (Mar. 1991) *JOM*, 21-25.
Shangguan, D., S. Ahuja, and D.M. Stefanescu, (1992) *Metall. Trans.*, **23A**, 669-680.
Shekhar, J.A., and R. Trivedi, (1989) *Mater. Sci. Eng.*, **A114**, 133-146.
Shekhar, J.A., and R. Trivedi (1990), in *Solidification of Metal Matrix Composites*, P. Rohatgi, ed., TMS, Warrendale, PA, pp. 23.
Spowart, J.E., B. Maruyama, and D.B. Miracle, (2001) *Mater. Sci. Eng.*, **A307**, 51-66.
Stefanescu, D.M., B.K. Dhindaw, A.S. Kakar, and A. Mitra, (1988) *Metall. Trans.*, **19A**, 2847-2855.
Surappa, M.K., and P.K. Rohatgi, (1981) *J. Mater. Sci.*, **16,** 983-993.
Thadhani, N.N., (1988) *Adv. Mater. Manuf. Proc.*, **3**, 493-549.
Thadhani, N.N., N. Chawla, and W. Kibbe, (1991) *J. Mater. Sci.*, **26**, 232-240.
Tham, L.M., M. Gupta, and L. Cheng, (2002) *Mater. Sci. Eng.*, **326**, 355-363.
Thomas, D.G., (1965) *J. Colloid. Sci.*, **20**, 267.
Trivedi, R., S.H. Han, and J.A. Shekhar, (1990) in *Solidification of Metal Matrix Composites*, P. Rohatgi, ed., TMS, Warrendale, PA, p. 23.

Tsunekawa, Y., M. Okumiya, I. Niimi, and K. Yoneyama, (1988) *J. Mater. Sci. Lett.*, **7**, 830-832.

Uhlmann, D.R., B. Chalmers, and K.A. Jackson, (1964) *J. Appl. Phys.*, **35**, 2986-2993.

Williams, J.J., G. Piotrowski, R. Saha, and N. Chawla, (2002) *Metall. Mater. Trans.*, **33A**, 3861-3869.

Yang, N., J. Boselli, and I. Sinclair, (2001) *J. Micros.*, **201**, 189-200.

Chapter 5

INTERFACE

Interface is a very general term used in various fields of science and technology to denote the location where two entities meet. The term in composites refers to a bounding surface between the reinforcement and matrix across which there is a discontinuity in chemical composition, elastic modulus, coefficient of thermal expansion, and/or thermodynamic properties such as chemical potential. The interface (fiber/matrix or particle/matrix) is very important in all kinds of composites. This is because in most composites, the interfacial area per unit volume is very large. Also, in most metal matrix composite systems, the reinforcement and the matrix will not be in thermodynamic equilibrium, i.e., a thermodynamic driving force will be present for an interfacial reaction that will reduce the energy of the system. All these items make the interface have a very important influence on the properties of the composite.

Instead of a bidimensional boundary, generally, we have an interfacial zone with finite thickness, possibly consisting of multiple layers. The multilayer boundary zone will be in equilibrium at the high temperature at which the components were originally brought together. At any other temperature, a complex stress field exists in the boundary zone because of mismatch in coefficient of thermal expansion of the various layers. These stresses will be proportional to the difference in the elastic moduli of the components, difference in the coefficients of thermal expansion, and of course the temperature difference between the equilibrium (or initial) temperature and final temperature. Thermodynamically speaking, the phases in the boundary zone will tend to change such that the free energy of the system is minimized. This may involve generation of dislocations, grain boundary migration, crack nucleation and/or propagation. An ideal interface in a metal matrix composite should promote wetting and bond the reinforcement and the matrix to a desirable degree. The interface should protect the ceramic reinforcement and allow load transfer from the soft metallic matrix to the strong reinforcement.

Besides composition, we also need to take into account other parameters that characterize the interfacial zone such as: Geometry and dimensions; microstructure and morphology; and mechanical, physical, chemical, and thermal characteristics of the different phases present in the interfacial zone.

The components of a composite system are chosen on the basis of their mechanical and physical characteristics in isolation. When one puts together two components to make a composite, the system will rarely be in thermodynamic equilibrium. More often than not, a driving force will be present for some kind of interfacial reaction(s) between the two components leading to a state of thermodynamic equilibrium. Of course, thermodynamic information such as phase diagrams can help predict the final equilibrium state of the composite. Data regarding reaction kinetics, for example, diffusivities of one constituent in another can provide information about the rate at which the system will tend to equilibrium. In the absence of thermodynamic and kinetic data, experimental studies need to be conducted to determine compatibility of the components.

In this chapter, we first describe some important concepts related to interfaces in metal matrix composites. Types of bonding in MMCs are provided with examples from various systems, followed by a description of some tests to determine the mechanical properties of interfaces.

5.1 CRYSTALLOGRAPHIC NATURE OF THE INTERFACE

In crystallographic terms, one can describe an interface between two crystalline phases as coherent, semi-coherent, or incoherent. A coherent interface implies one-to-one correspondence between lattice planes on the two sides of the interface. The lattice constants of any two phases are not likely to be the same. Thus, in order to provide continuity of lattice planes across an interface, i.e., a coherent interface, there will be some coherency strains associated with the interface because of straining of the lattice planes in the two phases. Commonly, such coherent interfaces are observed between some precipitates and the metallic matrix where the mismatch in the lattice parameter of the two phases is quite small. For example in the Al-Li system, Al_3Li precipitate is coherent with the aluminum matrix. An incoherent interface, on the other hand, consists of such severe atomic disorder that no matching of lattice planes occurs across the boundary, i.e., no continuity of lattice planes is maintained across the interface. Thus, an incoherent boundary or interface will have no coherency strains but the energy associated with the interfacial boundary increases because of severe atomic disorder. A situation that is intermediate between coherent and incoherent can exist, i.e., we can have a semi-coherent interface. A semi-coherent interface has some lattice mismatch between the phases which can be accommodated by the introduction of dislocations at the interface. Crystallographically, most of the interfaces that one encounters in fiber, whisker, or particle reinforced metal matrix composites are incoherent and high energy interfaces. Accordingly, they can act as efficient vacancy sinks, and provide rapid diffusion paths, segregation sites, sites of heterogeneous precipitation, as well as sites for precipitate-free zones. Among the possible exceptions to this are eutectic composites and XD-type particulate composites, in which the interfaces can be semi-coherent.

5.2 WETTABILITY

Wettability is defined as the ability of a liquid to spread on a solid surface. Figure 5.1 shows two conditions of a liquid drop resting on a solid substrate. There are three specific energy (energy per unit area) terms: γ_{SV}, the energy of the solid/vapor interface; γ_{LS}, the energy of the liquid/solid interface; and γ_{LV}, the energy of the liquid/vapor interface. It should be mentioned that the term surface tension is also used to denote surface energy, although rigorously speaking, surface energy is a more appropriate term for solids. When we put a liquid drop on a solid substrate, we replace a portion of the solid/vapor interface by a liquid/solid and a liquid/vapor interface. Thermodynamically, spreading of the liquid will occur if this results in a decrease in the free energy of the system, i.e.:

$$\gamma_{SL} + \gamma_{LV} < \gamma_{SV}$$

Figure 5.1 defines these terms. An important parameter with respect to wettability is the contact angle, θ, which is a measure of wettability for a system. From the equilibrium of forces in the horizontal direction, we can write:

$$\gamma_{SL} + \gamma_{LV} \cos\theta = \gamma_{SV}$$
$$\theta = \cos^{-1}(\gamma_{SV} - \gamma_{SL}) / \gamma_{LV}$$

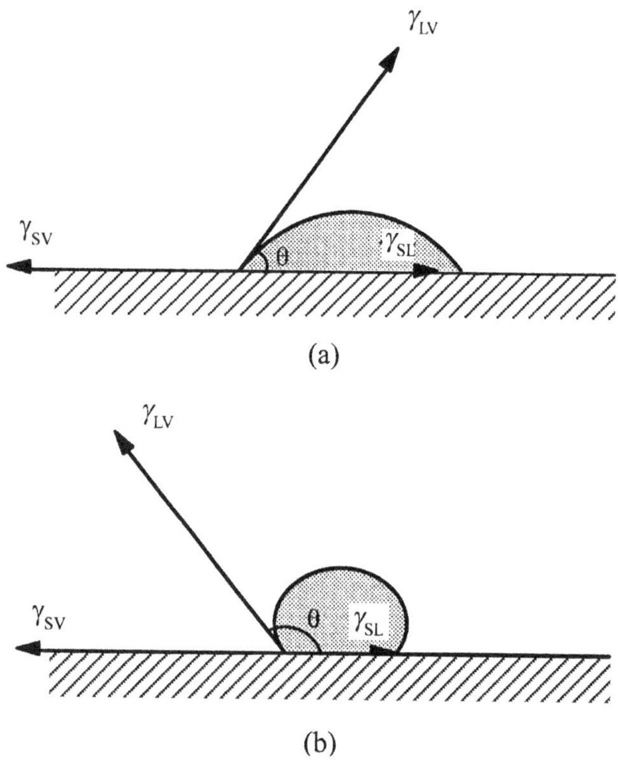

Fig. 5.1 Contact angle, θ, a measure of wettability for a system is defined by interaction among three surface energies: Solid-liquid surface energy, γ_{SL}, solid-vapor surface energy, γ_{SV}, and liquid-vapor surface energy, γ_{LV}.

From this equation, we see that for $\theta = 0^{\circ}$, we have perfect wetting, while for $\theta = 180^{\circ}$, we have no wetting. For $0^{\circ} < \theta < 180^{\circ}$, there will be partial wetting. It should be pointed out that the contact angle for a given system can vary with temperature, stoichiometry, hold time, interfacial reactions, presence of any adsorbed gases, roughness and geometry of the substrate, etc.

It is worth emphasizing that wettability only describes the extent of intimate contact between a liquid and a solid. It does not necessarily mean a strong bond at the interface. One can have excellent wettability but only a weak physical, low energy bond. A low contact angle, implying good wettability, is a necessary but not sufficient condition for strong bonding. Wettability of the ceramic reinforcement by the molten metal is very important for liquid state processing of MMCs. One can modify the contact angle by changing the composition of the liquid matrix, which changes the value of its surface energy. The effect of roughness of the solid substrate can be evaluated in terms of a ratio, r, defined as:

$$r = \frac{\text{true surface area}}{\text{mean plane surface area}}$$

In general, if $\theta < 90^{\circ}$, wettability is enhanced by roughness while if $\theta > 90^{\circ}$, wettability is reduced by roughness.

5.3 TYPES OF BONDING

There are two important types of bonding at an interface in a metal matrix composite:

- Mechanical bonding
- Chemical bonding

A brief description of mechanical and chemical bonding is given below with examples from various MMC systems.

5.3.1 Mechanical Bonding

Most fibers have a characteristic surface roughness or texture resulting from the fabrication process (techniques for measuring fiber roughness are shown in the inset.) This in turn imparts a roughness to the interface when the fibers are incorporated in a matrix to make a composite. Mechanical keying of the matrix into the hills and valleys of the surface of the reinforcement (due to roughness) is an important but generally neglected aspect of the interface in all composites. It turns out that interface roughness-induced mechanical bonding is quite important in all kinds of composites. Surface roughness can contribute to bonding only if the liquid matrix wets the reinforcement surface. If the liquid matrix (polymer, metal, or ceramic) is unable to penetrate the asperities on the fiber surface, then the matrix will leave interfacial voids on solidification. In a mechanical bond, the degree of interfacial roughness is a very important parameter, which, in turn, is controlled by the fiber surface roughness. In polymer matrix composites (PMCs) and metal matrix composites (MMCs), one would like to have mechanical bonding in addition to chemical bonding. In ceramic matrix composites (CMCs), one would like to have mechanical bonding rather than chemical bonding. In fiber reinforced composites mechanical bonding will be effective mostly in the longitudinal or fiber direction. Transverse to the fiber, it will provide only a minor effect.

Techniques for Measuring Roughness

Most fibers have surface markings or *striations* that originate from the edges of the surface of the spinneret or the edges of a die, in the case of wire drawing. Thus, the origin of surface roughness lies in processing. Such roughness is characterized by rather close and irregular changes in surface height, resulting in peaks and valleys about an imaginary mean surface line. A whole range of techniques is available for measuring surface roughness. In a mechanical profilometer, a diamond stylus is run over the surface and the up and down movement of the stylus is recorded. Atomic Force Microscopy (AFM) can offer a greater spatial resolution than any of the conventional techniques. The best mechanical profilometer can resolve variations in surface heights down to 0.1 μm, while the AFM can resolve heights as small as 1 nm. Scanning tunneling microscope (STM) can be used to characterize an electrically conducting surface. STM detects changes in the quantum-mechanical tunneling current of electrons between the surface and an electrode (W or Pt) as the electrode scans the surface. The magnitude of tunneling current decreases with distance between the electrode and the surface. This can be used to provide information about the topography of the surface. AFM can characterize most any material, including an electrically insulating material such as a glass or a ceramic. Commonly a silicon or silicon nitride tip is used at the end of a microcantilevered arm. Atomic forces (van der Waals or electrostatic) between the tip and the surface deflect the arm, with the deflection falling off with distance from the surface. The deflection is higher for peaks and lower for valleys. A laser beam from a laser diode is reflected off the back of the cantilever with an optical-lever detection system. Conventionally, a stylus-based profilometer provides height information in one direction. In this regard, AFM can be regarded as a high magnification profilometer that has a superior vertical and lateral resolution. AFM is controlled by a microprocessor that provides a variety of computer-based topographical images of the surface. For example, gray scale (and/or color images) can be obtained, in which x and y data form the horizontal and vertical axes while the z-data are used to give the gray scale. The gray scale is a linear scale in which the brightness of a point is in proportion to its height. A brighter point corresponds to a higher height while a darker point corresponds to a lower height on the surface. One can then draw a line across the gray scale image and display the surface topographical profile corresponding to the line. The figure on the next page shows an AFM picture of the surface of a Nicalon fiber and the roughness profile corresponding to the line between the arrows in the AFM picture.

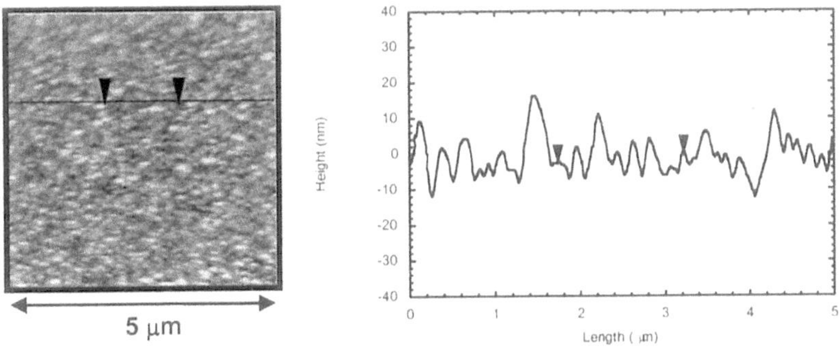

Another possibility is a three-dimensional representation of the surface wherein the height is indicated by superimposing the gray scale along the perpendicular axis and rotating the display to a convenient viewing angle. A three-dimensional picture of the surface of Nicalon fiber is shown in figure below (Chawla et al., 1995). It is possible to further enhance the three-dimensional aspect by adding a computer-generated light source to cast shadows. This is called illumination mode, not shown here.

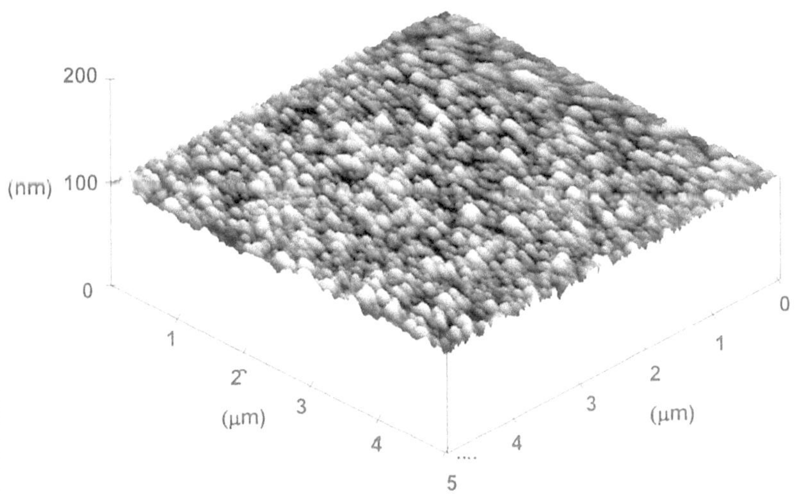

Mechanical keying between two surfaces can lead to bonding between them and can be quite important in metal matrix composites. An ideal, smooth interface is only an idealization. Interfaces in real composites are invariably rough, which allows for interlocking. The degree of interfacial roughness can be controlled to some extent, but there is always some mechanical bonding present. Consider the situation of an MMC made of a ceramic reinforcement and a metallic matrix. Metals generally have a higher coefficient of thermal expansion than ceramics. Thus, the metallic matrix in the composite will shrink radially more than the ceramic fiber on cooling from a high temperature. This will lead to mechanical gripping of the fiber by the matrix even in the absence of any chemical bonding. The matrix penetrating the crevices on the fiber surface, by liquid flow or high temperature diffusion, can also lead to some mechanical bonding. The radial gripping stress, σ_r, can be related to the interfacial shear strength, τ_i, by the following expression:

$$\tau_i = \mu\sigma_r$$

where μ is the coefficient of friction, generally, between 0.1 and 0.6. In general, a mechanical bond is a low energy bond vis-à-vis a chemical bond. Examples of mechanical bonding include alumina fibers in a variety of metals, e.g., Al_2O_3 fiber in aluminum. Mechanical bonding with carbon fibers is improved by chemically treating the surface with nitric acid. The oxidation of the fiber surface results in increased specific surface area of the fibers.

We cite two examples showing the importance of mechanical gripping effects in MMCs. Hill et al. (1969) confirmed this experimentally for tungsten filaments in an aluminum matrix while Chawla and Metzger (1978) observed mechanical gripping effects at Al_2O_3/Al interfaces. Hill et al. (1969) etched tungsten wires along a portion of their length to produce a rough interface. In the first work, tungsten filaments were incorporated into an aluminum matrix by liquid metal infiltration technique in vacuum. They evaluated three interface conditions by longitudinal tensile tests of composites. In the case of a smooth interface, a chemical bond formed between the aluminum matrix and tungsten fiber, which resulted in a high strength composite. In the case of a smooth fiber surface with a graphite layer, the graphite barrier layer prevented the reaction from taking place, that is, there was no chemical bonding, and because the interface was smooth (little or no roughness), there was also very little mechanical bonding. The resultant strength of the composite was, therefore, very low. In the third case, they etched the tungsten filaments and applied a graphite layer. In this case, there was no reaction bonding, but there was a mechanical keying effect because of the rough surface produced by etching. The result was that the mechanical bonding restored the strength of the composite to the level achieved with chemical reaction at the interface. Chawla and Metzger (1978) compared the load transfer from aluminum to alumina as a function of interfacial roughness. They used a polished aluminum surface and roughened aluminum surface produced by etch-pitting. Alumina was formed by electrolytically anodizing the aluminum. When this composite was loaded in tension, the crack appeared in the alumina perpendicular to the loading direction. Figure 5.2 shows these results in the form of linear crack density (number of cracks per unit length) in alumina as a function of strain in the alumina/aluminum composite for different degrees of interface roughness. Solid circles represent the polished substrate or smooth interface while filled squares represent steep sided pits (10^6 pits/cm^2) or rough interface. Cracks in the alumina film first appeared at about the same strain. Initially the crack density rose at about the same rate for the rough and smooth interfaces. For a smooth interface, however, the crack density remained constant above 8% strain in the matrix while for a rough interface, the crack density continued to increase beyond this value, i.e., in the case of a rough interface the load transfer from aluminum to alumina continued to higher strain values than in the case of a smooth interface. Thus, high degree of mechanical bonding at the rough interface was responsible for more efficient load transfer from the soft aluminum matrix to hard alumina. Similar mechanical bonding effects due to interface roughness have been observed in babbitt (Sn with small amounts of Cu and Sb)-bronze composites (Liaw et al., 1990).

5.3.2 Chemical Bonding

Ceramic/metal interfaces in metal matrix composites are generally formed at high temperatures. Diffusion and chemical reaction kinetics are faster at elevated temperatures. Knowledge of the chemical reaction products and, if possible, their properties are needed. It is therefore imperative to understand the thermodynamics and kinetics of reactions such that processing can be controlled and optimum properties obtained.

Chemical bonding in MMCs involves atomic transport by diffusion. Thus, chemical bonding includes solid solution and/or chemical compound formation at the interface. It may lead to the formation of an interfacial zone containing a solid solution and/or a reinforcement/matrix interfacial reaction zone.

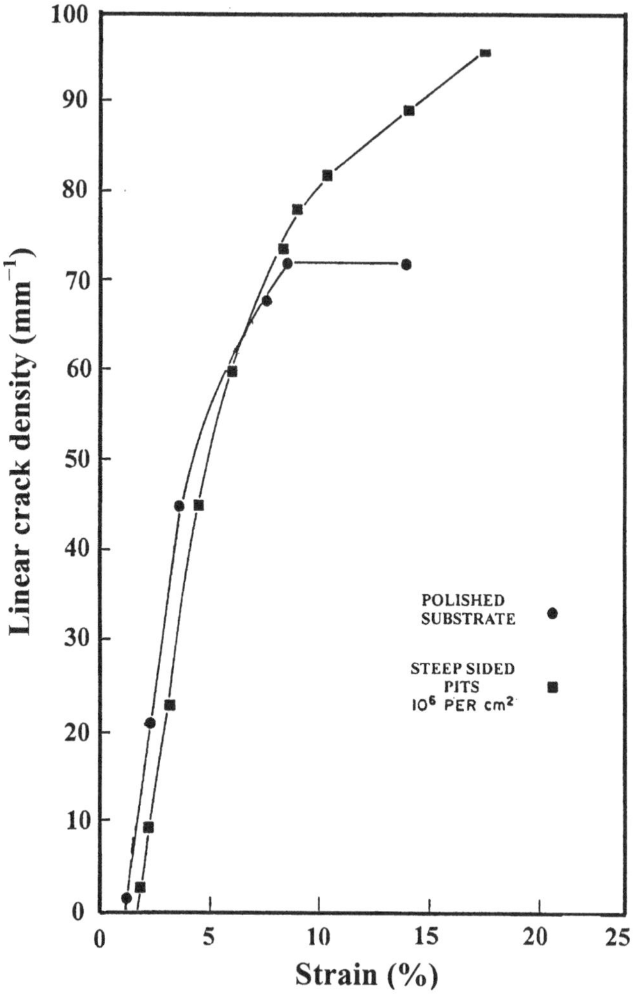

Fig. 5.2 Load transfer from aluminum to alumina as a function of interfacial roughness (after Chawla and Metzger, 1978). For a smooth interface, the crack density remained constant above 8% strain in the matrix while for a rough interface, the crack density continued to increase beyond this value, i.e., the load transfer from aluminum to alumina continued to higher strain values in the case of a rough interface.

For diffusion controlled growth in an infinite diffusion couple with a planar interface, we have the following important relationship:

$$x^2 \approx Dt$$

where x is the thickness of the reaction zone, D is the diffusivity, and t is time. The diffusivity, D, depends on the temperature in an exponential manner:

$$D = D_o \exp(-\Delta Q / kT)$$

where D_o is a pre-exponential constant, ΔQ is the activation energy for the rate controlling process, k is the Boltzmann's constant, and T is the temperature in kelvin. For a composite containing small diameter cylindrical fibers, the assumption of an infinite diffusion couple is not valid, i.e., the diffusion distances are quite small. However, to a first approximation, we can write:

$$x^2 \approx Bt$$

where B is a pseudo-diffusivity constant and has the dimensions of diffusivity, i.e., m^2s^{-1}. One may use this approximate relationship for composites where the thickness of the reaction zone is small compared to the interfiber spacing. Under these conditions, one can use an Arrhenius type relationship:

$$B = A \exp(-\Delta Q / kT)$$

where A is a pre-exponential constant. A plot of ln B vs. 1/T can then be used to obtain the activation energy, ΔQ, for a fiber/matrix reaction in a given temperature range. The dependence of the reaction zone thickness on the square root of time indicates the operation of volumetric diffusion. Figure 5.3 shows in a schematic manner the effect of temperature on the reaction zone thickness vs. $\sqrt{}$time. With increasing temperature, the slope of the line increases.

As we said above, most metal matrix composite systems are nonequilibrium systems in the thermodynamic sense; that is, a chemical potential gradient exists across the fiber/matrix interface. This means that given favorable kinetic conditions, which in practice means a high enough temperature or long enough time, diffusion and/or chemical reactions occur between the components. Prolonged contact between liquid metal and reinforcement can lead to a significant chemical reaction, which may adversely affect the

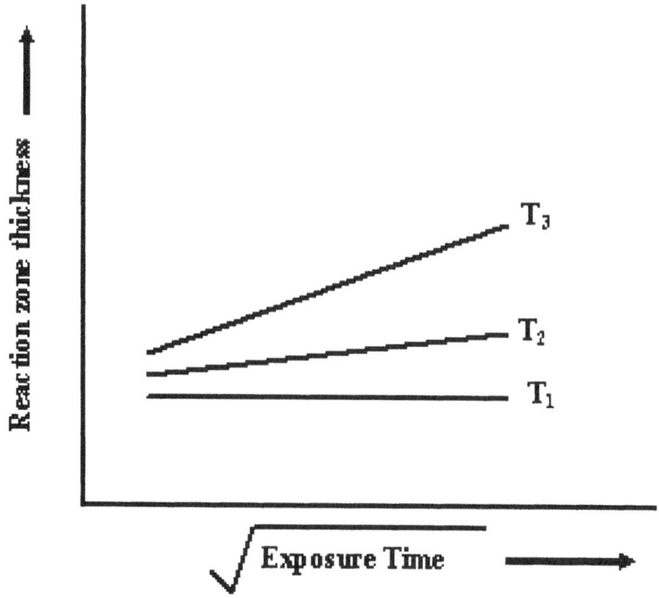

Fig. 5.3 Schematic representation of thickness of reaction zone between fiber and matrix as a function of square root of exposure time for three different temperatures, $T_3>T_2>T_1$.

behavior of the composite. For example, molten aluminum can react with carbon fiber to form Al_4C_3 and Si per the following reaction:

$$4Al + 3SiC \longleftrightarrow Al_4C_3 + 3Si$$

This reaction, as indicated by the two arrows, can go leftward or rightward. When it goes rightward, it injects silicon in the molten aluminum, which can have important consequences. Change of matrix alloy composition is one. Addition of silicon in aluminum also results in lowering of the melting point of the alloy; in fact it will result in a mushy range. Kinetics, i.e., time and temperature will control the Si level required to prevent the reaction. This reaction can be prevented from going rightward by using high levels of Si in the matrix, say ~ 10% Si. It is for this reason that only high silicon aluminum alloys are suitable for SiC particles for making composites by casting route. Al_2O_3 is stable in pure molten aluminum. It does, however, react with magnesium, a common alloying element in aluminum. The following reactions can occur between Mg and Al_2O_3:

$$3Mg + Al_2O_3 \longleftrightarrow 3MgO + 2Al$$

$$3Mg + 4Al_2O_3 \longleftrightarrow 3MgAl_2O_4 + 2Al$$

In alumina reinforced composites with a high level of Mg in the matrix, MgO is expected to form at the interface, while spinel forms at low levels of Mg (Pfeifer et al., 1990).

Figure 5.4 shows a transmission electron microscopy (TEM) micrograph of the reaction zone between alumina fiber and magnesium matrix made by liquid metal infiltration. The interface layer(s) formed generally have characteristics different from those of either one of the components. At times some controlled amount of reaction at the interface may be desirable for obtaining strong bonding between the fiber and the matrix; too thick an interaction zone, of course, adversely affects the composite properties.

The importance of matrix alloy composition cannot be over-emphasized. For example, in pure Mg matrix, SiC particle would be stable and no reaction is observed at the interface. In Mg alloys containing Al and Si as the main alloying elements, interfacial reaction can occur between Al and SiC as described above. Such a reaction will form the undesirable Al_4C_3 and inject Si into the matrix, which can react with Mg to form Mg_2Si. Another important composite system involves SCS-6 type silicon carbide fiber and a titanium alloy (Ti-6Al-4V) matrix. Interfacial reaction products in this case include TiC and Ti_xSi_y (Gabryel and McLeod, 1991).

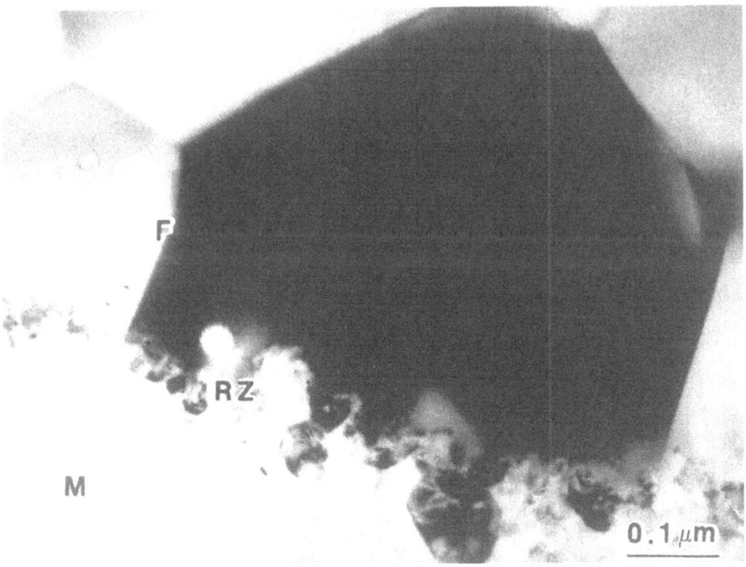

Fig. 5.4 Bright field (BF) transmission electron micrograph showing interface in a continuous fiber (F) α-Al_2O_3/Mg alloy (ZE41A) matrix (M). RZ indicates the interfacial reaction zone.

We provide a summary of interfacial reaction products in some important metal matrix composites in Table 5.1. It is important to have an understanding of the structure and properties of the reaction products that form in a given system because they will be the key in determining the final properties of the composite. For example, in carbon fiber reinforced aluminum processed at around 700°C, Al_4C_3 forms. It is also very sensitive to ambient moisture. It would be very desirable to avoid formation of such brittle phases at the interface because they lead to catastrophic failure of the composite.

5.3.3 Interactions at the Interface due to Thermal Mismatch

Thermal stresses arise in composite materials because of the mismatch between the coefficients of thermal expansion of the components making up the composite. Mismatch in elastic constants of the components exacerbates the problem. In a fiber reinforced composite, if the matrix thermal expansion coefficient is higher than that of the fiber, then on cooling from high temperature radial compression will result, i.e., the matrix will grip the fiber. Such radial gripping of the fiber by the matrix will increase the

Table 5.1: Interfacial reaction products in some important MMCs

Reinforcement	**Matrix**	**Reaction product(s)**
SiC	Ti alloy Aℓ alloy	TiC, Ti_5Si_3 $A\ell_4C_3$
$A\ell_2O_3$	Mg alloy	MgO, $MgA\ell_2O_4$ (spinel)
$A\ell_20_3$	Aℓ alloy	None
B	Aℓ alloy	$A\ell B_2$
ZrO_2	Aℓ alloy	$ZrA\ell_3$
C	Aℓ alloy	$A\ell_4C_3$
C	Cu	None
W	Cu	None
NbTi Nb_3Sn	Cu	None
Various Oxides	Ag	None

strength of the interface. This increases the tendency toward brittleness in the composite. This will make phenomena such as fiber debonding and pullout, desirable for enhanced toughness, difficult to operate.

Quite frequently, the very process of fabrication of a composite can involve interfacial interactions that can cause changes in the constituent properties and/or interface structure. For example, if the fabrication process involves cooling from high temperatures to ambient temperature, then the difference in the expansion coefficients of the two components can give rise to thermal stresses of such a magnitude that the softer component (generally the matrix) will deform plastically. Chawla and Metzger (1972) studied a tungsten reinforced single crystal copper matrix (non-reacting components). They observed that liquid copper infiltration of tungsten fibers at about 1100°C, followed by cooling to room temperature, resulted in a dislocation density in the copper matrix that was much higher in the vicinity of the interface than away from the interface. The high dislocation density in the matrix near the interface resulted from the plastic deformation of the matrix caused by high thermal stresses near the interface. Other researchers observed similar results in other systems such as SiC whiskers in an aluminum matrix (Arsenault and Fisher, 1983) and short alumina fibers in aluminum matrix (Dlouhy et al., 1993). If powder metallurgy fabrication techniques are used, the nature of the powder surface will influence the interfacial interactions. For example, an oxide film on the powder surface affects its chemical nature. Topographic characteristics of the components can also affect the degree of atomic contact obtainable between the components. This can result in geometrical irregularities (e.g., asperities and voids) at the interface, which can be source of stress concentrations.

5.4 MEASUREMENT OF INTERFACIAL BOND STRENGTH

Once the matrix and the reinforcement of a composite are chosen, it is the set of characteristics of the interface region that determines the final properties of the composite. In this regard, thorough characterization of the interface region assumes a great deal of importance. A variety of sophisticated techniques are available to mechanical characterization of the interface region. In particular, a quantitative measure of the strength of the interfacial bond between the matrix and reinforcement is of great importance. We describe below some of the important techniques to measure interfacial bond strength.

5.4.1 Bend Tests

Bend tests are easy to perform but they do not give a true measure of the interfacial strength. We describe some variants of bend tests below.

Transverse Bend Test

A three-point bend test configuration with fibers aligned perpendicular to the specimen length is called a transverse bend test. There are two possible arrangements of fibers, one with fibers parallel to the length of the specimen and the other with fibers transverse to the length of fibers. Under either one of these configurations, fracture will occur on the outermost surface of the specimen which is under maximum tensile stress. This will put the fiber/matrix interface under tension, which gives us a measure of tensile strength of the fiber/matrix interface. The transverse strength is given by:

$$\sigma = \frac{3PS}{2bh^2} \tag{5.1}$$

where P is the applied load, S is the load span, b is the specimen width, and h is the specimen height.

Longitudinal Bend Test or Short Beam Shear Test

This test is also known as the *I*nter*L*aminar *S*hear *S*trength (ILSS) test. In this test, the fibers are aligned parallel to the length of the three-point bend bar. In such a test, the maximum shear stress, τ, occurs at the midplane and is given by:

$$\tau = \frac{3P}{4bh} \tag{5.2}$$

where the symbols have the significance given above.

The maximum tensile stress occurs at the outermost surface and is given by Eq. 5.1. Dividing Eq. 5.2 by Eq. 5.1, we get:

$$\frac{\tau}{\sigma} = \frac{h}{2S} \tag{5.3}$$

Inspection of Eq. (5.3) shows that we can maximize the shear stress by making the load span, S, arbitrarily small. This ensures that the specimen fails predominantly under shear with a crack running along the midplane. The test becomes invalid if the fibers fail in tension before shear induced failure occurs. The test will also be invalid if shear and tensile failure occur simultaneously. It is advisable to make an examination of the fracture surface after the test and make sure that the crack is along the interface and not through the matrix.

5.4.2 Fiber Pullout and Pushout Tests

Single fiber pullout and pushout tests have been devised to measure interfacial characteristics. Such tests give load vs. displacement curves, with the peak load corresponding to fiber/matrix debonding and a frictional load corresponding to the fiber pullout from the matrix. The main simplification that is frequently made in the analysis of such tests involves averaging the load values over the entire interfacial surface area to get the interface debond strength and/or frictional strength. Analytical and finite element analyses show that the shear stress is a maximum close to the fiber end and falls rapidly within a distance of a few fiber diameters. Thus, one would expect the interface debonding to start near the fiber end and progressively propagate along the embedded length. Here we describe the salient features of these tests.

Single Fiber Pullout Tests

Single fiber pullout tests can provide useful information about the interface strength in *model* composite systems. They are not very helpful in the case of commercially available composites. One must also carefully avoid any fiber misalignment and introduction of bending moments. The mechanics of the single fiber pullout test is complicated (see, e.g., Penn and Lee, 1989; Marshall et al., 1992; Kerans and Parthasarthy, 1991). In all the variants of such test, the fiber is pulled out of the matrix in a tensile testing machine and a load vs. displacement record is obtained.

Figure 5.5 shows the experimental setup for such a test. A portion of fiber, length ℓ, is embedded in a matrix. We apply a pulling tensile force, as shown, and measure the stress required to pull the fiber out of the matrix as a function of the embedded fiber length. The stress required to pull the fiber out without breaking it increases with the embedded fiber length, up to a critical length, ℓ_c. This critical fiber length is the length of fiber that is used in load transfer form matrix to fiber. We discuss the concept of critical fiber

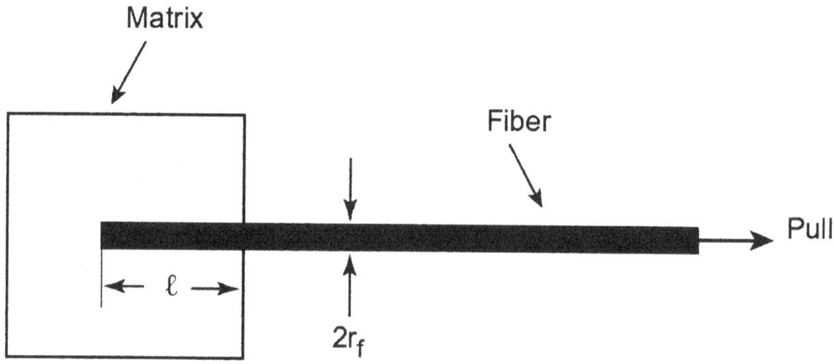

Fig. 5.5 A fiber of radius r embedded in a matrix and being pulled out. The longitudinal tensile force on the fiber generates shear at the fiber/matrix interface.

length in chapter 7. For an embedded fiber length, $\ell > \ell_c$, the fiber will fracture under the action of the tensile stress, σ, acting on the fiber. When we apply a tensile force on the fiber, a shear force at the fiber/matrix interface will result. A simple force balance along the embedded fiber length gives:

$$\sigma \pi r^2 = \tau 2 \pi r \ell$$

For $\ell < \ell_c$, the fiber is pulled out and the interfacial shear strength is given by:

$$\tau = \frac{\sigma r}{2\ell}$$

One measures the load, P, required to debond the interface as a function of the embedded fiber length. Then, we can write:

$$P = \tau 2 \pi r \ell$$

and the interfacial shear strength, τ, can be calculated from the slope of P vs. ℓ plot. For $\ell > \ell_c$, fiber failure rather than pullout occurs. The interfacial shear strength is a function of the coefficient of friction, μ, and any normal compressive stress at the interface, σ_r, i.e., $\tau_i = \mu\sigma_r$. As mentioned above, a common source of radial compressive stress is the shrinkage of the matrix during cooling from the processing temperature. The most doubtful assumption in this analysis is that the shear stress acting along the fiber/matrix interface is a constant.

The fabrication of the single fiber pullout test sample is often the most difficult part; it entails embedding a part of the single fiber in the matrix. The peak load corresponds to the initial debonding of the interface. This is followed by frictional sliding at the interface, and finally by fiber pullout from the matrix, during which a steady decrease in the load with displacement is observed. The steady decrease in the load is attributed to the decreasing area of the interface as the fiber is pulled out. Thus, the test simulates the fiber pullout that may occur in the actual composite, and more importantly, provides the bond strength and frictional stress values.

The effect of different Poisson contractions of fiber and matrix can lead to a radial tensile stress at the interface. The radial tensile stress will no doubt aid the fiber/matrix debonding process. The effect of Poisson contraction, together with the problem that the imposed shear stress is not constant along the interface, complicates the analysis of fiber pullout test.

Pushout or Indentation Tests

Interfacial frictional sliding is an important characteristic of the fiber/matrix itnerface. Many researchers (for example, Marshall, 1984; Doerner and Nix, 1986; Eldridge and Brindley, 1989; Ferber et al., 1993; Mandell et al., 1986; Mandell et al., 1987; Marshall and Oliver, 1987; Cranmer, 1991; Chawla et al., 2001) have used the technique of pressing an indenter on the cross-section of a fiber in a composite to measure the interfacial bond strength in a fiber/reinforced composite. An instrumented indentation system, the apparatus for which is sometimes called a *nanoindenter*, is available commercially. Such an instrument allows extremely small forces and displacements to be measured. A nanoindenter is essentially a computer-controlled depth-sensing indentation system. Indentation instruments have been in use for hardness measurement since early twentieth century, but "depth sensing" instruments having high resolution became available in the 1980s (Doerner and Nix, 1986; Weihs and Nix, 1991). Very small volumes of material can be studied and a very local characterization of microstructural variations is possible by mechanical means. Indenters with pointed, rounded, and flat tips can be used to displace a fiber aligned perpendicular to the composite surface. By measuring the applied force and the displacement, the interfacial stress can be obtained. One generally loads several fibers in a polished cross section of composite system. Most commercially available nanoindenter instruments are capable of accurately applying very small loads (mN) via a Berkovich pyramidal diamond indenter having the same depth-area ratio as a Vickers diamond tip indenter. A nanoindenter records the total penetration of an indenter into the sample. The position of the indenter is determined by a sensitive capacitance displacement gage. The capacitance gage can detect displacement changes less than one nm, while the applied force can be detected to less that 1 μN. The indenter can be moved toward the sample or away from the sample by means of a magnetic coil assembly.

In the pushout test, one pushes a fiber out, by means of an indenter, in a thin sample with the fibers aligned perpendicular to the viewing surface. Such a fiber pushout test gives us the frictional shear stress, τ, acting at the fiber/matrix interface. In a valid pushout test, a three-region curve is obtained, see Fig. 5.6. In the first region, the indenter is in the contact with the fiber. Following elastic loading, initial interface debonding takes place. Progressive debonding continues until the interface is fully debonded. This is followed by a horizontal region (region two) which consists of interfacial sliding of the fiber in the matrix. In the third region, the indenter comes in contact with the matrix. From the second region, we can determine the interfacial shear stress. The specimen thickness should be much greater that the fiber diameter for this relationship to be valid. In the horizontal region, the interfacial shear stress is given by:

$$\tau_i = \frac{P}{2\pi rt} \tag{5.4}$$

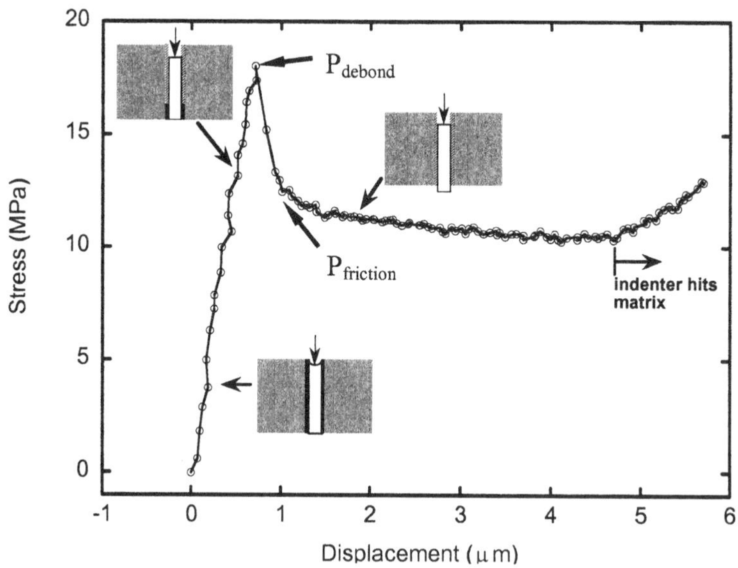

Fig. 5.6 Stress-displacement curve during fiber pushout. After elastic loading, progressive fiber debonding takes place, followed by interfacial sliding (Chawla et al., 2001).

where t is the specimen thickness. In the third region of Fig. 5.6 the indenter comes in contact with the matrix.

If the fiber is strongly bonded to the matrix, the energy at the crack tip at the interface will be high enough to fracture the fiber. In this case, the crack propagates straight through the interface, into the fiber, and the composite fails in a catastrophic manner. If the fiber/matrix interface is tailored, such that bonding between fiber and matrix is weak, a crack propagating normal to the interface will be deflected at the interface, causing it to lose energy. In this manner, debonding and sliding of the fiber with respect to the matrix acts as an energy-absorbing mechanism.

Several important factors control the fiber/matrix behavior, such as thermal residual stresses from processing, that can induce radial clamping stresses and axial stresses from the matrix to the fiber. Thermal mismatch in coefficient of thermal expansion (CTE) between fiber and matrix is responsible for residual stresses in all types of composites (see chapter 6). If the matrix shrinks more than the fiber then a compressive clamping stress will act at the interface

Fiber surface roughness can also contribute significantly to the radial clamping stresses, because the matrix becomes mechanically keyed to the fiber. A roughness induced strain arises because of this mechanical keying, and it can be estimated by the roughness amplitude between fiber and matrix. To relieve radial clamping stresses from CTE mismatch as well as from fiber roughness, a compliant fiber coating can be applied to the fiber. The effect of interfacial clamping stress, temperature, and environment on the debonding and sliding of the pushed fibers was examined by Eldridge and Ebihara (1994). They studied interfacial characteristics in the system SCS-6/Ti-24Al-11Nb (at.%), from ambient temperature to 1100°C. The fiber debond stress and frictional sliding stress decreased continuously at temperatures higher than 300°C, Fig. 5.7. The decrease in interfacial shear strength was attributed to relief of the residual compressive stresses on the fiber, and oxidation of the carbon coating at higher temperatures. Figure 5.8 shows the pushed out fiber at low and high temperatures, showing a relatively rough and smooth surface, respectively.

Eldridge and Ebihara (1994) also showed that the pushout characteristics were a function of environment. They postulated that the sensitivity of the degree of interfacial sliding to the environment can be attributed to the well-documented (Savage, 1948; Lancaster and Pritchard, 1981) to the lubricative properties of graphite, which exists on the outer shell of the

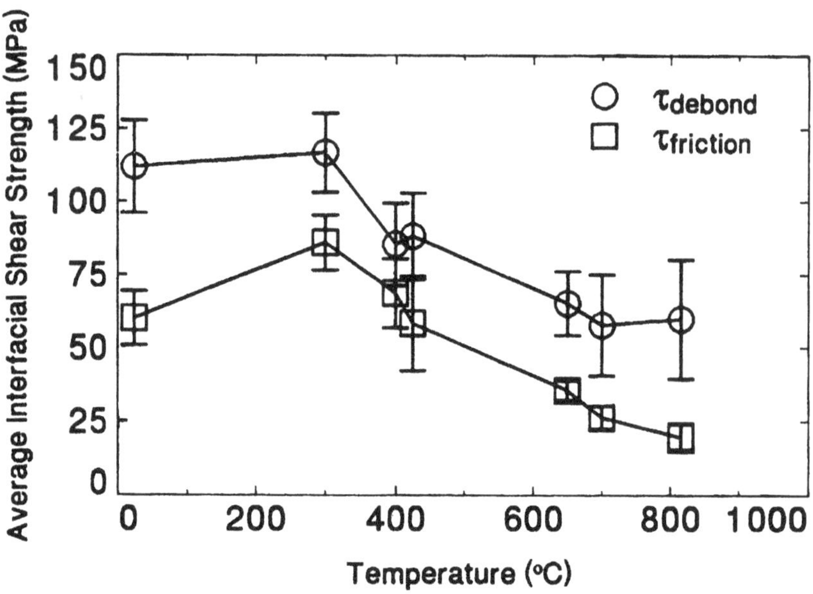

Fig. 5.7 Effect of temperature on fiber debond stress and frictional sliding stress in SCS-6/Ti-24Al-11Nb (Eldridge and Ebihara, 1994).

SCS-6 fiber and thus, acts as a weak interface between the fiber and the matrix. The low friction and wear at room temperature of graphite is attributed to water in the form of moisture, and secondarily oxygen adsorption (Savage, 1948). In the absence of adsorbed species (in vacuum), sliding graphite surfaces exhibited "dusting" wear, producing extremely fine debris. At room temperature in air, once the interface in SCS-6 fiber/matrix debonded, the gap created at the failed interface exposed both sides of the interface to the environment and the two surfaces could slide smoothly.

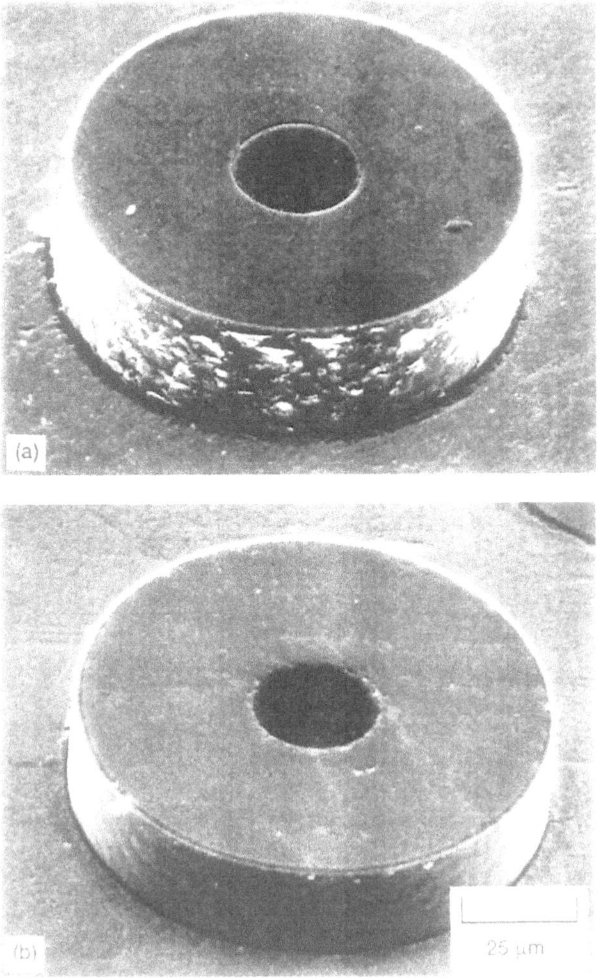

Fig. 5.8 SEM micrographs of pushed out fibers at (a) low temperature and (b) high temperature, showing a relatively rough and smooth surface, respectively (courtesy of J.I. Eldridge).

References

Arsenault, R.J., and R.M. Fisher, (1983) *Scripta Metall.*, **17**, 43.

Chawla, K.K., and Metzger, M., (1972) *J. Mater. Sci.*, **7**, 34.

Chawla, K.K. and M. Metzger, (1978) in *Advances in Research on Strength and Fracture of Materials*, Vol. 3, Pergaman Press, New York, p.1039.

Chawla, N., K.K. Chawla, M. Koopman, B. Patel, C. Coffin, and J.I. Eldridge, (2001) *Composites Sci. Tech.*, **61**, 1923.

Chawla, N., J.W. Holmes, and J.F. Mansfield, (1995) *Mater. Charac.*, **35** 199.

Cranmer, D.C., (1991) in *Ceramic and Metal Matrix Composites*, Pergamon Press, New York, p. 157.

Dlouhy, A., N. Merk, and G. Eggeler, (1993) *Acta Metall. Mater.*, **41**, 3245-3256.

Doerner, M.F., and W.D. Nix, (1986) *J. Mater. Res.*, **1**, 601.

Eldridge, J.I., and P.K. Brindley, (1989) *J. Mater. Sci. Lett.*, **8**, 1451.

Eldridge, J.I., and B.T. Ebihara, (1994) *J. Mater. Res.*, **9**, 1035.

Ferber, M.K., A.A. Wereszczsak, L. Riester, R.A. Lowden, and K.K. Chawla, (1993) *Ceram. Eng. Sci. Proc.*, **13**, 168.

Gabryel, C.M., and A.D. McLeod, (1991) *Metall. Trans*, **23A**, 1279.

Hill, R.G, R.P. Nelson, and C.L. Hellerich, (October, 1969) in *Proc. of the 16th Refractory Working Group Meeting*, Seattle, WA.

Kerans, R.J., and T.A. Parthasarathy, (1991) *J. Amer. Ceram. Soc.*, **74**, 1585.

Lancaster, J.K., and J.R. Pritchard, (1981) *J. Phys. D: Appl. Phys.*, **14** 747.

Liaw, P.K., Y. Ijiri, B.J. Taszarek, S. Frohlich, M.N. Gungor, and W.A. Logsdon, (1990) *Metall. Mater. Trans.*, **21**, 529-538.

Mandell, J.F., D.H. Grande, T.H. Tsiang, and F.J. McGarry, (1986) in *Composite Materials:Testing & Design*, ASTM STP 327, ASTM, Philadelphia, p. 87.

Mandell, J.F., K.C.C. Hong, and D.H. Grande, (1987) *Ceram. Eng. Sci. Proc.*, **8**, 937.

Marshall, D.B., (1984) *J. Amer. Ceram. Soc.*, **67**, C259.

Marshall, D.B., and W.C. Oliver, (1987) *J. Amer. Ceram. Soc.*, **70**, 542.

Marshall, D.B., M.C. Shaw, and W.L. Morris, (1992) *Acta Metall.*, **40**, 443.

Penn, L.S., and S.M. Lee, (1989) *J. Comp. Tech. & Res.*, **11**, 23.

Pfeifer, M., J.M. Rigsbee, and K.K. Chawla, (1990) *J. Mater. Sci.*, **25**, 1563.

Savage, R.H., (1948) *J. Appl. Phys.*, **19** 1.

Weihs, T.P., and W.D. Nix, (1991) *J. Amer. Ceram. Soc.*, **74**, 524.

Chapter 6

MICROMECHANICS

In this chapter, we examine a variety of means of calculating elastic and physical properties of metal matrix composites, given the same constants for the individual components and the arrangement of components in the composite; and thermal stresses generated because of mismatch in the coefficient of thermal expansion (CTE) of the components. In fact, most of the material discussed in this chapter is applicable to all kind of composites. Specifically, we provide a micromechanical description of physical properties such as density, thermal expansion coefficients, thermal and electrical conductivity, and various elastic constants. Of particular interest are methods or expressions that predict elastic constants of composites because of the generally high anisotropy found in composites. A description of conventional and microstructure-based finite element techniques to predict the elastic and thermal constants is also provided.

We briefly review the theory of elasticity and the concept of number of independent elastic constants required for a fiber reinforced composite. We then provide expressions for the elastic constants as per different schemes available. This is followed by a description of physical properties and thermal stresses in composites.

6.1 ELASTIC CONSTANTS OF A FIBER REINFORCED COMPOSITE

Hooke's law relates the second rank **stress** tensor, σ_{ij}, and the second rank **strain** tensor, $\varepsilon_{k\ell}$, and can be written as:

$$\sigma_{ij} = C_{ijkl}\varepsilon_{k\ell} \tag{6.1}$$

where C_{ijkl} is a fourth-rank tensor called the elastic stiffness tensor and the indices i, j, k, and ℓ have values of 1, 2, and 3. The stress and strain tensors, being second rank, have $3^2 = 9$ components each. The stiffness tensor is a fourth rank tensor, it has $3^4 = 81$ components. Stress, strain, and stiffness are symmetric tensors, i.e., $\sigma_{ij} = \sigma_{ji}$, $\varepsilon_{k\ell} = \varepsilon_{\ell k}$, and $C_{ijk\ell} = C_{k\ell ji} = C_{ji\ell k} = C_{jik\ell}$. These symmetry relationships reduce the maximum number of independent elastic constants to 21. Hooke's law in (Eq. 6.1) can then be simplified and rewritten in a contracted notation more appropriate for matrix operations as follows:

$$\sigma_i = C_{ij}\varepsilon_j \qquad (6.2)$$

In a similar manner we can write Hooke's law in terms of the elastic compliance matrix, S_{ij}, as:

$$\varepsilon_i = S_{ij}\sigma_j \qquad (6.3)$$

Both C_{ij} and S_{ij} are 6 x 6 symmetric matrices, i.e., $C_{ij} = C_{ji}$ and $S_{ij} = S_{ji}$. Also, the stiffness and the compliance matrices are inverse of each other, i.e.:

$$[C][S] = [I]$$

where $[I]$ is the identity matrix. The total number of independent constants required to describe the elastic behavior of a material completely decreases with increasing symmetry elements present. Table 6.1 presents a summary.

Table 6.1: Independent elastic constants for systems of different symmetry.

Symmetry	**Number of Independent Constants**	**Constants**
Orthorhombic	9	$C_{11},C_{12},C_{13},C_{22},C_{23},C_{33},C_{44},C_{55},C_{66}$
Trigonal	6	$C_{11},C_{12},C_{13},C_{14},C_{33},C_{44}$
Tetragonal	6	$C_{11},C_{12},C_{13},C_{33},C_{44},C_{66}$
Hexagonal	5	$C_{11},C_{12},C_{13},C_{33},C_{44}$
Cubic	3	C_{11},C_{12},C_{44}
Isotropic	2	C_{11},C_{12}

For an isotropic material, there are only two independent elastic constants. For a material with cubic symmetry, we need three independent elastic constants. The most general situation is that of the triclinic system (not shown in Table 6.1) in which there are no symmetry elements present; it requires 21 independent elastic constants. For a material with cubic symmetry, we can write Hooke's law in an extended form as follows:

$$\begin{bmatrix} \sigma_1 \\ \sigma_2 \\ \sigma_3 \\ \sigma_4 \\ \sigma_5 \\ \sigma_6 \end{bmatrix} = \begin{bmatrix} C_{11} & C_{12} & C_{12} & 0 & 0 & 0 \\ & C_{11} & C_{12} & 0 & 0 & 0 \\ & & C_{11} & 0 & 0 & 0 \\ & & & C_{44} & 0 & 0 \\ & & & & C_{44} & 0 \\ & & & & & C_{44} \end{bmatrix} \begin{bmatrix} \varepsilon_1 \\ \varepsilon_2 \\ \varepsilon_3 \\ \varepsilon_4 \\ \varepsilon_5 \\ \varepsilon_6 \end{bmatrix}$$

Note that there are only three independent constants. For an isotropic material, the number of independent constants is reduced from 3 to 2 because of the following relationship:

$$C_{11} - C_{22} = 2C_{44}$$

In engineering usage, for an isotropic material, any two of the following four constants will suffice: Young's modulus (E), Poisson's ratio (ν), shear modulus (G), and bulk modulus (K). This is because of the following relationships among the four constants:

$$E = 3K(1 - 2\nu)$$

$$G = \frac{E}{2(1 + \nu)}$$

$$K = \frac{E}{3(1 - 2\nu)}$$

$$\text{and} \quad \nu = \frac{E}{2G} - 1$$

Thus, only two of the four constants for an isotropic material are independent.

A unidirectionally reinforced fiber reinforced composite with fibers arranged in a random manner in the transverse section is said to be transversely isotropic, i.e., there is no preferred direction in the 2-3 plane, see Fig. 6.1. Such an arrangement of fibers gives us the same elements as the hexagonal crystal, i.e., five independent elastic constants are needed to fully describe the elastic behavior of such a composite. Note that this follows from the disposition of long fibers in a matrix, even though the two components, fiber and matrix, may individually be isotropic in nature. On the other hand, a particle or whisker/short fiber reinforced composite, with no preferential alignment of reinforcement, can be treated as an isotropic material. We describe a variety of approaches to obtain the elastic constants of the composite below, given the elastic constants of the individual components.

6.1.1 Strength of Materials Approach

We can get some quick estimates of elastic constants of a composite by using the strength of materials approach. Here we make simplifying assumptions of uniform strain or uniform stress in the constituents of the composite. The results are satisfactory for E_{11} and ν_{12}, but underestimate E_{22} and G_{12}. We use $E_{c\ell}$ and E_{11} interchangeably to indicate longitudinal Young's modulus in the fiber direction. Similarly, E_{ct} and E_{22} indicate Young's modulus in the direction transverse to the fiber axis. Two simple

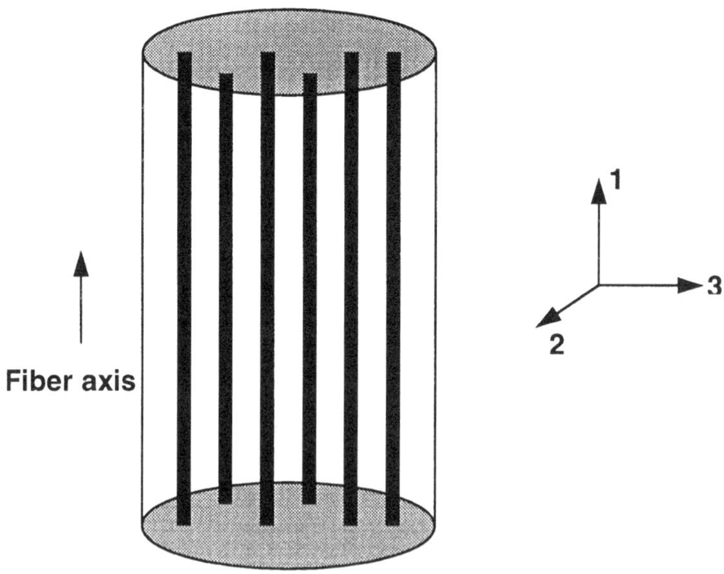

Fig. 6.1 A unidirectionally reinforced fiber reinforced composite with fibers arranged in a random in the transverse (2-3) plane. Such a composite is called transversely isotropic.

cases are *isostrain* and *isostress* models for Young's modulus of a unidirectionally aligned, fiber reinforced composite, see Fig. 6.2. We shall also derive expressions for the principal shear modulus and principal Poisson's ratio.

Longitudinal Young's Modulus

If we apply the *isostrain* condition to a unidirectional, fiber reinforced composite loaded along the fiber direction, we get the longitudinal Young's modulus, E_{cl} or E_{11}, of the composite. The isostrain condition says that the strains in the fiber, matrix, and composite are identical, see Fig. 6.2; i.e.:

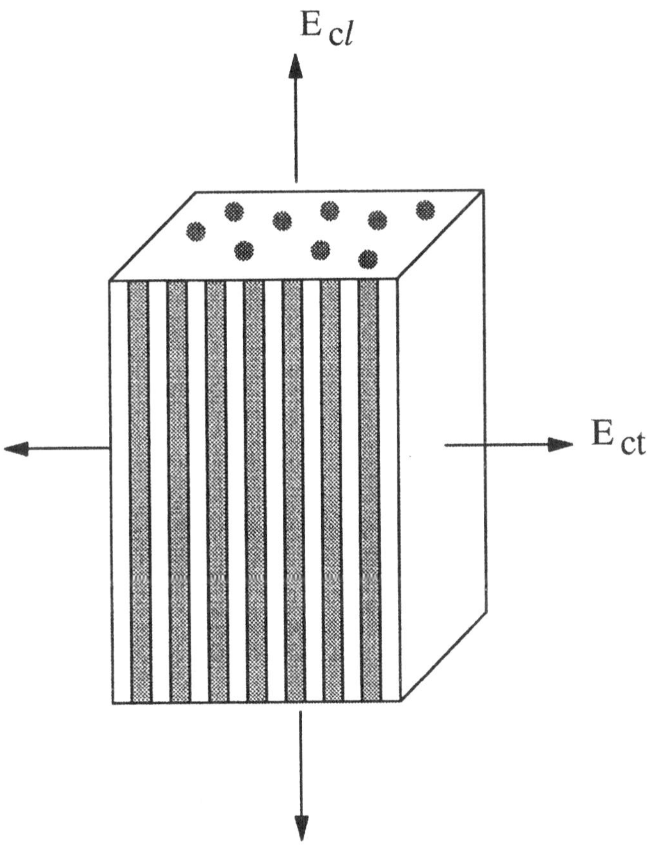

Fig. 6.2 Longitudinal ($E_{c\ell}$) and transverse (E_{ct}) Young's moduli for a unidirectionally reinforced fiber reinforced composite.

$$\varepsilon_f = \varepsilon_m = \varepsilon_{c\ell} = \frac{\Delta\ell}{\ell_o} \tag{6.4}$$

where ε is the strain, $\Delta\ell$ is the change in length, ℓ_o is the original length, and the subscripts f, m, and $c\ell$ indicate fiber, matrix, and composite in the longitudinal direction, respectively.

For components behaving elastically, we can use Hooke's law for a uniaxial stress acting on the fiber and the matrix:

$$\sigma_f = E_f \varepsilon_{c\ell} \quad \text{and} \quad \sigma_m = E_m \varepsilon_{c\ell}$$

where σ is the stress, E is the Young's modulus, and the subscripts have the meanings given above.

The applied load on the composite, P_c, is partitioned between the fiber and the matrix, i.e.:

$$P_c = P_f + P_m$$

where P_f and P_m indicate the load on the fiber and matrix, respectively. Converting loads into stresses, we can write:

$$\begin{aligned} \sigma_{c\ell} A_c &= \sigma_f A_f + \sigma_m A_m \\ &= E_f A_f \varepsilon_{c\ell} + E_m A_m \varepsilon_{c\ell} \\ &= (E_f A_f + E_m A_m)\, \varepsilon_{c\ell} \\ \sigma_{c\ell} &= E_{c\ell} \varepsilon_{c\ell} = (E_f A_f + E_m A_m)\, \varepsilon_{c\ell} \end{aligned}$$

From this we get the following expressions:

$$E_{c\ell} = E_f V_f + E_m V_m \qquad (6.5)$$

$$\sigma_{c\ell} = E_{c\ell} \varepsilon_{c\ell} = (E_f A_f + E_m A_m)\, \varepsilon_{c\ell} \qquad (6.6)$$

It turns out that predictions given by Eq. 6.5 are quite reasonable. As an example, Fig. 6.3 shows the linear dependence of longitudinal Young's modulus of a tungsten fiber reinforced copper as a function of fiber volume fraction (McDanels et al., 1965).

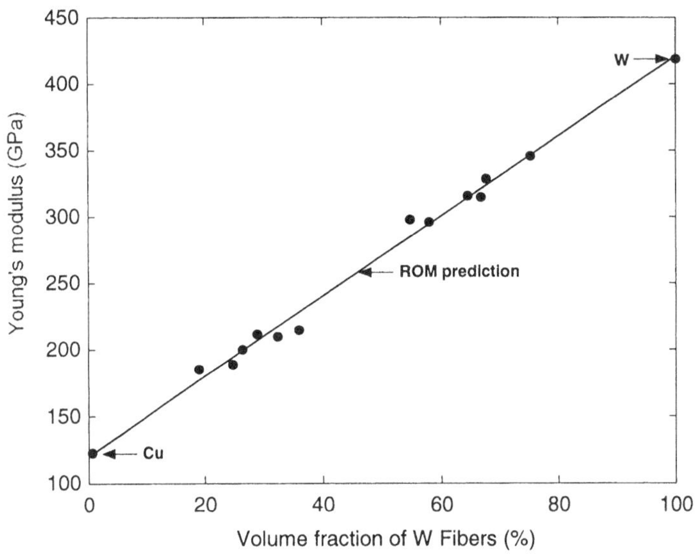

Fig. 6.3 Linear dependence of longitudinal Young's modulus of tungsten fiber reinforced copper as a function of fiber volume fraction (after McDanels et al., 1965).

Transverse Young's Modulus

The transverse modulus, E_{ct} or E_{22}, can be estimated by applying an isostress condition, i.e., the fiber, matrix, and composite experience the same stress, Fig. 6.2. Thus:

$$\sigma_f = \sigma_m = \sigma_{ct} \qquad (6.7)$$

where the subscript ct denotes the composite in the transverse direction, and the other subscripts have the significance given earlier.

The total displacement of the composite in the thickness direction, Δt_c, is the sum of the displacement in the fiber, Δt_f, and that in the matrix, Δt_m. We can then write the following relationship for the displacement in the thickness direction:

$$\Delta t_c = \Delta t_m + \Delta t_f$$

Let the original thickness of the composite be t_c. Dividing throughout by t_c, the gage length, we get the strain in the transverse direction:

$$\varepsilon_{ct} = \frac{\Delta t_c}{t_c} = \frac{\Delta t_m}{t_m}\frac{t_m}{t_c} + \frac{\Delta t_f}{t_f}\frac{t_f}{t_c}$$

$$\varepsilon_{ct} = \varepsilon_m \frac{\Delta t_m}{t_c} + \varepsilon_f \frac{t_f}{t_c} \qquad (6.8)$$

Also, we can write the volume fractions of fiber and matrix as:

$$V_m = \frac{t_m}{t_c} \text{ and } V_f = \frac{t_f}{t_c}$$

Using Hooke's Law, we can rewrite Eq. (6.8) as:

$$\frac{\sigma_{ct}}{E_{ct}} = \frac{\sigma_{ct} V_m}{E_m} + \frac{\sigma_{ct} V_f}{E_{ft}}$$

or

$$\frac{1}{E_{ct}} = \frac{V_m}{E_m} + \frac{V_f}{E_{ft}} \qquad (6.9)$$

We should point out that when computing Young's modulus in the transverse direction, we have chosen the transverse modulus of the fiber, E_{ft}. This is especially important when the fiber is anisotropic, for example, carbon fiber.

Poisson's Ratio

Consider a composite containing unidirectionally aligned fibers loaded in tension parallel to the fibers, see Fig. 6.4. The composite will extend longitudinally by ε_1 (direction 1) and contract transversely by ε_2 (direction 2). The contraction in direction 2 due to all the fibers will be $-\varepsilon_1 V_f \nu_f$, where V_f is the fiber volume fraction and ν_f is the Poisson's ratio of the fiber. The contraction due to the matrix will be $-\varepsilon_1 V_m \nu_m$, where V_m is the matrix volume fraction and ν_m is the Poisson's ratio of the matrix.

The total contraction of the composite in direction 2 is then given by:

$$\varepsilon_2 = -\varepsilon_1 V_f \nu_f - \varepsilon_1 V_m \nu_m$$

$$= -\varepsilon_1 (V_f \nu_f + \varepsilon_1 V_m \nu_m)$$

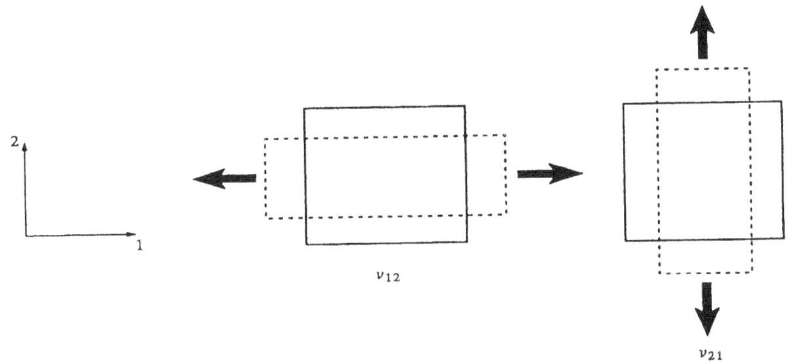

Fig. 6.4 The two Poisson's ratios in an anisotropic material. The principal Poisson's ratio (ν_{12}) gives the transverse strain caused by an axial stress. The secondary Poisson's ratio (ν_{21}) gives the axial strain caused by a transversely applied stress.

Defining the principal Poisson's ratio of the composite as $\nu_{12} = -\varepsilon_2/\varepsilon_1$ we get:

$$\nu_{12} = V_f \nu_f + V_m \nu_m$$

Shear Modulus

In the case of longitudinal or principal shear modulus, both the fiber and matrix are subjected to the same shear stress as shown in Fig. 6.5. Note that the fibers have been bundled together to show total shear in all the fibers. We can write the shear strain in the matrix and fiber as follows:

$$\gamma_m = \frac{\tau_m}{G_m}$$

and

$$\gamma_f = \frac{\tau_f}{G_f}$$

The total shear displacement in the composite, Δ, is given by:

$$\Delta = \gamma t$$

where γ is the average shear strain in the composite and t is the thickness of the composite. We can then write the total shear displacement as the sum of the shear displacements of the components:

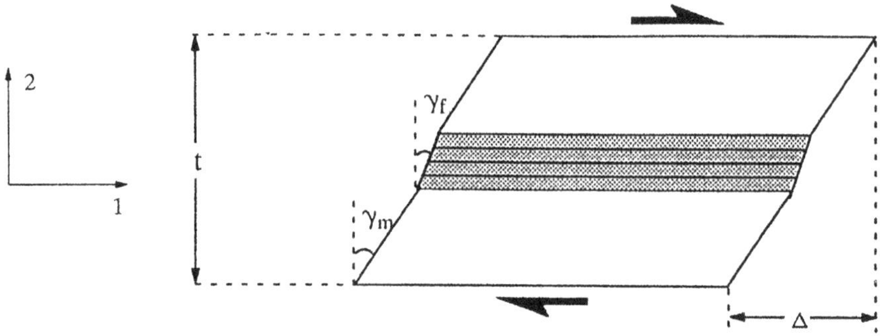

Fig. 6.5 A unidirectional fiber reinforced composite loaded in shear parallel to the fibers.

$$\Delta = \Delta_f + \Delta_m$$

or,
$$\Delta = \gamma_f V_f t_f + \gamma_m V_m t_m$$

$$\gamma = \frac{\Delta}{t} = \gamma_f V_f + \gamma_m V_m$$

Shear strain, γ, is nothing but shear stress divided by shear modulus, G_{12}, i.e.,:

$$\tau/G_{12} = V_m \tau/G_m + V_f \tau/G_f$$

or $$1/G_{12} = V_m/G_m + V_f/G_f$$

6.1.2 Micromechanical Approaches

In this section we include the following important techniques:

- Self-consistent field methods
- Variational calculus methods
- Numerical methods

Brief descriptions of these as well as some key results are presented below.

Self-Consistent Field Methods: The strength of materials approach described above involves gross simplification of uniform stress (isostress) or uniform strain (isostrain). Self-Consistent (SC) models improve on the internal stress and strain fields by introducing a simplified geometry of the phases. It should be pointed out, though, that in these techniques one still makes approximations of microstructure, i.e., the actual microstructure is not used. In one version, the phase geometry is represented by one single fiber embedded in a matrix cylinder. This outer cylinder is embedded in an infinite, homogeneous material whose properties are taken to be the average properties of the composite. In the second version, a three-cylinder model is used. The intermediate cylinder surrounding the fiber has the properties of the matrix. The outermost cylinder has average properties of the composite. The radii of the cylinders are dictated by the fiber volume fraction. A uniform load, applied at infinity, introduces a uniform strain field in the fiber. Elastic constants are then obtained from this strain field. The results obtained are independent of fiber arrangements in the matrix and, in general, are reliable at low fiber volume fractions (V_f), reasonable at intermediate V_f, and unreliable at high V_f (Hill, 1964).

Variational Calculus or Energy Methods: Energy methods involving variational calculus can be used to obtain bounds on a property of the composite. These techniques are also called bounding methods. Effective elastic constants (or compliances) of a composite (a heterogeneous system) are obtained from those of a homogeneous system with the same free energy. In simple terms, we can describe the rationale behind these methods as follows. Consider a linear elastic solid under deformation. We can express the strain energy stored in this solid in terms of a strain field:

$$U_s = \frac{1}{2} S_{ij} \sigma_j$$

and in terms of stress field:

$$U_c = \frac{1}{2} C_{ij} \varepsilon_j$$

These two expressions are equivalent for a homogeneous material but not for a heterogeneous material. The difference between these two expressions for a heterogeneous material (i.e., a composite) can be exploited to obtain the upper and lower bounds. Specifically, the theorem of least work gives the lower bound while the theorem of minimum potential energy provides the upper bound. There is a considerable amount of literature available on the subject; see, for example, Paul (1960); Hermans (1967); Hashin and Rosen (1964); Whitney and Riley (1966).

These bounding methods do not predict properties exactly, but give upper and lower bounds on elastic constants. The property is determined exactly only if the upper and lower bounds coincide. More often than not, the upper and lower bounds are not very close. Only when these bounds are close enough can we safely use them as indicators of the material behavior. It turns out that this is the case for longitudinal properties of a unidirectionally aligned fiber reinforced composite such as longitudinal constants (E_{11}, ν_{12}), but they can be far apart in case of transverse and shear properties (E_{22} and G_{12}).

Hill (1965) derived bounds for the elastic constants. In particular, he put rigorous bounds on the longitudinal Young's modulus, E, in terms of bulk modulus in plane strain (k_p), Poisson's ratio (ν), and the shear modulus (G) of the two phases. No restrictions were made on fiber form or packing geometry. The bulk modulus in plane strain, k_p, is the modulus for lateral dilation with zero longitudinal strain and is given by:

$$k_p = \frac{E}{2(1-2\nu)(1+\nu)}$$

The bounds on the longitudinal modulus, $E_{c\ell}$, are:

$$4V_fV_m(\nu_f-\nu_m)^2 / (V_f/k_{pm} + V_m/k_{pf} + 1/G_m)$$

$$\leq E_{c\ell} - E_fV_f - E_mV_m \leq$$

$$4V_fV_m(\nu_f-\nu_m)^2 / (V_f/k_{pm} + V_m/k_{pf} + 1/G_f) \qquad (6.10)$$

We can readily verify, from Eq. 6.10, that deviations from the rule of mixtures for $E_{c\ell}$ are quite small. If we substitute some values of practical composites, such as silicon carbide fibers in an aluminum matrix, we find that the deviations in $E_{c\ell}$ from the rule of mixtures are < 2%. Note that the deviation from the rule-of-mixtures value comes in from the $(\nu_f - \nu_m)^2$ factor. For $\nu_f = \nu_m$, we have $E_{c\ell}$ given precisely by the rule-of-mixtures expression. Numerical simulations confirm that Hill's bounds are the best possible general bounds for linear elastic behavior of unidirectionally aligned fiber reinforced composites under axial loading (Rossoll et al., 2005)

For Poisson's ratio of a unidirectionally aligned fiber composite, Hill also showed that

$$\nu_{12} > \nu_fV_f + \nu_mV_m \text{ for } (\nu_f - \nu_m)(k_{pf} - k_{pm}) > 0,$$

$$\text{and} \quad \nu_{12} < \nu_f V_f + \nu_m V_m \text{ for } (\nu_f - \nu_m)(k_{pf} - k_{pm}) < 0 \qquad (6.11)$$

If $\nu_f < \nu_m$ and $E_f >> E_m$, then, ν_{12} will be less than that predicted by the rule of mixtures ($= \nu_f V_f + \nu_m V_m$). It is easy to see that the bounds on ν_{12} are not as close as the ones on $E_{c\ell}$. This is because $(\nu_f - \nu_m)$ appears in the case of ν_{12} (Eq. 6.13), while $(\nu_f - \nu_m)^2$ appears in the case of $E_{c\ell}$ (Eq. 6.10). If $(\nu_f - \nu_m)$ is very small, the bounds will be close enough to allow us to write:

$$\nu_{12} \approx \nu_f V_f + \nu_m V_m \qquad (6.12)$$

We can summarize the results of Hashin and Rosen (1964) and Hill (1965) as follows. For a transversely isotropic composite, with fibers along direction 1 and the 2-3 plane being the transverse (isotropic) plane, the equations for five independent moduli are given below.

Plane-strain bulk modulus, k_{23}:

$$k_{23} = \frac{k_f k_m + G_m(v_f k_f + v_m k_m)}{G_f + v_m k_f + v_f k_m}$$

where k_m and k_f are the plane-strain bulk moduli of the matrix and fiber, respectively. The plain-strain bulk modulus is defined as:

$$k = K + \frac{1}{3}G$$

In-plane shear modulus, G_{12}:

$$G_{12} = G_m \frac{G_f(1+V_f) + G_m V_m}{G_f V_m + G_m(1+V_f)}$$

Longitudinal Young's modulus, E_{11}:

$$E_{11} = E_f V_f + E_m V_m + \left[4V_f V_m(\nu_m - \nu_f)^2\right] / \left[(V_f / k_m) + (V_m / k_f) + 1/G_m\right]$$

For most practical purposes the last term is negligible in the above expression.

Longitudinal Poisson's ratio, ν_{12}:

$$\nu_{12} = \nu_f V_f + \nu_m V_m +$$
$$\left[V_f V_m(\nu_f - \nu_m)\right]\left[(1/k_m) - (1/k_f)/(V_f/k_m) + (V_m/k_f) + 1/G_m\right]$$

Transverse plane shear modulus, G_{23}:

In this case, the upper and lower bounds are not coincident. The lower bound is:

$$G_{23L} = G_m + \frac{V_f}{\dfrac{1}{G_f - G_m} + (K_m + 2G_m)\dfrac{V_m}{2G_m(k_m + G_m)}}$$

The upper bound is:

$$G_{23U} = G_m + \frac{(\alpha + \beta_m V_f)(1 + \rho v_f^3) - 3v_f v_m^2 \beta_m^2}{(\alpha - v_f)(1 + \rho v_f^3) - 3v_f v_m^2 \beta_m^2}$$

where

$$\alpha = \frac{\gamma + \beta_m}{\gamma - 1} \qquad \rho = \frac{\beta_m - \gamma\beta_f}{1 + \gamma\beta_f}$$

$$\gamma = \frac{G_f}{G_m} \qquad \beta_m = \frac{1}{3 - 4V_m} \qquad \beta_f = \frac{1}{3 - 4V_f}$$

For particulate composites, Hashin (1962) proposed a composite sphere assembly model wherein the composite is made up of units consisting of a spherical particle and its surrounding matrix shell. In each spherical unit, the volume fractions of the particle and matrix are the same, but the spherical units can be of any size. Figure 6.6 shows such an assembly. Hashin analyzed this model by variational calculus methods and obtained a closed-form solution for the bulk modulus and upper and lower bounds for the effective shear modulus. The bulk modulus, K, is given by:

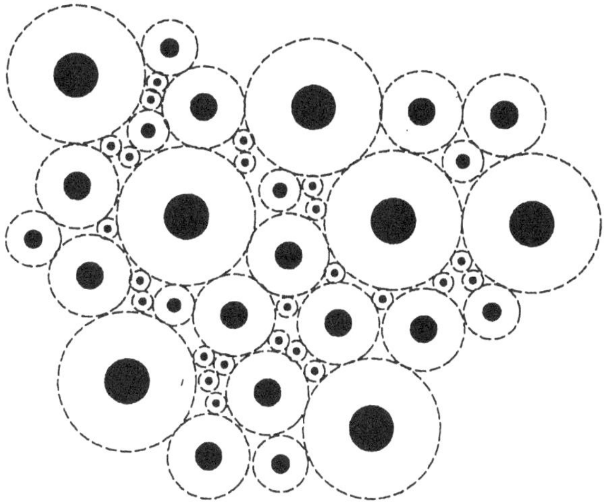

Fig. 6.6 A composite made up of an assemblage of spherical particles surrounded by their respective matrix shells. The volume fractions of phases are maintained constant in each unit; individual units can be of any size.

$$K = \frac{K_p K_m + \frac{4}{3} G_m (V_p K_p + V_m K_m)}{V_p K_m + V_m K_p + \frac{4}{3} G_m}$$

where K, G, and V indicate the bulk modulus, shear modulus, and volume fraction, respectively, and the subscripts p and m refer to the particle and the matrix, respectively. Hashin and Shtrikman (1963) and Rosen (1973) have also analyzed macroscopically isotropic, particulate composite with an arbitrary internal phase geometry and only phase volume fractions being specified. The bounds of Hashin and Shtrikman on the bulk modulus, K and shear modulus, G are given below:

$$K_{upper} = K_p + (1 - V_p) \left[\frac{1}{K_m - K_p} + \frac{3V_p}{3K_p + 4G_p} \right]^{-1}$$

$$K_{lower} = K_m + V_p \left[\frac{1}{K_p - K_m} + \frac{3(1 - V_p)}{3K_m + 4G_m} \right]^{-1}$$

$$G_{upper} = G_p + (1 - V_p) \left[\frac{1}{G_m - G_p} + \frac{6V_p (K_p + 2G_p)}{5G_p (3K_p + 4G_p)} \right]^{-1}$$

$$G_{lower} = G_m + V_p \left[\frac{1}{G_p - G_m} + \frac{6(1 - V_p)(K_m + 2G_m)}{5G_m (3K_m + 4G_m)} \right]^{-1}$$

where $K_m > K_p$ and $G_p > G_m$. For $K_m < K_p$ and $G_p < G_m$ the inequalities will be reversed. Treating the particulate composite as an isotropic material, we can obtain the bounds on the Young's modulus of the composite, E_c, by using the following relationship:

$$E = \frac{9K}{1 + 3K/G}$$

We can easily use this relationship to obtain the bounds on Young's modulus. For $0.5 < E_p/E_m < 3$, the bounds are close enough to give us a value within ~10 % of the true modulus.

6.1.3 Semi-Empirical Expressions

Halpin and Tsai (1967) and Halpin and Kardos (1976) used an empirical approach to obtain some generalized equations which give quite satisfactory results for unidirectional composites compared to the complicated micromechanical equations. These equations contain adjustable fitting parameters and work quite well at low fiber volume fractions. They can also provide useful estimates of properties of composites containing oriented short fibers or whiskers. The adjustable parameters must be obtained from experimental data or must conform to some analytical solution. One uses a single equation of the form:

$$\frac{p}{p_m} = \frac{1 + \xi\eta V_f}{1 - \eta V_f}$$

$$\eta = \frac{\dfrac{p_f}{p_m} - 1}{\dfrac{p_f}{p_m} + \xi} \qquad (6.13)$$

where p represents one of the various moduli of the composite, e.g., E_{11}, E_{22}, G_{12} or G_{23}; p_f and p_m are the corresponding matrix and fiber moduli, respectively; V_f is the fiber volume fraction; and η is a measure of reinforcement which depends on the boundary conditions (fiber geometry, fiber distribution, and loading conditions). The term ξ is a fitting parameter that is used to make Eq. (6.13) conform to the experimental data. The function ξ in Eq. (6.13) is constructed in such a way that we get the two extreme values of property corresponding to $V_f = 0$ and 1, i.e., when $V_f = 0$, we have $p = p_m$ and when $V_f = 1$, we get $p = p_f$. Furthermore, the form of η is such that:

$$p = p_f V_f + p_m V_m \quad \text{for } \xi \to \infty$$

$$1/p = V_m/p_m + V_f/p_f \quad \text{for } \xi \to 0$$

These two extremes provide the bounds, although not necessarily tight bounds, on the properties of the composite. For example, for the transverse modulus of the composite E_{ct} or E_{22}, we can write:

$$\frac{E_{ct}}{E_m} = \frac{1 + \xi\eta V_f}{1 - \eta V_f}$$

$$\eta = \frac{\frac{E_f}{E_m} - 1}{\frac{E_f}{E_m} + \xi}$$

Comparing these expressions with the exact elasticity solutions, one can obtain a value of ξ. Note that, in principle, ξ can range form 0 to ∞. ξ can be thought of as a measure of the "effectiveness" of fiber reinforcement. For small values of ξ, fibers are less effective than for large values of ξ. For computing E_{ct}, $\xi = 1$ or 2 works for hexagonal or square array of fibers, respectively. One can also point out some limiting values of η. The term ηV_f is akin to reduced fiber volume fraction. For a homogeneous material, $\eta = 0$. For very stiff fibers, $\eta = 1$ while for very compliant inclusions (e.g., voids) $\eta = -1/\xi$. Figure 6.7 shows plots of Young's modulus, E, as function of volume fraction of SiC particles in 2080 Al matrix, as per lower and upper bounds of Hashin-Shtrikman expressions, Halpin-Tsai expression, and the experimental values measured in the longitudinal and transverse directions (Ganesh and Chawla, 2005). Note that Hashin-Shtrikman and Halpin-Tsai expressions treat the composite to be an isotropic material. The experimental results show that because of particle alignment caused by extrusion, the longitudinal values of modulus are higher than the transverse values. The experimental values fall somewhere between the Hashin-Shtrikman upper and lower bounds. Figure 6.8 shows similar plots for a composite where the particles are distributed in a homogenous manner, a WC/Co composite (Koopman et al., 2002).

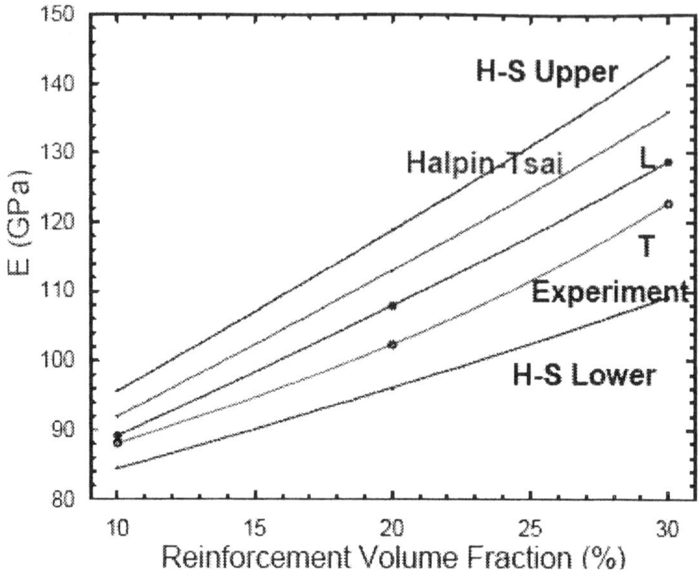

Fig. 6.7 Young's modulus, E as a function of SiC particle reinforcement showing lower and upper bounds of Hashin-Shtrikman, Halpin-Tsai expression, and the experimental values measured in the longitudinal (L) and transverse (T) directions (after Ganesh and Chawla, 2005). The experimental results show difference between longitudinal and transverse directions because of particle alignment caused by extrusion. Hashin-Shtrikman, Halpin-Tsai expressions treat the composite to be isotropic which is not quite true in this case.

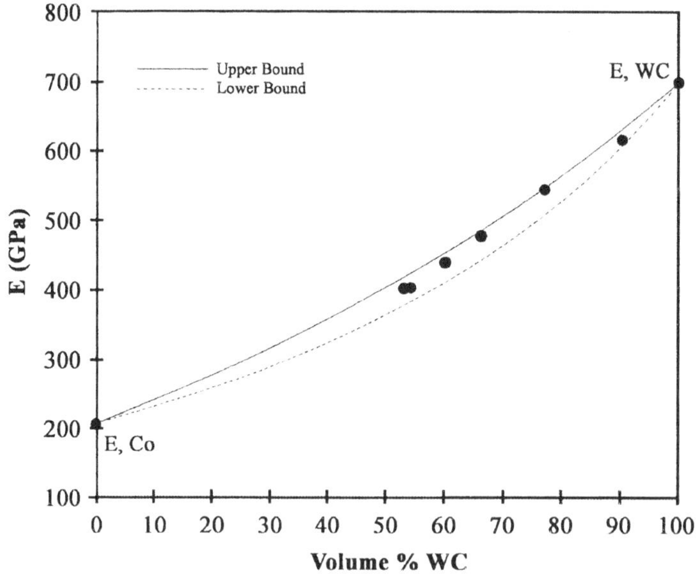

Fig. 6.8 Young's modulus, E as a function of volume fraction of WC in a WC/Co composite (after Koopman et al., 2002)

6.1.4 Eshelby Method or Equivalent Homogeneous Inclusion Technique

Eshelby (1957) formulated the problem of an elastic ellipsoidal inclusion embedded in an elastic matrix subjected to a displacement, with uniform strain field, an infinite distance away from the inclusion. The Eshelby technique provides the solution to a general problem of an infinite medium containing an inclusion. It involves replacing the real inclusion by an inclusion made of matrix (hence the use of expression *equivalent homogeneous inclusion*) but with a proper misfit strain, called *equivalent transformation strain*. The resultant stress field is the same as that for the real inclusions. In very simple terms, we can outline the technique as follows:

- Cut a piece of material from the unstressed, homogenous matrix leaving a hole.
- Allow the cut piece of material to undergo a shape change that corresponds to a transformation strain, ε^t.
- Insert the transformed inclusion back into the original hole, preventing interfacial sliding, and then releasing the tractions. This requires the application of surface tractions so that the transformed shape of the inclusion can be fit back into the original hole.

A constrained strain, ε^c is developed in the inclusion relative to the initial shape. The Eshelby tensor, S, relates the constrained strain, ε^c to the transformation strain, ε^t, by:

$$\varepsilon^c = S\varepsilon^t$$

The Eshelby tensor is a function of the aspect ratio of the inclusion (or reinforcement) and the Poisson's ratio of the matrix. The reinforcement (short fibers, whiskers, or particles) are represented as prolate spheroids. The equation for such an inclusion in terms of Cartesian coordinates is:

$$\frac{x^2 + y^2}{a^2} + \frac{z^2}{c^2} = 1$$

where c/a represents the aspect ratio of the inclusion and $c/a > 1$.

The important general result of the Eshelby analysis is that for an ellipsoidal inclusion the resulting stress and strain in the inclusion are homogeneous. The strains in the matrix and in the inclusion are therefore related algebraically. We can write for the stress in the inclusion, σ_i, in terms of stiffness of the material, C_m, and the elastic strain in the inclusion as:

$$\sigma_i = C_m (\varepsilon^c - \varepsilon^t)$$

In a composite, the ellipsoid inclusion or reinforcement will generally be stiffer than the matrix, i.e., the reinforcement has elastic constants, C_i. The inclusion being a different material than the matrix, we need to replace ε^t with an equivalent strain. With appropriate rearrangement, the resulting stress in the inclusion is

$$\sigma_i = C_i (\varepsilon^c - \varepsilon^t_{eq})$$

Now, $\varepsilon^c \neq S\varepsilon^t_{eq}$, since $C_i \neq C_m$. However, we can find an equivalent homogenous transformation strain, ε^t, such that the equivalent inclusion resembles the inhomogeneity, i.e.:

$$C_i (\varepsilon^c - \varepsilon^t_{eq}) = C_m (\varepsilon^c - \varepsilon^t)$$

$$\text{or} \quad C_i(S\varepsilon^t - \varepsilon^t_{eq}) = C_m(S - I)\varepsilon^t \qquad (6.14)$$

From Eq. (6.14), we can obtain the transformation strain for any shape change, ε^t_{eq} and stiffness mismatch $(C_i - C_m)$ as:

$$\varepsilon^t = [(C_i - C_m)S + C_m]^{-1} C_i\varepsilon^t$$

This then gives the stress in the reinforcement as:

$$\sigma_i = C_m(S - I)[(C_i - C_m)S + C_m]^{-1} C_i\varepsilon^t \qquad (6.15)$$

This expression allows us to calculate the internal stresses in the inclusion. It is valid when the concentration of the inclusion phase is small. If this is not so, the interaction of the fields from the other inclusions would affect the average fields in the matrix and the reinforcement. The main advantage of the Eshelby tensor is that it allows us to determine the stress and strain in the inclusion without worrying about the complex stress field in the matrix. Determining the full field values can be an extremely difficult proposition depending on the geometry and the selected load case.

The Eshelby method would work nicely if the composite had ellipsoidal inclusions of arbitrary aspect ratios. The reinforcements, however, are generally of any shape causing singularities and problems in the local fields.

The averaging schemes, in general, break down for some reinforcement shapes and if there occurs a clustering of particles or fibers. Last but not least, the effect of shape and spatial distribution of particles can become quite marked when constrained plastic deformation occurs in one of the phases, for example, in a WC/Co composite.

6.1.5 Numerical Methods

As an alternative to analytical analysis, numerical techniques such as finite element method (FEM) have become very popular, especially when the geometry of the component and thermomechanical history of the composite can be incorporated into the simulation of the properties. One common approach is to use a unit-cell model, where one or more reinforcement, fibers or particles, are embedded within the matrix, to simulate a composite material with a periodic array of reinforcement. It should be noted that in actual composites the particles commonly contain sharp corners, so spherical particles are not necessarily a realistic choice for simulation. Microstructure-based finite element techniques have been employed that are able to incorporate the "true" microstructures, that take into consideration particle morphology and clustering of particles, as a basis for analysis using finite element techniques (Chawla et al., 2003, Chawla et al., 2004). Figure 6.9 shows a microstructure-based model and two unit-cell models.

Chawla et al. (2003) applied an object oriented finite element technique to particulate reinforced metal matrix composites from images (micrographs) of real systems. In particular, the effect of reinforcement particle volume fraction and alignment (with respect to the loading axis) on the anisotropy of elastic properties was investigated. Chawla et al. (2004) obtained numerical

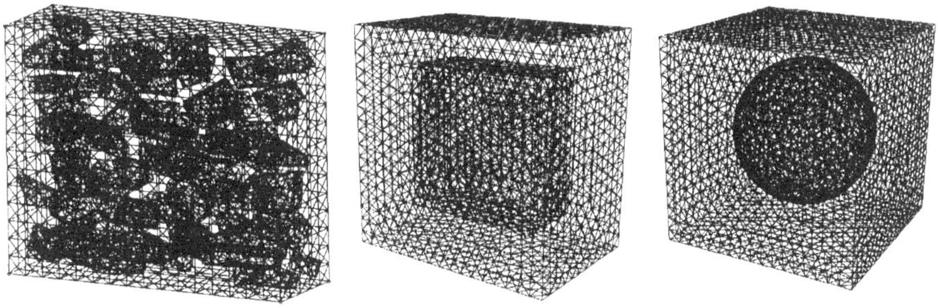

Fig. 6.9 Comparison of different types of numerical models: microstructure-based, unit cell rectangular prism, and unit cell sphere (after Chawla et al., 2004)

predictions of the effective Young's modulus for an Al/SiC_p composite, which are shown in Table 6.2. The unit cylinder particle has a greater stiffening effect than the spherical particle. Clearly, load transfer by a shear-lag type of mechanism (see chapter 7) is more effective across a planar interface than a spherical interface.

The advantage of using a three-dimensional (3D) microstructure-based model is shown by a comparison of the predicted Young's modulus of the composite, versus typical prismatic (rectangular) and spherical unit cell models, Table 6.2. The highest and lowest simulated moduli and strength were obtained by the prismatic rectangle unit cell and sphere unit cell, respectively. This can be attributed to the highest degree of load transfer for the prismatic, rectangular particle than for the spherical particles. The 3D microstructural models, from two different regions of the material, exhibited a higher degree of strengthening, since the actual microstructure incorporated the inherent aspect ratio and alignment of the SiC particles along the loading direction. A comparison of all predicted moduli with experimental tensile data on the same composite, from Chawla et al. (2004), shows that the result from the 3D microstructure model correlates very well with the experimentally determined Young's modulus value of 108 GPa (Ganesh and Chawla, 2004).

6.2 PHYSICAL PROPERTIES

In this section we provide expressions for some important physical properties such as density, coefficients of thermal expansion, and electrical and thermal conductivity.

6.2.1 Density

Density of a material is mass per unit volume. Density is one property for which the rule-of-mixtures works for any composite irrespective of the

Table 6.2 Young's Modulus Predicted by Various Finite Element Models and Comparison to Experiment (Chawla et al., 2004)

Method	Young's Modulus (GPa)
Unit Cell – Rectangular Prism	113
Unit Cell – Sphere	100
3D Microstructure	107.4 ± 0.4
Experiment (Ganesh and Chawla, 2004)	107.9 ± 0.7

distribution of the constituents. The mass of a composite is the sum of the masses of the constituents. Thus:

$$m_c = m_m + m_r \qquad (6.16)$$

where m represents the mass and the subscripts c, m, and r indicate composite, matrix, and reinforcement, respectively. The above equation is valid even in the presence of voids in the composite. For the volume of the composite, we can write:

$$v_c = v_m + v_r + v_v \qquad (6.17)$$

where v denotes the volume of a component while the subscripts c, m, r, and v indicate composite, matrix, reinforcement, and void, respectively. We can then write in terms of mass fractions and volume fractions of the composite the following two expressions:

$$M_m + M_r = 1$$

$$V_m + V_r + V_v = 1$$

For the density of the composite, we can write:

$$\rho_c = \frac{m_c}{v_c} = \frac{m_r + m_m}{v_c} = \frac{\rho_r v_r + \rho_m v_m}{v_c} = \rho_r V_r + \rho_m V_m \qquad (6.18)$$

An experimental measurement of density of a composite is frequently used to test for the presence of any porosity in the composite. It is worth pointing out that should there be any interfacial reaction between the components of a composite, the reaction product(s) should be treated as additional component(s) of the composite.

6.2.2 Coefficient of Thermal Expansion (CTE), α

Most materials, with very few exceptions, expand when heated, i.e., their volume increases. This of course stems from the atomic or molecular vibrations that materials undergo at all temperatures. The amplitude of these vibrations increases with temperature. We can characterize this in terms of a *volumetric* coefficient of thermal expansion, β, defined as:

$$\beta_{ij} = \frac{1}{V}\left(\frac{\delta V}{\delta T}\right)$$

where V is the volume of the material, T is its temperature and the two indices of the volumetric coefficient of thermal expansion indicate that it is a second rank tensor.

We can also define a *linear* coefficient of thermal expansion, α, as:

$$\alpha_{ij} = \frac{\delta \varepsilon_{ij}}{\delta T}$$

where ε is the strain and α is another second-rank symmetric tensor. The coefficient of thermal expansion, α, does not have a constant value over a very large range of temperature. Over a small temperature range, ΔT, we write:

$$\varepsilon_{ij} = \alpha_{ij}\Delta T$$

It is instructive to recall that the sum of the diagonal terms of the strain tensor represent the volume change, i.e.,

$$\beta = \varepsilon_{11} + \varepsilon_{22} + \varepsilon_{33} = 3\alpha$$

$$\alpha = \frac{1}{3}\left[\varepsilon_{11} + \varepsilon_{22} + \varepsilon_{33}\right]$$

For small strains,

$$\beta_{ij} = 3\alpha_{ij}$$

In matrix notation, we can write:

$$\begin{vmatrix} \varepsilon_{11} & \varepsilon_{12} & \varepsilon_{13} \\ & \varepsilon_{22} & \varepsilon_{23} \\ & & \varepsilon_{33} \end{vmatrix} = \begin{vmatrix} \alpha_{11} & \alpha_{12} & \alpha_{13} \\ & \alpha_{22} & \alpha_{23} \\ & & \alpha_{33} \end{vmatrix} \Delta T$$

In contracted notation, the above expression takes the following form:

$$\begin{vmatrix} \varepsilon_1 \\ \varepsilon_2 \\ \varepsilon_3 \\ \varepsilon_4 \\ \varepsilon_5 \\ \varepsilon_6 \end{vmatrix} = \begin{vmatrix} \alpha_1 \\ \alpha_2 \\ \alpha_3 \\ \alpha_4 \\ \alpha_5 \\ \alpha_6 \end{vmatrix} \Delta T$$

If we are interested in finding the CTE in any arbitrary direction [hkl] in the plane of a fiber reinforced composite, we need to use the direction cosines formula. Let α_{hkl} be the CTE along a direction [hkl] that has direction cosines n_1, n_2, and n_3.

$$\alpha_{hkl} = n_1^2\alpha_l + (n_2^2 + n_3^2)\alpha_l$$

$$\alpha_{hkl} = \alpha_{\perp}\sin^2\theta + \alpha_{||}\cos^2\theta$$

$$\alpha_{hkl} = \alpha_{\perp} + (\alpha_{||} - \alpha_{\perp})\cos^2\theta$$

where we have taken $\alpha_l = \alpha_{||}$, parallel to the fiber axis and $\alpha_2 = \alpha_3 = \alpha_{\perp}$, perpendicular to the fiber axis. Also, we have made use of the relationship $n_1^2 + n_2^2 + n_3^2 = 1$. θ is the angle between direction [hkl] and the fiber axis.

Expressions for CTE of Composites

In general, the CTE of a composite is different from that given by a simple rule of mixtures ($\alpha_f V_f + \alpha_m V_m$). This is because the presence of reinforcement, with an expansion coefficient less than that of the matrix, introduces a mechanical constraint on the matrix. The degree of constraint is also dependent on the nature of the reinforcement, e.g., a fiber will cause a greater constraint on the matrix than a particle.

Many models have been proposed to predict the coefficients of thermal expansion (CTE) of fibrous and particulate composites. Experimental determination of these coefficients and analysis of the general thermal expansion characteristics of a variety of composites are available in the literature; see for example, Turner (1946); Kerner (1956); Rosen and Hashin (1970); Schapery (1968); Marom and Weinberg (1975); Vaidya and Chawla (1994).

Unidirectionally Aligned Fibrous Composites

In fiber reinforced composites, two thermal expansion coefficients are needed: α_{cl} in the longitudinal direction and α_{ct} in the transverse direction. Fibers generally have a lower expansion coefficient than that of the matrix. Thus, the fibers mechanically constrain the matrix resulting in α_{cl} being generally smaller than α_{ct} for the composite. At low V_f, it is possible that the transverse expansion coefficient of a fibrous composite, α_{ct}, may be greater than the CTE of the matrix by itself. Long, stiff fibers prevent the matrix from expanding in the longitudinal direction, which forces the matrix to expand more than usual in the transverse direction. Schapery (1968) put bounds on the coefficient of thermal expansion of a unidirectional fiber reinforced composite. These bounds are quite narrow in the longitudinal direction. The following assumptions are made in this analysis:

(a) The bonding between the fiber and matrix is perfect and mechanical in nature, i.e., no chemical interaction is allowed.
(b) The fibers are continuous and perfectly aligned.
(c) The properties of the constituents do not change with temperature.
(d) Poisson's ratios of the components are not very different.

Expressions for expansion coefficient of a fibrous composite are as follows. The longitudinal expansion coefficient for the composite is:

$$\alpha_{cl} = \frac{\alpha_m E_m V_m + \alpha_f E_f V_f}{E_m V_m + E_f V_f} \qquad (6.19)$$

The transverse expansion coefficient is:

$$\alpha_{ct} \cong (1 + \nu_m)\alpha_m V_m + (1 + \nu_f)\alpha_f V_f - \alpha_{cl}\bar{\nu} \qquad (6.20)$$

$$\bar{\nu} = \nu_f V_f + \nu_m V_m$$

For low fiber volume fractions, $V_f > 0.2$ or 0.3, α_{ct} can be approximated by

$$\alpha_{ct} \cong (1 + V_m)\alpha_m V_m + \alpha_f V_f$$

Anisotropy in expansion can be reduced if the composite contains randomly oriented short fibers or whiskers in three dimensions. Fore such a composite, the isotropic thermal expansion coefficient is given by:

$$\alpha \cong \frac{\alpha_{cl} + 2\alpha_{ct}}{3}$$

Particle Reinforced Composites

One can regard a particulate composite as a homogeneous material in a statistical sense, i.e., assuming a uniform distribution of the particles in the matrix, Fig. 6.10. Volumes fractions of the two phases are:

$$V_1 \text{ and } V_2 (= 1 - V_1)$$

Volumetric CTE of a composite consisting of spherical particles dispersed in a matrix (Kerner, 1956) is given by:

$$\beta_c = \beta_m V_p + \beta_p V_p - (\beta_m - \beta_p) V_p \left[\frac{\dfrac{1}{K_m} - \dfrac{1}{K_p}}{\dfrac{V_m}{K_p} + \dfrac{V_p}{K_m} + \dfrac{0.75}{G_m}} \right]$$

Kerner's expression does not differ significantly from the rule of mixtures because the particle constraints the matrix a lot less than fibers. The

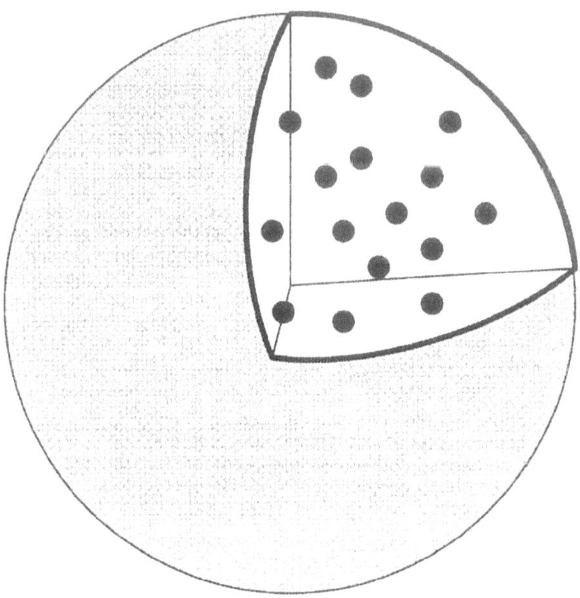

Fig. 6.10 Schematic of a particulate composite.

Schapery solution (Schapery, 1968), gives upper and lower bounds for the thermal expansion coefficient. The upper bound is given by:

$$\alpha_c = V_p\alpha_p + V_m\alpha_m + \left[\frac{4G_m}{K_c}\right]\left[\frac{(K_c - K_p)(\alpha_m - \alpha_p)V_p}{4G_m + 3K_p}\right]$$

and the lower bound is:

$$\alpha_c = V_p\alpha_p + V_m\alpha_m + \left[\frac{4G_p}{K_c}\right]\left[\frac{(K_c - K_m)(\alpha_p - \alpha_m)V_m}{4G_p + 3K_m}\right]$$

where K_c is the bulk modulus of the composite and is obtained from the Hashin-Shtrikman analytical model (Hashin and Shtrikman, 1963) described above.

The expression of the linear CTE of a particulate composite has also been given by Turner (1946):

$$\alpha_c = \frac{\alpha_m V_m K_m + \alpha_p V_p K_p}{V_p K_p + V_m K_m}$$

Turner's expression, generally, gives an expansion coefficient much lower than the rule-of-mixtures value.

The models described above have been used extensively to predict the experimental behavior of particle reinforced MMCs. Elomari et al. (1997) studied the CTE behavior of SiC particle reinforced pure Al matrix composites, in the volume fraction range of 47-55% particles. The SiC particles were oxidized prior to matrix infiltration to minimize interfacial reaction. In their study the rule-of-mixtures predicted the highest CTE, followed by the predictions of Kerner, Schapery, and Turner, which was the lowest. The experimental data were in between the broad range between Kerner and Schapery predictions. Similar trends were observed by Sadanandam et al. (1992) for composites of 2124 Al reinforced with SiC (10,20%), Al_2O_3 (20%), and TiC (20%). Vaidya and Chawla (1994) measured CTE in several particle reinforced composites, 2014/SiC/17_p, 2014/Al_2O_3/17_p, 8090 (Al-Li)/SiC/15_p, 6061/SiC/15_p, 6061/B_4C/15_p. Kerner's model was closest to experiment, although the Turner model predictions were much lower than the experiment. They attributed this to the fact that the Kerner's model is close to the rule of mixtures approximation when the constraint term is small, i.e., since the reinforcement is in particulate form, the constraint is not as large as that from fiber reinforcement.

Chawla et al. (2005) studied the thermal expansion behavior of extruded SiC particle reinforced Al matrix composites, using 2D microstructures as the basis for the models, Fig. 6.11(a). The stress state in the composite was very much influenced by the morphology and distribution of the SiC particles, Fig. 6.11(b). The stress was highest at the particle/matrix interface, while the matrix between closely-spaced particles was also under high thermal stress. "Networks" of high stress regions, linking particles in the composite, were also observed. The CTE of the composites was measured in three different

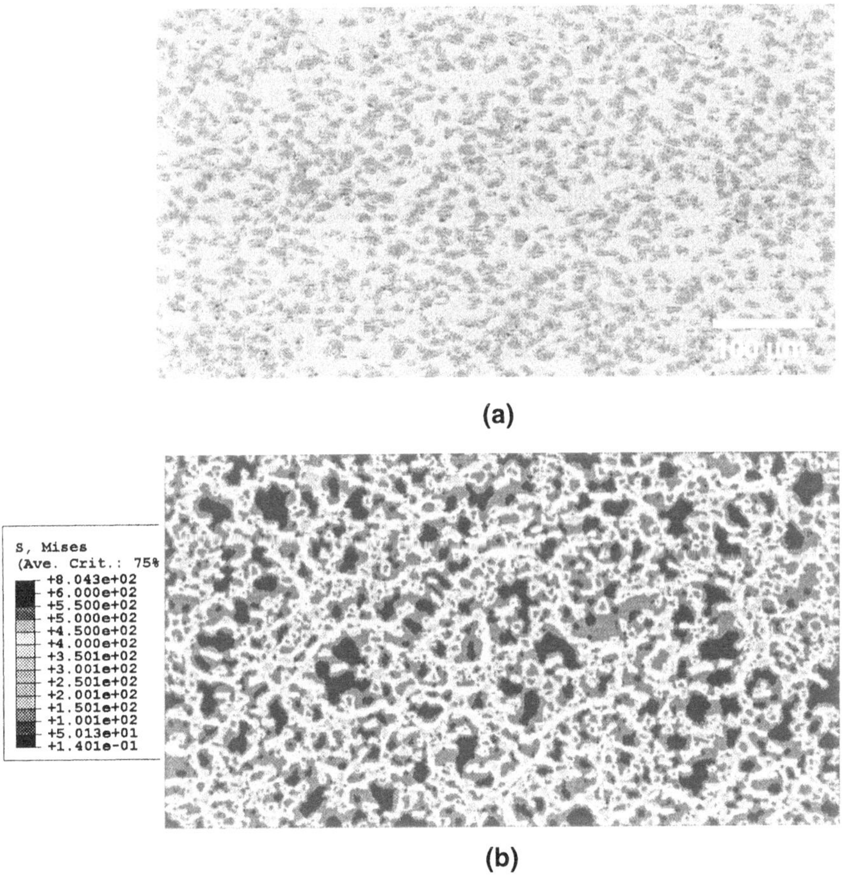

Fig. 6.11 (a) 2D microstructure used as the basis for finite element modeling of thermal expansion behavior in extruded SiC particle reinforced Al matrix composites (b) von Mises stress distribution after thermal expansion. The stress state in the composite was very much influenced by the morphology and distribution of the SiC particles (after Chawla et al., 2005).

orientations: longitudinal (parallel to the extrusion axis), transverse (normal to the extrusion axis but in the extrusion plane), and short transverse (normal to the extrusion axis, and out of the plane). They also showed that the Turner prediction was below the experimental values, while the Schapery bounds were somewhat higher than the experiments, Fig. 6.12(a). Using the microstructure-based models, better agreement was obtained with experiments than with the analytical models above. Because of the 2D nature of the model, however, the predictions were slightly higher than the experimental values, Fig. 6.12(b).

Effect of Interface on CTE of a Composite

As we have described above, most models predict a value of thermal expansion coefficient of a composite that is less than that given by a simple rule of mixtures (ROM). The CTE of the composite is generally less than the ROM value because the presence of ceramic particles (usually of low α) introduces a constraint on the expansion of the metallic matrix (usually of high α). In this regard, it is important to point out that interface can exert some influence on the value of CTE, especially for very small particle size.

Xu et al. (1994) examined the effect of particle size on the thermal expansion of Al/TiC_p composites, in which the TiC particle size was varied. In this system, the interface is semi-coherent (Mitra et al., 1993). The particle volume fraction was maintained constant at about 15% while two particle sizes were used, 0.7 μm and 4 μm. Figure 6.13 shows the values of coefficient of thermal expansion (CTE) for the two composites (circles and squares). Also plotted in Fig. 6.13 are the CTE values for pure aluminum and titanium carbide. The 4 μm particle composite showed, consistently, a higher CTE than the 0.7 μm particle composite. Note that the CTE of the 4 μm particle composite is very close to the rule of mixture (ROM) value, while that of the 0.7 μm particle composite is not, implying that more constraint on the aluminum matrix is caused by the 0.7 μm particle than by the 4.0 μm particle. A very good interfacial bond between the TiC particle and matrix exists. Although no chemical compound formation was observed in the interfacial area, lattice distortion in the interfacial area was observed. The thickness of this region of strain localization varied between 10-50 nm (Mitra et al., 1993). Such a lattice distortion at the interface will affect the CTE value of the composites. Since the interfacial area is related to the particle size, the volume of such a lattice distortion layer will depend on the particle size, and, accordingly, the effect of the particle size on the CTE of the composites will vary with the particle size as well as shape. For a given particle volume fraction, the smaller particle size, the greater the volume

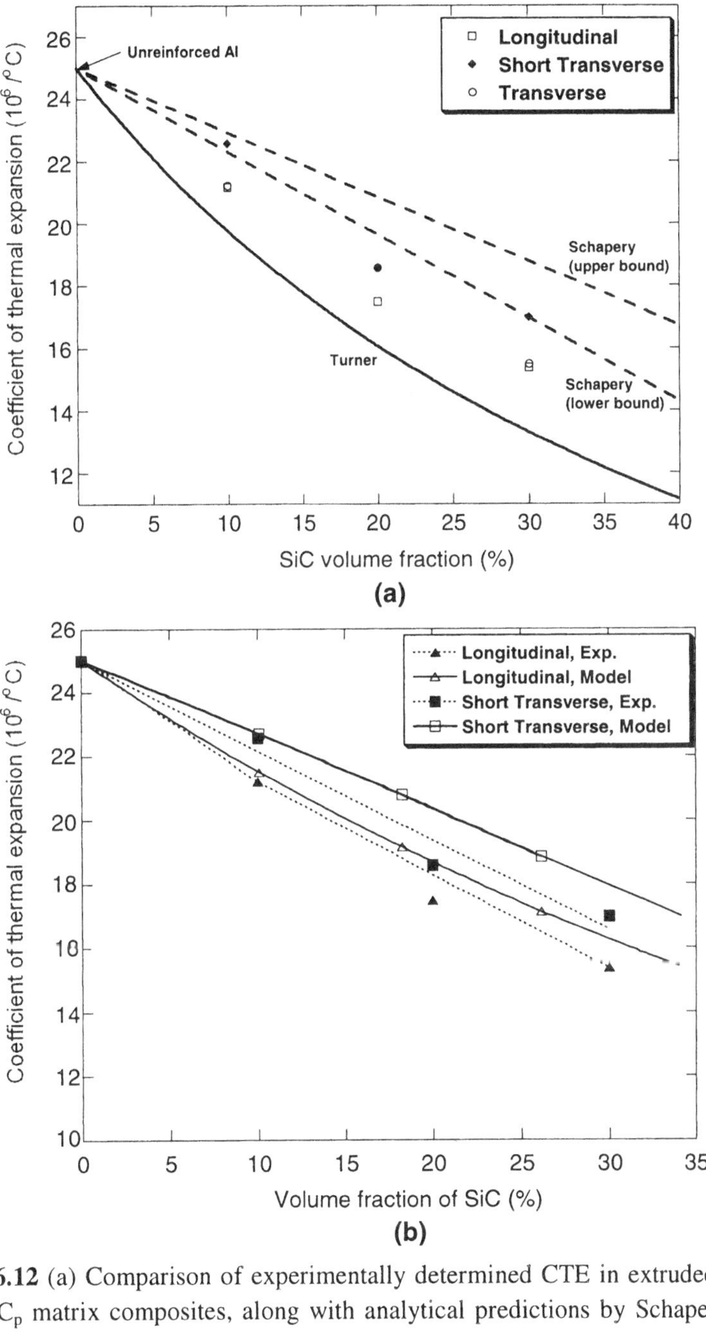

Fig. 6.12 (a) Comparison of experimentally determined CTE in extruded 2080 Al/SiC_p matrix composites, along with analytical predictions by Schapery and Turner, and (b) 2D microstructure-based FEM predictions. The effect of CTE anisotropy is more adequately captured by the microstructure-based model (after Chawla et al., 2005).

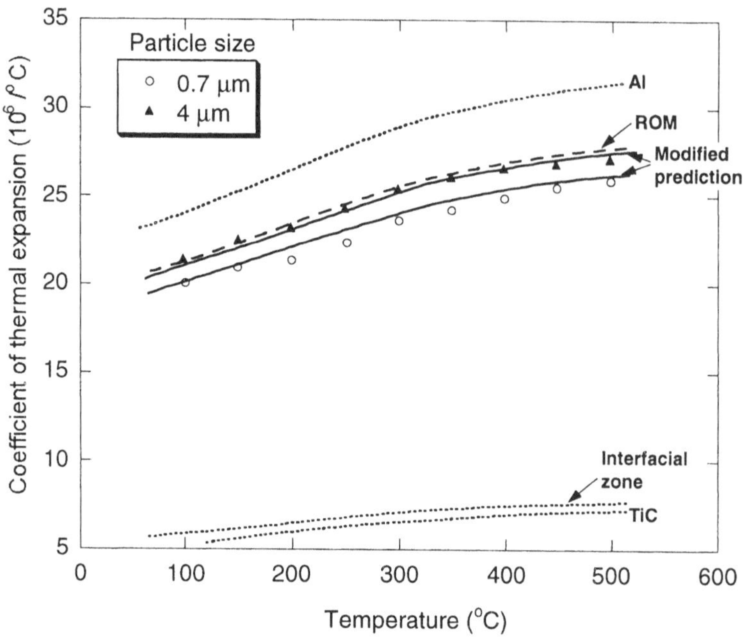

Fig. 6.13 CTE versus temperature for TiC particle reinforced Al composites with different particle size (after Xu et al., 1994). The 4 μm particle composite showed, consistently, a higher CTE than the 0.7 μm particle composite. The modified rule of mixtures (ROM) prediction, which includes the CTE of the interface, predicted the experimental behavior very well.

fraction of interfacial zone. Figure 6.14(a) illustrates the effect of particle size on the interfacial zone size for a given volume fraction of particle. A quantitative comparison of volume fraction of interfacial zone, V_i, between two particle size composites is shown in Fig 6.14(b). If the thickness of the distortion layer is assumed to be 25 nm, V_i can be 3%, which is quite significant. This interfacial zone will be constrained to expand or contract with the particle, and thus, CTE of this zone will be closer to that of the particle than that of the metal matrix. In order to explain the effect of the particle size on the CTE of the particle reinforced metal matrix composite, consider that the composite consists of a matrix phase, a particulate reinforcement, and an interfacial zone to which a CTE different from that of the matrix can be ascribed. In this approximation, the constraint on the matrix expansion is thus embedded in the interfacial zone term. We can then write for the CTE of a three-component composite as follows:

$$\alpha_c = \alpha_p V_p + \alpha_m V_m + \alpha_i V_i \qquad (6.21)$$

where the subscripts, c, p, m, and i, denote the composite, particle, matrix, and interfacial zone, respectively. The interfacial zone volume fraction, V_i, was calculated by the interfacial area and the thickness of the lattice distortion layer. The best fit CTE curves using Equation (6.21) for the 4.0 μm and 0.7 μm particle composites are obtained for the same assumed value α_i but different V_i, see the solid lines with data points in Fig. 6.13. The

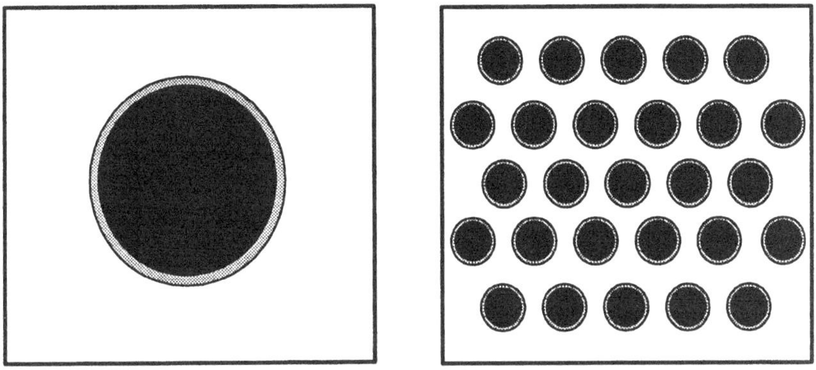

(a)

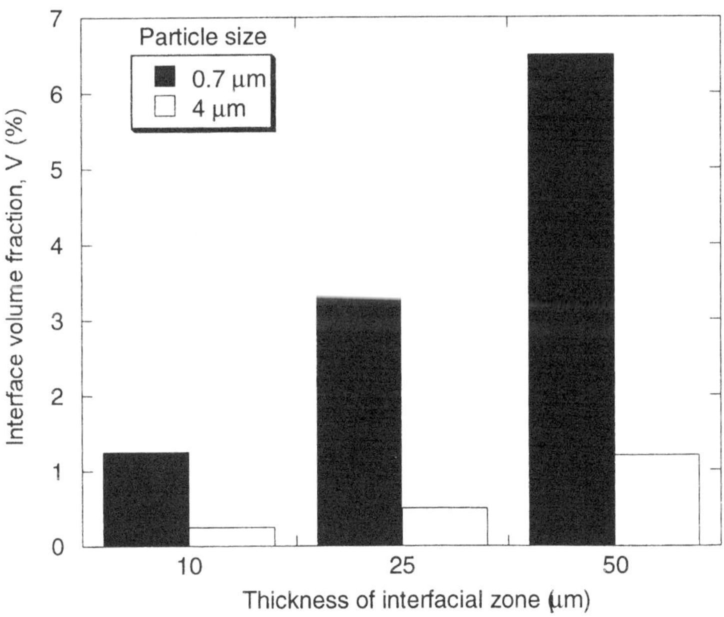

(b)

Fig. 6.14 (a) Schematic of the effect of particle size on fraction of interfacial zone. A smaller particle size results in a larger interfacial area. (b) Quantitative measurements of interfacial zone in Al/TiC_p composites. A larger interfacial zone will have a stronger effect on the CTE of the composite (after Xu et al., 1994).

apparent CTE values for the interfacial zone are also shown in Fig. 6.13. The agreement between the prediction (Eq. 6.21) and the experimental results confirms the effect of particle size on the CTE of this particle reinforced composite.

A lower CTE may also result from an interfacial phase between particle and matrix. Elomari et al. (1997) studied the CTE behavior of SiC particle reinforced pure Al matrix composites. The SiC particles were oxidized prior to matrix infiltration to minimize interfacial reaction. A thin SiO_2 coating formed on the particles resulted in a lower CTE than that of unoxidized materials. Furthermore, a decrease in particles size (i.e., more surface area for oxides to form) also resulted in reduced CTE.

Other Factors Influencing CTE of Composites

In addition to the interface, the CTE of particle reinforced MMCs is affected by several other factors. These include: Plasticity due to CTE mismatch between reinforcement and matrix, during heating or cooling; reinforcement fracture, residual stress, and local stresses at points of contact between reinforcement. Lee et al. (1991), for example, noted that the experimentally measured CTE in Al alloy/SiC_p composites was lower than that of model predictions by Paul (1960). They postulated that the irregular shape of the SiC particles may cause different thermal stress fields relative to perfectly spherical particles. Furthermore, an incompressible plastic layer may form at the particle/matrix interface, and result in lower expansion of the matrix, and thus, the overall composite.

The CTE may also be affected the particle fracture. Elomari et al. (1996) examined the CTE in particle reinforced MMCs, 6061/Al_2O_3/10_p and 20_p, with increasing particle fracture. An increase in particle fracture resulted in increased CTE. Residual stress also has an important effect on CTE. Chang et al. (2000) studied the CTE behavior of 40 vol.% Al_2O_3 particle reinforced Al matrix composites. Water-quenched simples had a lower CTE than furnace cooled simples. This was attributed to thermal residual stresses present in the water-quenched samples. At temperatures above 100°C the CTE increased, and were closer to the prediction by Kerner. This was attributed to a relief in the residual stresses with increasing temperature. Shen et al. (1994) used a unit-cell FEM approach to model the thermal expansion behavior of SiC particle reinforced Al matrix composites. They showed that residual stress increased the apparent CTE of the composite, since the residual stress in the metal matrix has a tensile component. Shen et al. (1994) also showed that the CTE is much less sensitive to particle distribution (clustering) than tensile stiffness. This was attributed to the fact that for an elastically deforming, isotropic composite, the effective CTE is given as (per Hashin-Shtrikman and Schapery, for example):

$$\alpha = \alpha_2 + (\alpha_1 - \alpha_2)\frac{1/K - 1/K_2}{1/K_1 - 1/K_2}$$

where K is the bulk modulus and the subscripts 1 and 2 refer to ductile and brittle phases. From this equation one can see that the CTE is highly dependent on K, which is relatively insensitive to the distribution of particles in the matrix of the composite. Interestingly, the contiguity of a phase had a significant effect on CTE. For example, when the brittle phase is continuous, an additional constraint is placed on the expansion of the ductile phase. Microvoid nucleation and propagation at points where particles touch or at pre-existing processing-induced voids may also cause a deviation in strain versus temperature, with slight increase in CTE taking place (Balch et al., 1996).

6.2.3 Thermal Conductivity

Thermal conductivity is another thermal property of great importance. Some general background on thermal conductivity is given in the inset. A composite consisting of unidirectionally aligned carbon fibers in an aluminum matrix can show extremely high thermal conductivity along the fiber direction, especially if pitch-based carbon fibers are used. This is because pitch-based carbon fibers have very high thermal conductivity along the fiber axis. Thus, depending on the volume fraction of carbon fiber used, the thermal conductivity of a C/Al composite along the fiber direction can be greater than that of even copper. The thermal conductivity transverse to the fibers in such a composite will be about two-thirds that of aluminum. Such a C/Al composite can find applications in heat transfer applications where weight reduction is an important consideration. Examples of such applications might include high-density, high-speed integrated-circuit packages for computers and base plates for electronic equipment (see the chapter on applications, Chapter 11). Another possibility is to use this composite to dissipate heat from the leading edges of wings in high-speed airplanes. Tungsten carbide/cobalt composites are used extensively in operations such as metal cutting or rock drilling where they experience very high temperatures. The ability to dissipate heat away from the wear surface, and thus avoid any localized temperature-induced softening, depends on the thermal conductivity of the composite. In WC/Co composite, both phonons (quanta of lattice waves) and electrons contribute to thermal conductivity. In ceramics, only phonon (lattice) thermal conductivity contributes to the total thermal conductivity.

The heat flow in a material is proportional to the temperature gradient, and the constant of proportionality is called the thermal conductivity. In indicial notation, the equation for heat transfer is

$$q_i = -k_{ij}\, dT/dx_j$$

where q_i is the heat flux along x_i axis, dT/dx_j is the temperature gradient across a surface that is perpendicular to the x_j axis, and the coefficient, k_{ij} is called thermal conductivity. Thermal conductivity is also a second-rank tensor, note the two indexes. Similar to thermal expansion coefficient, the second rank, symmetric thermal conductivity tensor, k_{ij}, in an isotropic material reduces to a scalar number, k. In an orthotropic material, we need three constants along the three principal axes: k_{11}, k_{22}, and k_{33}. In a transversely isotropic material such as a unidirectionally reinforced fibrous composite, there will be two constants: k_{cl}, in the axial direction and k_{ct} in the transverse direction. The thermal conductivity in the axial direction, k_{cl} can be predicted by a rule-of-mixtures type expression (Behrens, 1968)

$$k_l = k_{cl} = k_{fl} V_f + k_m V_m$$

where k_{fl} is the thermal conductivity of the fiber in the axial direction and k_m is the thermal conductivity of the isotropic matrix. This relationship is similar to the rule-of-mixtures in the longitudinal direction for the Young's modulus obtained by action-in-parallel.

Thermal conductivity of a unidirectionally aligned fiber composite (i.e., transversely isotropic) in the transverse direction can be approximated by the action-in-series model. This would give:

$$k_{ct} = k_2 = k_{f2}\, k_m/(k_{f2} V_f + k_m V_m)$$

Halpin-Tsai-Kardos equations can be used to obtain the following expression for the transverse thermal conductivity of a composite containing unidirectionally aligned fibers:

$$k_{c2} = k_{c3} = k = (1 + \eta V_f)/(1 - \eta V_f) k_m$$
$$\text{and } \eta = [(k_{f2}/k_m) - 1]/[(k_{f2}/k_{fm}) + 1]$$

Transport Properties

Electrical and thermal conductivities represent two very important transport properties. Electrical resistivity, ρ, is the inverse of electrical conductivity, σ. Electrical resistance of a piece of metal of length, L, and cross-sectional area, A, is given by:

$$R = \frac{L\rho}{A}$$

Ohm's law relates the voltage drop, V and current, I as V= IR. Another form of Ohm's laws is as follows:

$$J_i = \sigma_{ij} E_j \qquad \text{(A)}$$

where J is the current density (current per unit area) I/A (Am^{-2}), E is the electric field (Vm^{-1}), and σ is the electrical conductivity ($\Omega^{-1}m^{-1}$), which is equal to 1/ρ. Metals are generally good conductors with σ on the order of 10^7 $\Omega^{-1}m^{-1}$. In general, impurities, solid solution alloying, plastic deformation reduce the electrical conductivity of metals.

The heat flow in a material is proportional to the temperature gradient, and the constant of proportionality is called the thermal conductivity. The basic equation has the form:

$$q_i = -k_{ij}\, dT/dx_j$$

where q_i is the heat flux along x_i axis, dT/dx_j is the temperature gradient across a surface that is perpendicular to the x_j axis, and k_{ij} is the thermal conductivity. Thermal conductivity is also second-rank tensor, note the two indexes. Another form of heat transfer equation is

$$q_i = -k_{ij}\, \Delta T \qquad \text{(B)}$$

where q_i is the heat flux, ΔT is the temperature gradient and k_{ij} the thermal conductivity tensor (W/mK). Equations A and B are mathematically similar. In these equations, the quantity on the left hand side is the response to the stimulus on the right hand side. The two are related via a constant of proportionality (σ and k) which is a tensor of rank two. For cubic symmetry, the tensor reduces to a scalar.

where we have taken ξ equal to 1.

For a unidirectionally reinforced fibrous composite, if we know the thermal conductivity in directions 1 and 2, we can find the thermal conductivity k_x and k_y in any arbitrary directions, x and y, respectively, by the following equations:

$$k_x = k_1 \cos^2\theta + k_2 \sin^2\theta$$

$$k_y = k_1 \sin^2\theta + k_2 \cos^2\theta$$

$$k_{xy} = (k_2 - k_1) \sin\theta \cos\theta$$

where θ is the angle measured from the x-axis to the 1-axis and k_{xy} can be considered to be a thermal coupling coefficient.

6.2.4 Electrical Conductivity

In terms of electrical conductivity, most metal matrix composites are mixtures of good electrical conductors (e.g., Cu, Al, Ti, etc.) and insulators (e.g., B, C, SiC, Al_2O_3, etc.). There are exceptions such as W/Cu composites. Some general background on electrical conductivity is given in the inset. The conductivity (thermal or electrical) of a composite will depend on the conductivity characteristics of the matrix, reinforcement, volume fraction and aspect ratio and shape of the reinforcement, and, of course, the interfacial characteristics (Weber et al., 2003a, 2003b; Weber, 2005). In particular, interfacial resistance will vary with the form and size of reinforcement and the connectivity of the phases. Weber (2005) analyzed the influence of reinforcement size on the electrical conductivity of a metal matrix composite. He explained the observed size effect in terms of additional scattering of the conduction electrons at the metal/ceramic particle interface. The scattering effect of dislocations, generated by plastic deformation in the metal matrix during cooling from the processing temperature, on electrical conductivity was not significant.

For the simple case of a fiber reinforced composite, assuming no significant interfacial effect, we can write for the resistivity of the composite in the axial direction, in a manner similar to other rule-of-mixtures type relationships,

$$\rho_{c\ell} = \rho_1 V_1 + \rho_2 V_2$$

where ρ is the electrical resistivity, V is volume fraction of a phase, and the subscripts $c\ell$, 1, and 2 designate the composite in the longitudinal direction and the two components of the composite, respectively. Or in terms of electrical conductivity, σ (unfortunately, it is customary to use the same symbol as that for strength), we can write:

$$\sigma_{c\ell} = \sigma_1 V_1 + \sigma_2 V_2$$

In the transverse direction, we have:

$$\frac{1}{\sigma_{ct}} = \frac{V_1}{\sigma_1} + \frac{V_2}{\sigma_2}$$

where the symbols have the significance given earlier. There is a logarithmic expression also:

$$\log \sigma_c = V_1 \log\sigma_1 + V_2 \log\sigma_2$$

Of course, one can generalize these expressions to more than two composites in the composite, should that be the case. One can also use self-consistent models to obtain the electrical conductivity of a composite, see for example, Hale (1976).

An important point that needs to be emphasized here is that the electrical resistivity (and therefore electrical conductivity) of the metal matrix in a composite is likely to be different from that of the unreinforced metal. This is because of possible plastic deformation during processing which will introduce dislocations due to thermal mismatch between the matrix and the reinforcement. The dislocations in turn will increase the resistivity of the matrix. Unlike the unreinforced metal, one cannot recover electrical conductivity by resorting to an annealing treatment because the annealing treatment will again cause plastic deformation of the matrix because of the thermal mismatch between the components.

6.3 Thermal Stresses in Composites

Thermal stresses are internal stresses that arise when a constraint on free dimensional change of a body exists (Chawla, 1973a, 1973b). In the absence of this constraint, the body can experience free thermal strains without any accompanying thermal stresses. In a composite material, this constraint comes from the fact that it is made of dissimilar materials, i.e., materials having different coefficients of thermal expansion. Thermal stresses result in a composite because of the ever present mismatch in the coefficients of thermal expansion of the reinforcement and the matrix ($\Delta\alpha = \alpha_r - \alpha_m$). Thermal strain, in the absence of any temperature gradients, is given by $\Delta\alpha\Delta T$, where ΔT is the amplitude of temperature change. During cooling from the processing temperatures, a large magnitude of thermal stresses can result because of the thermal mismatch between the reinforcement (particle, short fiber or long fiber) and the matrix. Thermal stresses generated will depend on the reinforcement volume fraction, reinforcement geometry, thermal mismatch, temperature interval ($T_{final} - T_{initial}$), and modulus ratio, E_r/E_m, where the subscript r denotes the reinforcement. Generally, $\alpha_m > \alpha_r$, i.e., on cooling from $T_{initial}$ to T_{final} ($T_{initial} > T_{final}$), the matrix would tend to contract more than the reinforcement, putting the reinforcement in compression and the matrix in tension. That is the explanation in simple terms. In reality, it is complex three-dimensional situation. We derive below the analytical expressions for (three-dimensional) thermal stress components in two types of composites: a central particle surrounded by its associated spherical matrix shell and a central fiber surrounded by its cylindrical shell of matrix. The following assumptions are made in this analysis:

- Matrix and reinforcement obey Hooke's law.
- Elastic reinforcement is embedded in an elastic continuum of matrix.
- No chemical reactions between reinforcement and matrix.

In what follows, we need to use the following sets of relationships to solve problems in elasticity involving a complex state of stress:

- Equations of equilibrium of forces
- Equations of compatibility
- Constitutive equations (stress-strain relations)
- Boundary conditions

6.3.1 Thermal Stresses in Particulate Composites

Consider a particulate composite consisting of small ceramic particles distributed in a ceramic matrix. If we regard this composite as an assembly of elastic spheres of uniform size embedded in an infinite elastic continuum, then it can be shown from the theory of elasticity (Timoshenko and Goodier, 1951; Brooksbank and Andrews, 1970) that an axially symmetrical stress distribution will result around each particle. Figure 6.15 shows a schematic of such a particle reinforced composite. Let us say that each particle has a radius, a while the surrounding matrix sphere has the outer radius, b. This spherical symmetry problem calls for the use of spherical coordinates, r, θ,

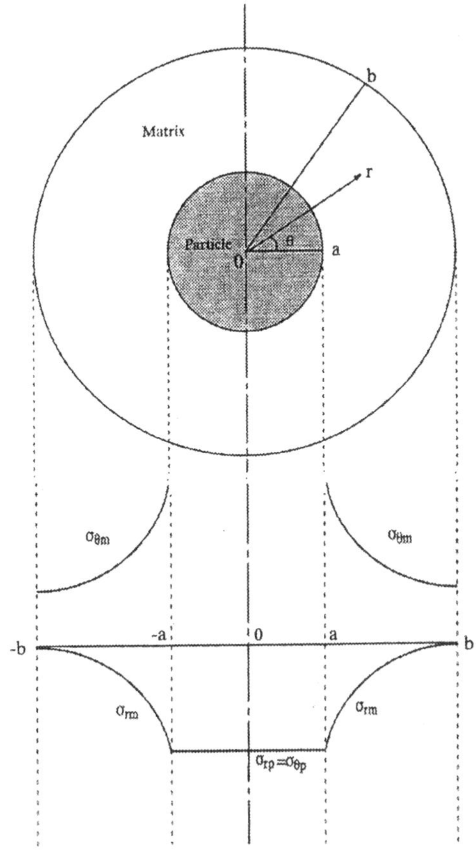

Fig. 6.15 Thermal stress distribution, on cooling, in a particulate composite that has CTE of matrix, α_m > particle CTE, α_p. The particle is under a uniform pressure, P, while the matrix has radial and tangential stress components, radial and tangential components in the matrix vary with distance as $1/r^3$.

and φ as indicated in Fig. 6.15. We have the following stress, strain, and displacement components:

$$\sigma_r,\ \sigma_\theta = \sigma_\varphi$$

$$\varepsilon_r,\ \varepsilon_\theta = \varepsilon_\varphi$$

$$u_r = u, \text{ independent of } \theta \text{ or } \varphi$$

The stress-strain and strain-displacement relationships are:

$$\varepsilon_\theta = \frac{u}{r} = \frac{1}{E}\left[\sigma_\theta(1-\upsilon) - \upsilon\sigma_r\right] + \alpha\Delta T = \varepsilon_\varphi$$
$$\varepsilon_r = \frac{du}{dr} = \frac{1}{E}\left[\sigma_r(1-\upsilon) - \upsilon\sigma_\theta\right] + \alpha\Delta T \quad (6.22)$$

The equilibrium equation for this spherical problem is

$$\frac{d\sigma_r}{dr} + \frac{2(\sigma_r - \sigma_\theta)}{r} = 0 \quad (6.23)$$

From Eqs. (6.22) and (6.23), we get the following governing differential equation:

$$\frac{d^2u}{dr^2} + \frac{2}{r}\frac{du}{dr} - \frac{2}{r^2}u = 0 \quad (6.24)$$

The solution to this differential equation is:

$$u = Ar + \frac{C}{r^2} \quad (6.25)$$

This is the general solution that gives the displacement as a function of radial distance. We can obtain our stress components by applying the following boundary conditions relevant to our problem:

(i) stress vanishes at the free surface (i.e., at r = b)

(ii) the radial stress at the interface (r = a) is the interfacial pressure, P, i.e., $\sigma_r(a) = -P$.

When we apply these boundary conditions, we find that the stresses in the particle are:

$$\sigma_{rp} = \sigma_{\theta p} = -P$$

while the stresses in the matrix are:

$$\sigma_{rm} = \frac{P}{1-V_p}\left[\frac{a^3}{r^3} - V_p\right] \text{ and } \sigma_{\theta m} = -\frac{P}{1-V_p}\left[\frac{1}{2}\frac{a^3}{r^3} + V_p\right]$$

The expression for P becomes:

$$P = \frac{(\alpha_m - \alpha_p)\Delta T}{\left[\dfrac{0.5(1+\nu_m)+(1-2\nu_m)V_p}{E_m(1-V_p)} + \dfrac{1-2\nu_p}{E_p}\right]} \tag{6.26}$$

$$V_p = \left(\frac{a}{b}\right)^3$$

In the above expressions, V_p is the particle volume fraction, a is the particle radius, b is the matrix outer radius, and other symbols have the significance given earlier. A schematic of the stress distribution in the particulate composite is shown in the bottom half of Fig. 6.15 for case of cooling for a particulate composite that has CTE of matrix, α_m > particle CTE, α_p. We can summarize the main results for a particulate composite as follows:

- The particle is under a uniform pressure, P.
- The matrix has radial and tangential stress components that vary as $1/r^3$.
- The radial component goes to zero at the free surface, r = b, as per our boundary conditions. The tangential component has a nonzero value at the free surface.

6.3.2 Thermal Stresses in Fiber Reinforced Composites

This important problem has been analyzed by a number of researchers (Poritsky, 1934; Hull and Berger, 1934; Chawla and Metzger, 1972; Scherer, 1986; Herrmann and Wang, 1991; Hsueh et al., 1988). In the case of a unidirectionally reinforced fibrous composite, it is convenient to use polar coordinates because of the inherent cylindrical symmetry. Figure 6.16 shows the basic fiber/matrix unit consisting of a central fiber (radius, a) surrounded by its sleeve of matrix (outer radius, b). The matrix outer radius, b, will depend on the volume fraction of the matrix. Such a simple axisymmetric model can be used to obtain the three-dimensional state of thermal stress up to a moderate fiber volume fraction.

The axial symmetry means that we can treat the problem in terms of the principal stresses that are independent of θ. We derive expressions for the thermal stresses in a two-element cylindrical composite. For the axial symmetric case under consideration, the radial and circumferential stresses will have expressions of the following form (Poritsky, 1934; Hull and Berger, 1934; Chawla and Metzger, 1972):

$$\varepsilon_r = \frac{\partial u}{\partial r} = \frac{1}{E}\left[\sigma_r - \upsilon\left(\sigma_\theta + \sigma_2\right)\right] + \alpha\Delta T$$

$$\varepsilon_\theta = \frac{u}{r} = \frac{1}{E}\left[\sigma_\theta\left(1-\upsilon\right) - \upsilon\left(\sigma_1 + \sigma_2\right)\right] + \alpha\Delta T \qquad (6.27)$$

$$\varepsilon_z = \frac{\partial w}{\partial z} = \frac{1}{E}\left[\sigma_z - \upsilon\left(\sigma_r - \sigma_\theta\right)\right] + \alpha\Delta T$$

The only equilibrium equation for this rotationally symmetric problem is:

$$\frac{d\sigma_r}{dr} + \frac{\left(\sigma_r - \sigma_\theta\right)}{r} = 0 \qquad (6.28)$$

From Eqs. (6.27) and (6.28), we get the governing differential equation for our problem:

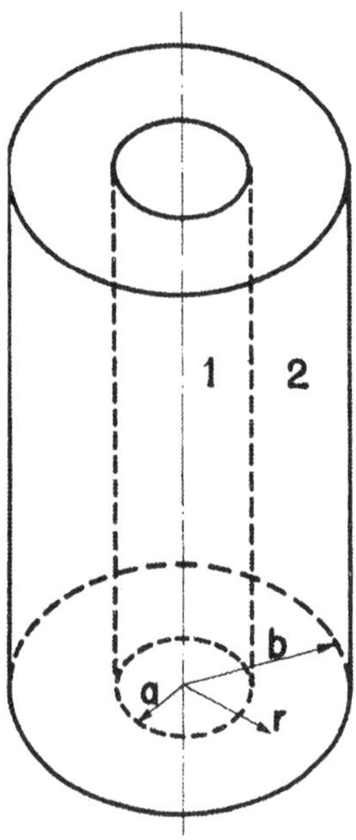

$$\frac{d^2u}{dr^2}+\frac{1}{r}\frac{du}{dr}-\frac{u}{r^2}=0 \tag{6.29}$$

The general solution to this differential equation is:

$$u = Ar+\frac{C}{r} \tag{6.30}$$

This relationship between displacement, u and radial distance, r needs to be solved for each component, central fiber and outer matrix sleeve by applying boundary conditions. Thus, in component 1 or fiber, we have:

$$u_1 = A_1 r+\frac{C_1}{r} \tag{6.31}$$

and in component 2 or matrix, we have:

$$u_2 = A_2 r+\frac{C_2}{r} \tag{6.32}$$

The constants of integrations in these expressions need to be obtained by applying our boundary conditions. The boundary conditions are:

(i) Stress vanishes at the free surface, i.e., $\sigma_{r2} = 0$ at $r = b$.
(ii) Continuity of displacement at the fiber/matrix interface, i.e., $u_1 = u_2$ at $r = a$.
(iii) Continuity of radial stress at the fiber/matrix interface, i.e., $\sigma_{r1} = \sigma_{r2}$ at $r = a$.
(iv) The radial displacement, u must vanish at the axis of symmetry, i.e., $u_1 = 0$ at $r = 0$.

By using these boundary conditions, we can determine the constant of integration in Eqs. (6.31) and (6.32). The last condition gives C_2 to be identically zero because if that were not so, u_1 will become infinity at $r = 0$. That leaves us with three equations and three unknowns. Since ceramic fibers used to reinforce metals remain elastic until fracture but the metal can undergo plastic deformation in response to the thermal stresses generated, we give below full expressions for the three stress components in the matrix. These are:

$$\sigma_r = A\left(1-\frac{b^2}{r^2}\right) \quad \sigma_\theta = A\left(1+\frac{b^2}{r^2}\right) \quad \sigma_z = B \qquad (6.33)$$

where we have omitted the subscript m, and the constants A and B have the following expressions:

$$A = -\left(\frac{E_m(\alpha_m - \alpha_f)\Delta T(a/b)^2}{1+(a/b)^2(1-2)[(b/a)^2-1](E_m/E_f)}\right)$$

$$B = \frac{A}{(a/b)^2}\left(2\nu\left(\frac{a}{b}\right)^2 \frac{1+(a/b)^2(1-2\nu\nu+(a/b)^2(1-2\nu)[(b/a)^2-1](E_m/E_f)}{1+[(b/a)^2-1](E_m/E_f)}\right)$$

$$\text{and } \nu_m = \nu_f = \nu$$

This thermoelastic solution can provide information about the magnitude of the elastic stresses involved and if the elastic range will be exceeded or not. In the case of a metallic matrix, it is likely to deform plastically in response to these thermal stresses (Chawla, 1973a; Chawla, 1973b; Arsenault and Fisher, 1983; Christman and Suresh, 1988). We can draw some important conclusions from Figure 6.17:

(i) Axial stress is uniform in fiber and matrix, although its magnitude is different in the two and depends on the respective elastic constants.

(ii) In the fiber, i.e., the central component f, σ_{rf} and $\sigma_{\theta f}$ are equal in magnitude and sense. In the matrix, i.e., the sleeve, σ_{rm} and $\sigma_{\theta m}$ vary as [1 – (b5/r5)] and [1 + (b5/r5)], respectively.

(iii) When either the temperature difference or the expansion coefficient difference goes to zero, the thermal stresses vanish, as expected.

As an illustration, the variation of thermal stresses generated in the magnesium matrix in an alumina fiber/magnesium matrix as a function of radial distance from the interface is plotted in Fig. 6.18. The stress values correspond a cooling of –1°C. Normalized radial distance is used, i.e., r/a = 1 corresponds to the interface. Note the constant value of the axial stress. Also to be noted is the compressive radial stress, which is highest at the fiber/matrix interface and goes to zero at the free surface. The tangential stress is tensile in nature, highest at the interface and drops to a positive value at the surface.

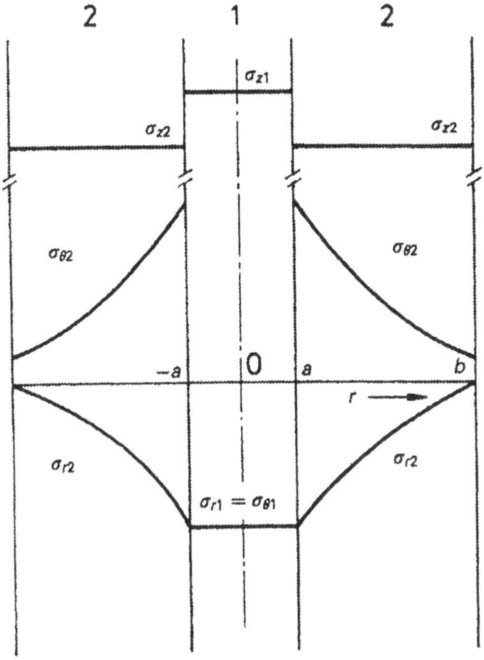

Fig. 6.17 Schematic of three-dimensional stress distribution in a fiber reinforced composite. 1 and 2 indicate fiber and matrix, respectively.

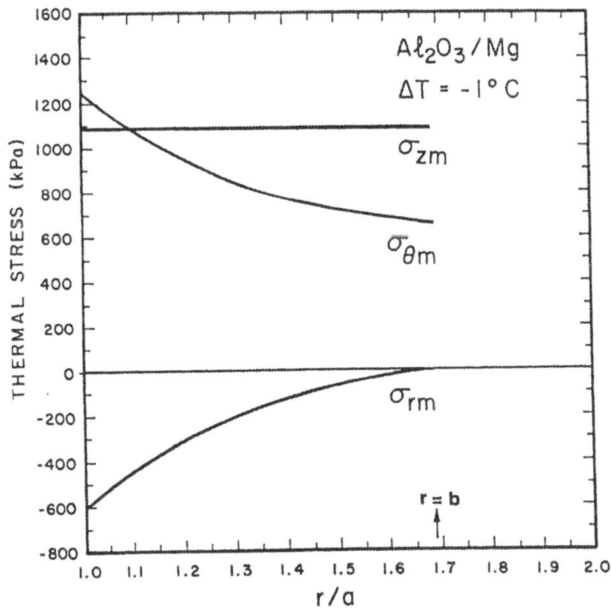

Fig. 6.18 Thermal stress distribution in magnesium matrix in an alumina fiber/magnesium matrix composite as a function of radial distance from the interface. The stress values correspond a cooling of −1°C. Normalized radial distance is used, i.e., r/a = 1 corresponds to the interface.

References

Arsenault, R.J., and R.M. Fisher, (1983) *Scripta Metall.*, **17,** 67.
Balch, D.K., T.J. Fitzgerald, V.J. Michaud, A. Mortensen, Y.-L. Shen, and S. Suresh, (1996) *Metall. Mater. Trans.*, **27A**, 3700-3717.
Behrens, E., (1968) *J. Composite Mater.*, **2**, 2.
Brooksbank, D., and K.W Andrews, (1970) *J. Iron and Steel Inst.*, **208,** 582.
Chang, S.-Y., S.-J. Lin, and M.C. Flemings, (2000) *Metall. Mater. Trans.*, **31A**, 291-298.
Chawla, K.K., (1973a) *Phil. Mag.*, **28**, 401.
Chawla, K.K., (1973b) *Metallography*, **6**, 55.
Chawla, K.K., and M. Metzger, (1972) *J. Mater. Sci.*, **7,** 34.
Chawla, N., B.V. Patel, M. Koopman, K.K. Chawla, R. Saha, B.R. Patterson, E.R. Fuller, and S.A. Langer, (2003) *Mater. Charac.*, **49,** 395-407.
Chawla, N., V.V. Ganesh, and B. Wunsch, (2004) *Scripta Mater.*, **51,** 161-165.
Chawla, N., X. Deng, and D.R.M. Schnell, (2005) *Mater. Sci. Eng.*, submitted.
Christman, T., and S. Suresh, (1988) *Acta Metall.*, **36**, 1691.
Elomari, S., R. Boukhili, and D.J. Lloyd, (1996) *Acta Mater.*, **44**, 1873-1882.
Elomari, S., R. Boukhili, C. San Marchi, A. Mortensen, and D.J. Lloyd, (1997) *J. Mater. Sci.*, **32**, 2131-2140.
Eshelby, J. D., (1957) *Proc. Royal Soc.*, **A241**, 376-396.
Ganesh, V.V., and N. Chawla, (2004) *Metall. Mater. Trans.*, **35A**, 53-62.
Ganesh, V.V., and N. Chawla, (2005) *Mater. Sci. Eng.*, **A391**, 342-353.
Hale, D. K., (1976) *J. Mater. Sci.*, **11**, 2105.
Halpin, J.C., and J.L. Kardos, (1976) *Polymer Eng. Sci.*, **16**, 344.
Halpin, J.C., and S.W. Tsai, (1967) "Environmental Factors Estimation in Composite Materials Design," AFML TR 67-423.
Hermans, J.J., (1967) *Proc. K. Ned.Akad. Wet.*, **B70** (1), 1.
Herrman, K.P., and Y.Q. Wang, (1991) in *Inelastic Deformation of Composite Materials*, Springer-Verlag, New York, p. 445.
Hashin, Z., (1962) *J. Appl. Mech.*, **29**, 143-53.
Hashin Z., and B.W. Rosen, (1964) *J. Appl. Mech.*, **31**, 233.
Hashin, Z., and S. Shtrikman, (1963) *J. Mech. Phys. Solids*, **11**, 127-140.
Hill, R., (1964) *J. Mech. Phys. Solids*, **12**, 199.
Hill, R., (1965) *J. Mech. Phys. Solids*, **13**, 189.
Hsueh, C.-H., P.F. Becker, and P. Angelini, (1988) *J. Am. Ceram. Soc.*, **71**, 929-933.
Hull, A.W., and E.E. Burger, (1934) *Physics*, **5**, 384.
Kerner, E.H., (1956) *Proc. Phys. Soc.*, **B69**, 808.

Koopman, M., K.K. Chawla, C. Coffin, B.R. Patterson, X. Deng, B.V. Patel, Z. Fang, and G. Lockwood, (2002) *Adv. Eng. Mater.*, **4**, 37.
Lee, Y.S., M.N. Gungor, T.J. Batt, and P.K. Liaw, (1991) *Mater. Sci. Eng.*, **A145**, 37-46.
Marom, G.D., and A. Weinberg, (1975) *J. Mater. Sci.*, **10**, 1005.
Mitra, R., W.A. Chiou, M.E. Fine, and J.R. Weertman, (1993) *J. Mater. Res.*, **8**, 2300.
McDanels, D.L., R.W. Jech, and J.W. Weeton, (1965) *Trans. TMS-AIME*, **233**, 636.
Paul, B., (1960) *Trans. AIME*, **218**, 36.
Poritsky, H., (1934) *Physics*, **5**, 406.
Rosen, B.W., (1973) *Composites*, **4**, 16.
Rosen, B.W., and Z. Hashin, (1970) *Int. J. Eng. Sci.*, **8,** 15.
Rossoll, A., B. Moser, and A. Mortensen, (2005) *Mech. Mater.*, **37**, 1.
Sadanandam, J., G. Bhikshamaiah, B. Gopalakrishna, S.V. Suryanarayana, Y.R. Mahajan, and M.K. Jain, (1992) *J. Mater. Sci. Lett.*, **11**, 1518-1520.
Schapery, R.A., (1968) *J. Comp. Mater.*, **2**, 380.
Scherer, G., (1986) *Relaxation in Glass and Composites*, Wiley, New York.
Shen, Y.-L., A. Needleman, and S. Suresh, (1994) *Metall. Mater. Trans.*, **25A**, 839-850.
Timoshenko, S., and J.N. Goodier, (1951) *Theory of Elasticity*, McGraw-Hill, New York, p. 416.
Turner, P.S., (1946) *J. Res. Natl. Bur. Stand.*, **37**, 239.
Vaidya, R.U., and K.K. Chawla, (1994) *Composites Sci. Tech.*, **50**, 13.
Weber, L., (2005) *Acta Mater,* **53**, 1945.
Weber, L., J. Dorn, and A. Mortensen, (2003a) *Acta Mater.*, **51**, 3199
Weber, L., C. Fischer, and A. Mortensen, (2003b) *Acta Mater.*, **51**, 495.
Xu, Z.R., K.K. Chawla, R. Mitra, and M.E. Fine, (1994) *Scripta Metall. Mater.*, **31**, 1525.
Whitney, J.M., and M.B. Riley, (1966) *AIAA Journal*, **4**, 1537.

Chapter 7

MONOTONIC BEHAVIOR

In this chapter we discuss the monotonic strengthening and fracture mechanisms of continuous fiber and discontinuously reinforced metal matrix composites. Cyclic fatigue and creep of MMCs are discussed in chapters 8 and 9, respectively.

7.1 STRENGTHENING MECHANISMS

The monotonic strength and stiffness of MMCs is usually much higher than that of the unreinforced metal. Figure 7.1 shows a general schematic of the evolution of damage in a MMC during monotonic loading. Since the reinforcing phase typically is much stiffer than the matrix, a significant fraction of the stress is initially borne by the reinforcement. Microplasticity then takes place, at a fairly low stress, which corresponds to the original deviation from linearity in the stress-strain curve. This point is termed the proportional limit stress. Microplasticity in the composites has been attributed to stress concentrations in the matrix at the sharp ends of fibers, whiskers, and particles; or at the poles of the reinforcement (Goodier, 1933; Corbin and Wilkinson, 1994; Chawla et al., 1998b; Chawla et al., 2003). With increasing strain, microplasticity increases in magnitude to global plasticity in the matrix. The incorporation of reinforcement results in an increase in work hardening rate of the material, relative to the unreinforced matrix. The higher observed work hardening rate, relative to the unreinforced material, is a simple function of lower matrix volume (by incorporation of the reinforcement), and not necessarily due to a change in work hardening mechanisms. When the matrix is significantly work hardened, the matrix is placed under great constraint (i.e., a triaxial tensile stress develops) with an inability for strain relaxation to take place. This causes the onset of void nucleation and propagation, which take place at a lower far field applied strain than that observed in the unreinforced material. With the onset of void growth in the matrix, the stress in the reinforcement

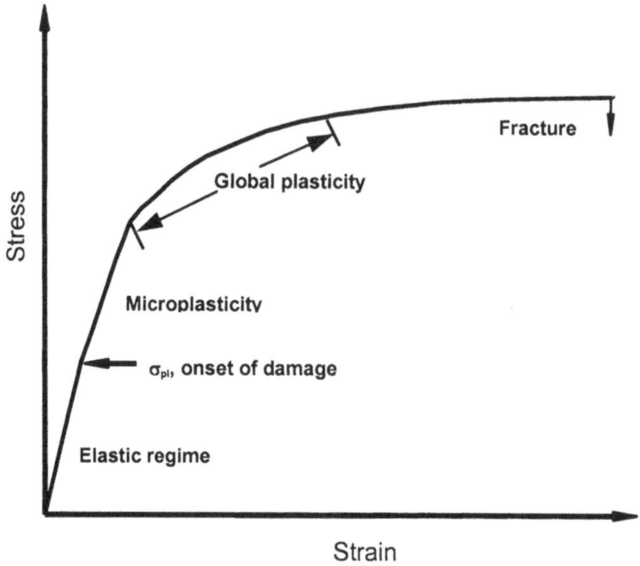

Fig. 7.1 Schematic of damage evolution in MMCs under tensile loading.

approaches its fracture stress, which results in particle fracture and fracture of the composite. It should be pointed out that the damage evolution will also be very much influenced by the strength and nature of the reinforcement/matrix interface, as discussed in sections 7.2 and 7.3 below.

7.1.1 Direct Strengthening

The strengthening mechanisms observed in MMCs may be divided into two broad categories, direct and indirect strengthening. Direct strengthening takes place primarily in continuous fiber reinforced composites, but also takes place in discontinuously reinforced composites. Under an applied load, the load is transferred from the weaker matrix, across the matrix/reinforcement interface, to the typically higher stiffness reinforcement (Cox, 1952; Kelly and Lilholt, 1969; Cheskis and Heckel, 1970; Kelly, 1973; Chawla, 1997; Chawla and Shen, 2001). In this manner, strengthening takes place by the reinforcement "carrying" much of the applied load. This is shown schematically in Fig. 7.2. Let's assume a single high stiffness fiber embedded in a lower modulus matrix. The composite is loaded without direct loading of the fiber itself. If a set of parallel imaginary lines are drawn on the composite, after loading the lines will become distorted because of the shear stresses generated by differing axial displacements in fiber and matrix. Thus, load transfer to the fiber occurs by means of shear strains in the matrix.

Let us consider the case of a purely elastic fiber in an elastic matrix mathematically, Fig. 7.3. Let u be displacement in the matrix in the presence of the fiber, at a distance x from one end, and let v be the displacement at x in the absence of the fiber. If P_f is the normal load on the fiber, then we can write the following expression for load transfer from the matrix to the fiber:

$$\frac{dP_f}{dx} = B(u - v) \qquad (7.1)$$

where B is a constant that is a function of fiber arrangement and matrix and

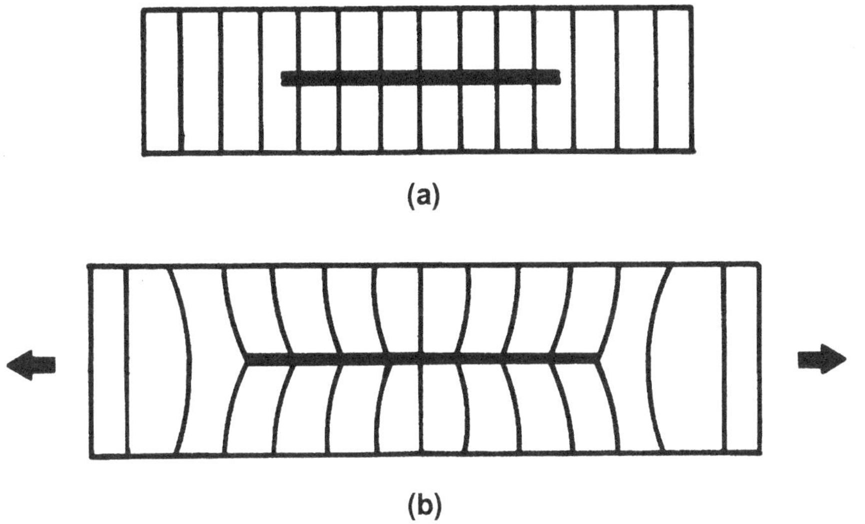

Fig. 7.2 A single fiber embedded in a matrix with lower modulus: (a) unstressed state and (b) stressed state. The imaginary vertical lines become distorted, indicating shear strains at the interface and load transfer to the fiber.

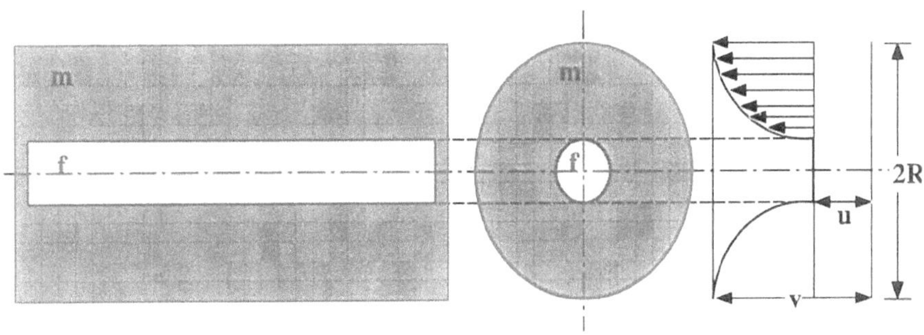

Fig. 7.3 A single fiber embedded in a cylindrical matrix shell subjected to an axial strain, e, which causes displacements in the fiber (u) and in the matrix (v).

fiber properties. The exact expression for B is described later in this section. Differentiating eq. 7.1, we get:

$$\frac{d^2P_f}{dx^2} = B\left(\frac{du}{dx} - \frac{dv}{dx}\right) \tag{7.2}$$

where $\frac{du}{dx}$ is the strain in fiber $= \frac{p_f}{E_f A_f}$ and

$\frac{dv}{dx}$ is the strain in the matrix away from the fiber = e, imposed strain

Equation 7.2 can be rewritten as:

$$\frac{d^2P_f}{dx^2} = B\left[\frac{P_f}{A_f E_f} - e\right]$$

This second-order differential equation has the following solution:

$$P_f = E_f A_f e + S\sinh\beta x + T\cosh\beta x$$

where $\beta = \sqrt{\frac{B}{A_f E_f}}$.

In order to evaluate the integration constants, S and T, we apply the boundary conditions, $P_f = 0$ at $x = 0$ and $x = \ell$. Using half-angle trigonometric relations we obtain the following result:

$$P_f = E_f A_f e\left[1 - \frac{\cosh\beta\left(\frac{\ell}{2} - x\right)}{\cosh\frac{\beta\ell}{2}}\right] \quad \text{for } 0 < x < \ell/2 \tag{7.3a}$$

or

$$\sigma_f = \frac{P_f}{A_f} = \frac{E_f e}{\ell}\left[1 - \frac{\cosh \beta\left(\frac{\ell}{2} - x\right)}{\cosh \frac{\beta\ell}{2}}\right] \quad \text{for } 0 < x < \ell/2 \qquad (7.3b)$$

The maximum possible value of strain in the fiber is the imposed strain, e, so the maximum stress in the fiber is eE_f. The parameter β is a measure of how fast the load is transferred from the matrix to the fiber from the two ends. A value of $\beta = 0.5$ indicates a linear dependence of load transfer. If we have a long enough fiber, Fig. 7.4, the stress in the fiber will increase from both ends to a maximum value of the ultimate tensile strength of the fiber, i.e., $\sigma_{fu} = E_f e$. Only a portion of the fiber (in the center) will be under the maximum stress. The *average* stress in the fiber, then, can be written as:

$$\overline{\sigma}_f = \frac{E_f e}{\ell}\int_0^\ell \left[1 - \frac{\cosh \beta\left(\frac{\ell}{2} - x\right)}{\cosh \frac{\beta\ell}{2}}\right] dx = E_f e\left[1 - \frac{\tanh \frac{\beta\ell}{2}}{\frac{\beta\ell}{2}}\right]$$

We can obtain the variation of shear stress, τ, along the fiber/matrix interface by considering the equilibrium of forces acting over an element of fiber (radius r_f). Thus, we can write, from Fig. 7.4:

$$\frac{dP_f}{dx} dx = 2\pi r_f dx \tau \qquad (7.4)$$

Let us now consider the variation of shear stress, τ, along the fiber/matrix interface. We can obtain an expression for the interfacial shear stress by considering the equilibrium of forces over an element of fiber. The tensile load on the fiber, P_f, is equal to $P_f = \sigma_f \pi r_f^2$. Substituting this into eq. 7.4 we get:

$$\tau = \frac{1}{2\pi r_f}\frac{dP_f}{d\ell} = \frac{r_f}{2}\frac{d\sigma_f}{dx} \qquad (7.5)$$

From eqs. 7.3(b) and 7.5 we obtain:

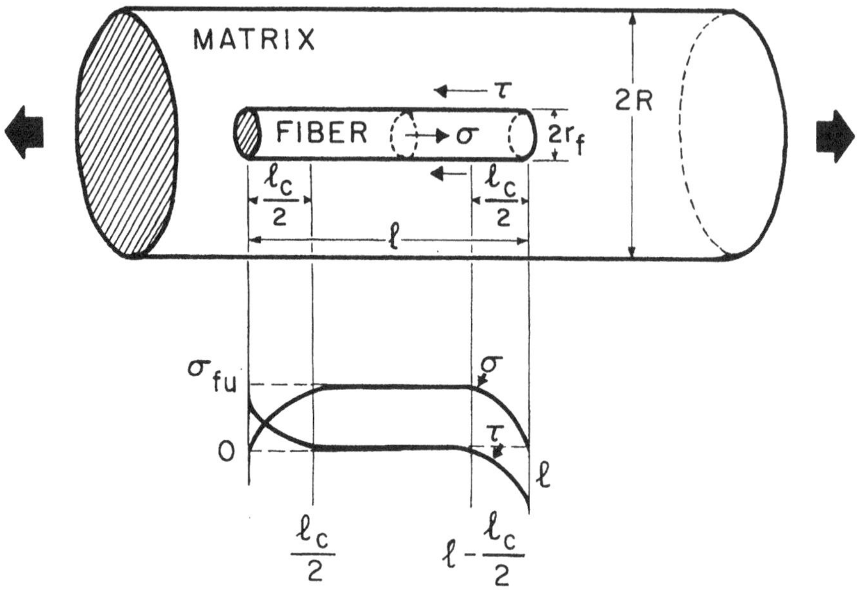

Fig. 7.4 Tensile stress (σ) and shear stress (τ) distribution in a single fiber embedded in a cylindrical matrix shell.

$$\tau = \frac{E_f r_f e\beta}{2} \frac{\sinh\beta(\ell/2 - x)}{\cosh\beta\ell/2}$$

Figure 7.4 shows the variation of τ and σ with distance x. At the ends of the fiber, the axial stress is zero. It increases until it reaches σ_{fu} in the center of the fiber (assuming a sufficiently long fiber that will enable the stress to build to σ_{fu}). The shear stress is a maximum at the fiber ends, and a minimum in the center of the fiber. Such a stress distribution has also been confirmed by finite difference technique (Termonia, 1987) and by micro-Raman spectroscopy for polymer matrix composites (Young, 1994).

We now return to the expression for the constant B, which is a function of fiber packing geometry. In Fig. 7.4 we note that 2R is the average center-to-center fiber spacing. Let τ(r) represent the shear stress at a distance r from the fiber axis. Then at the fiber surface we can be written as:

$$\frac{dP_f}{dx} = -2\pi r_f \tau(r_f) = B(u - v)$$

Or

$$B = -\frac{2\pi r_f \tau(r_f)}{(u - v)} \tag{7.6}$$

Consider now the equilibrium of forces on volume of matrix material between r_f and R. We can write:

$$2\pi r\tau(r) = \text{constant} = 2\pi r_f \tau(r_f)$$

$$\tau(r) = \frac{r_f \tau(r_f)}{r}$$

The shear strain in the matrix at a distance r is given by:

$$\tau(r) = G_m \gamma$$

where G_m is the shear modulus of the matrix and γ is the shear strain in the matrix. The shear strain in the matrix is given by

$$\gamma = \frac{dw}{dr} = \frac{\tau(r)}{G_m} = \frac{\tau(r_f) r_f}{G_m r}$$

where w is the real displacement in the matrix at any distance r. Integrating the above expression between the surface of the fiber, r_f and the outer radius of the matrix, R, gives us the total displacement in the matrix:

$$\int_0^R dw = \Delta w = \frac{\tau(r_f) r_f}{G_m} \int_0^R \frac{1}{r} dr = \frac{\tau(r_f) r_f}{G_m} \ell n\left(\frac{R}{r_f}\right) \tag{7.7}$$

We can also write for the total displacement as:

$$\Delta w = v - u = -(u - v) \tag{7.8}$$

From Eqs. (7.7) and (7.8), we obtain the following relationship:

$$\frac{\tau(r_f)r_f}{u-v} = -\frac{G_m}{\ln(R/r_f)} \tag{7.9}$$

From Eqs. (7.6) and (7.9), we get:

$$B = \frac{2\pi G_m}{\ln(R/r_f)} \tag{7.10}$$

B is related to load transfer parameter β as follows:

$$\beta = \left(\frac{B}{E_f A_f}\right)^{\frac{1}{2}} = \left[\frac{2\pi G_m}{E_f A_f \ln(R/r_f)}\right]^{\frac{1}{2}} \tag{7.11}$$

The ratio R/r_f is a function fiber packing. For a square array and hexagonal array of fibers, we can write the following two expressions:

$$\ln\left(\frac{R}{r_f}\right) = \frac{1}{2}\ln\left(\frac{\pi}{V_f}\right) \quad \text{(square array)}$$

$$\ln\left(\frac{R}{r_f}\right) = \frac{1}{2}\ln\left(\frac{2\pi}{\sqrt{3}V_f}\right) \quad \text{(hexagonal array)}$$

A more general, maximum fiber packing factor, ϕ_{max}, can also be introduced into the above equation:

$$\ln\left(\frac{R}{r_f}\right) = \frac{1}{2}\ln\left(\frac{\phi_{max}}{V_f}\right)$$

Substituting into Eq. (7.11), we get:

$$\beta = \left[\frac{4\pi G_m}{E_f A_f \ln(\phi_{max}/V_f)}\right]^{\frac{1}{2}}$$

From the discussion above, it can be seen that in order to load the fibers to their ultimate tensile strength, the matrix shear strength must be relatively high. The maximum shear stress will be the smaller of the following two stresses: (a) yield stress of the matrix in shear and (b) shear strength of the fiber/matrix interface. In MMCs, the interface shear strength is quite high, so plastic yielding of the matrix will take place first. If we assume that the matrix does not work-harden, the matrix shear yield strength, τ_y will control the load transfer. Then, the equilibrium of forces over a fiber length $\ell/2$ (since the fiber is loaded from both ends) gives us the following relation:

$$\sigma_f \frac{\pi d^2}{4} \ell = \tau_y \pi d \frac{\ell}{2} \quad \text{or} \quad \frac{\ell}{d} = \frac{\sigma_f}{2\tau_y} \tag{7.12}$$

The term ℓ/d is called the aspect ratio of the fiber. Given a sufficiently long fiber, it should be possible to load the fiber to its ultimate tensile strength, σ_{fu}, by means of load transfer through the plastically deforming matrix. Thus, in order to load the fiber to σ_{fu}, a critical aspect ratio of fiber is required, $(\ell/d)_c$, which is obtained by rewriting Eq. (7.12):

$$\left(\frac{\ell}{d}\right)_c = \frac{\sigma_{fu}}{2\tau_y} \tag{7.13}$$

Thus, in order to load the fiber to σ_{fu} at a single point, ℓ must be equal to ℓ_c. In order to load a larger fraction of the fiber to σ_{fu}, ℓ should be much larger than ℓ_c. Thus, load transfer is more efficient in composites with large aspect ratio reinforcement, such as continuous fibers or whiskers. Due to the lower aspect ratio of particulate materials, load transfer is not as efficient as in the case of continuous fiber reinforcement, but is still significant in providing strengthening (Nardone and Prewo, 1986; Davis and Allison, 1993; Chawla et al., 1998a; Chawla et al. 2000).

Nardone and Prewo (1986) proposed a modified shear lag model for load transfer in particulate materials, Fig. 7.5. The model incorporates load transfer from the particle ends, which is neglected in fiber reinforced composites because of the large aspect ratio. The yield strength of the particulate composite, σ_{cy}, is increased over the matrix yield strength, σ_{my}:

$$\sigma_{cy} = \sigma_{my}\left[V_p\left(\frac{S+4}{4}\right) + V_m\right]$$

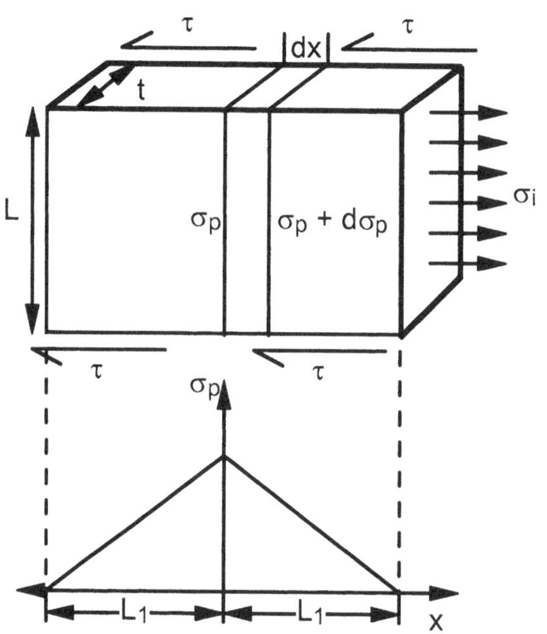

Fig. 7.5 Schematic of modified shear lag analysis used to quantify load transfer in low aspect ratio particles (after Nardone and Prewo, 1986).

where S is the aspect ratio of the particle (equal to the particle length, L, divided by the particle thickness, t, for a rectangular particle), V_p is the volume fraction of particles, and V_m is the volume fraction of matrix. Note that this relation does not account for particle size or matrix microstructure effects on load transfer.

7.1.2 Indirect Strengthening

Indirect strengthening arises from a change in matrix microstructure and properties due to the addition of reinforcement. In this section we describe some possible sources of indirect strengthening. Thermal expansion mismatch between reinforcement and matrix can result in internal stresses whenever there is a temperature change (e.g., during cool down from processing or during service). Such a mismatch is generally present in all kinds of composites. It is a very important feature of MMCs comprised as they are of a high coefficient of thermal expansion (CTE) metallic matrix and a low CTE ceramic reinforcement. If the thermal mismatch-induced stress is greater than the yield stress of the matrix, upon cooling, dislocations form at the reinforcement/matrix interface because plastic deformation in common metals occurs via dislocations. In this manner, thermally-induced dislocation punching results in "indirect strengthening" of the matrix

(Chawla and Metzger, 1972; Chawla, 1973a; Chawla, 1973b; Vogelsang et al., 1986; Arsenault and Shi, 1986; Dunand and Mortensen, 1991). Chawla and Metzger (1972) showed this effect in W fiber reinforced Cu single crystal matrix composites. The density of dislocations in the matrix was measured (by an etch pitting technique) as a function of distance from the fiber/matrix interface, for composites with various volume fractions of W fiber, Fig. 7.6(a). With increasing volume fraction, the dislocation density in the matrix increased. The dislocation density was also highest in the interface region because the thermal stresses between fiber and matrix in the interface region are the highest. The three-dimensional thermal stress model described in chapter 6 can be used to explain the distribution of dislocation density in the matrix. Fig. 7.6(b) shows thermally-induced slip lines in the polycrystalline Cu matrix in a W/Cu composite that was thermally cycled between room temperature and 450°C. Dunand and Mortensen (1991) used a model system of glass fiber reinforced AgCl to study dislocation punching due to thermal mismatch. The AgCl matrix is optically transparent and the dislocations can be seen emanating from the fiber, Fig. 7.7. The degree of thermally-induced dislocations can also be exacerbated by the degree of fiber roughness (Isaacs and Mortensen, 1992).

Arsenault and Shi (1986) developed a model to quantify the degree of dislocation punching that takes place due to CTE mismatch between a particle and matrix, Fig. 7.8. The dislocation density generated due to the mismatch is given by:

$$\rho_{CTE} = \frac{A\varepsilon V_p}{b(1 - V_p)d} \tag{7.14}$$

where A is a geometric constant, b is the Burgers vector, d is the diameter of the particle, V_p is the particle volume fraction, and ε is the thermal misfit strain equal to $\Delta\alpha\Delta T$. The incremental increase in strength due to dislocation punching, then, can be written as:

$$\Delta\sigma = BGb\rho_{CTE}^{1/2}$$

where B is a constant and G is the shear modulus of the matrix. Substituting in Eq. 7.14, we can write $\Delta\sigma$ as:

$$\Delta\sigma = BGb\left[\frac{A\varepsilon V_p}{b(1 - V_p)}\frac{S}{d}\right]^{\frac{1}{2}} \tag{7.15}$$

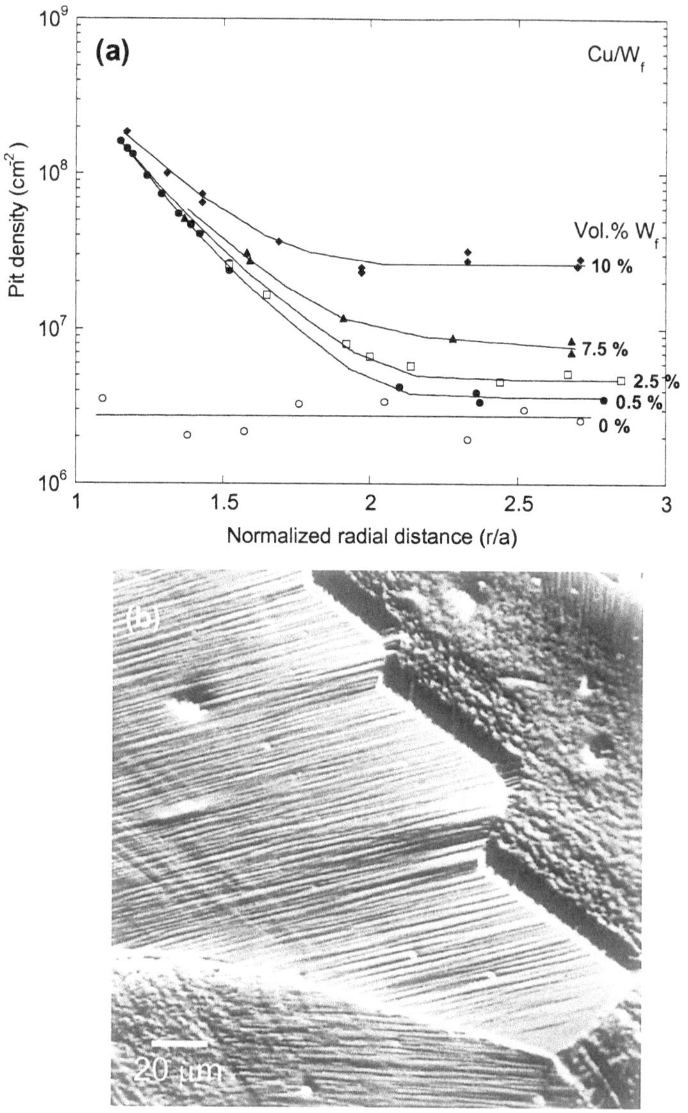

Fig. 7.6 (a) Dislocation density in W fiber reinforced Cu matrix composites, as a function of distance from the fiber/matrix interface. With increasing volume fraction of fiber, the thermal stress mismatch at the interface increased, resulting in a higher dislocation density (after Chawla and Metzger, 1972). (b) Thermally-induced slip lines in the Cu matrix.

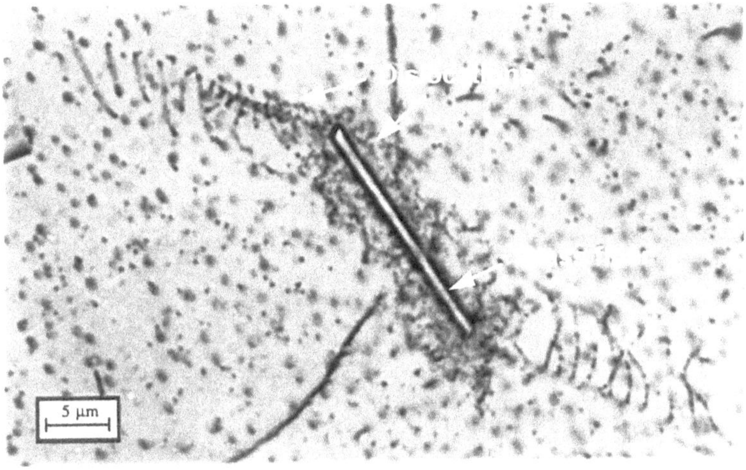

Fig. 7.7 Thermally-induced dislocation punching in a model glass fiber/AgCl system (Dunand and Mortensen, 1991). Note the high density of dislocations at the interface, and the punching of dislocation loops from the fiber ends (courtesy of A. Mortensen and reprinted with permission of Elsevier Ltd.).

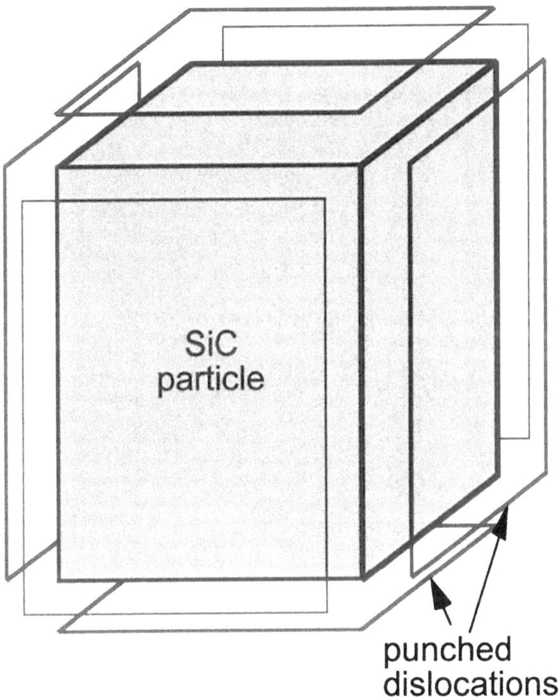

Fig. 7.8 Schematic of model used to quantify the degree of dislocation punching that takes place due to CTE mismatch between particle and matrix (after Arsenault and Shi, 1986).

where S is the aspect ratio of the particle and the other symbols have the significance given above. Inspection of Eq. 7.15 shows that the degree of indirect strengthening is directly proportional to volume fraction and inversely proportional to particle size. Thus, a larger degree of interfacial area (i.e., smaller particle size) will result in enhanced dislocation punching. This is shown schematically in Fig. 7.9. Note that for a constant volume fraction of 0.3, a significant enhancement in indirect strengthening is observed for particles of 3 μm diameter, vis-à-vis particles of 100 μm diameter

In age-hardenable matrix materials, the thermally-induced dislocations (formed upon quenching from the solution treatment) serve as heterogeneous nucleation sites for precipitate formation during the aging treatment (Suresh and Chawla, 1993). Not only is there a preferential distribution of precipitates in the particle/matrix interface region, but the higher density of dislocations also causes an acceleration in the time to peak-age compared to the unreinforced alloy of the same composition.

In composites processed by liquid phase routes, the matrix grain size can be much finer than that of the unreinforced alloy, due to pinning of grain boundaries by the particles or Zener pinning (Humphreys, 1977;

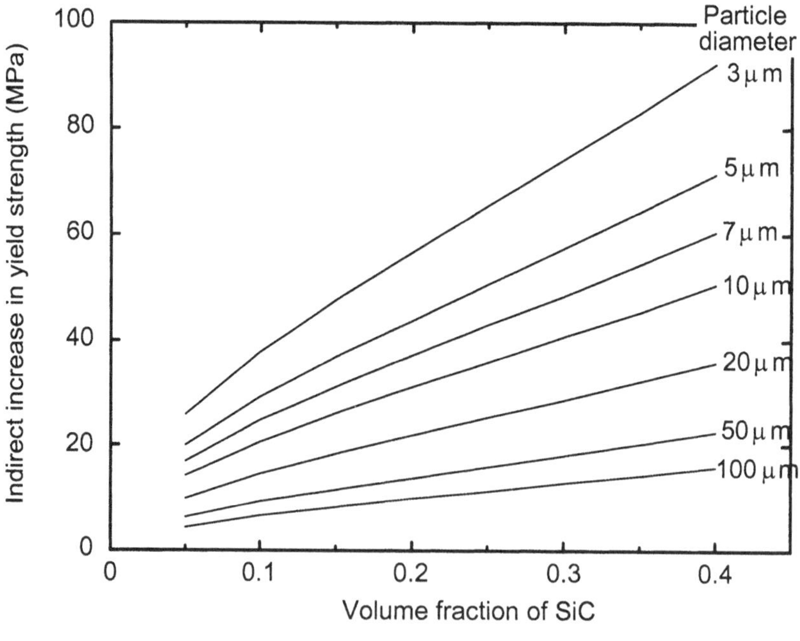

Fig. 7.9 Increase in yield strength due to dislocation punching, as predicted by the model of Arsenault and Shi (1986). The degree of strengthening is significantly influenced by particle size, for a given volume fraction.

Humphreys, 1991). Differences in matrix texture may also result by the incorporation of reinforcement, for example, in deformation processed materials (see chapter 4).

Separating and quantifying the contributions of direct and indirect strengthening, to the overall composite strength is a challenge. The extent of indirect strengthening is more difficult to quantify than the contribution from direct strengthening. One way to separate the two types of strengthening is to process composites such that the matrix microstructure is similar to that of the unreinforced alloy. Krajewski et al. (1993), used a thermomechanical treatment in Al 2080/SiC_p, consisting of solution treating, rolling, followed by aging (T8 treatment) to provide a homogeneous distribution of dislocations (and subsequently precipitates) in both the matrix of the composite and the unreinforced alloy. In such a situation, the difference in strengthening between unreinforced alloy and composite was attributed primarily to load transfer to the reinforcement (Chawla et al., 1998a). This is shown in Fig. 7.10. Chawla et al., (1998a) showed that in T8 matrix

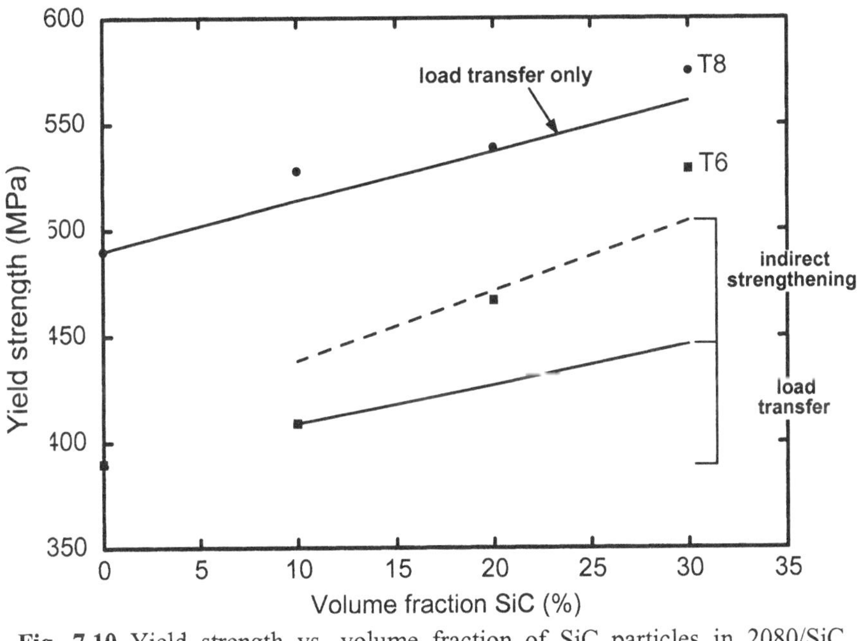

Fig. 7.10 Yield strength vs. volume fraction of SiC particles in 2080/SiC_p composite (after Chawla et al., 1998a). The rolled and aged materials (T8 matrix composites) have similar microstructure to the unreinforced T8 alloy. T6 matrix composites are aged only, so their microstructure is different from 2080-T6. The experimental increase in yield strength in T8 materials correlated well with predictions from the modified shear lag model. In T6 matrix composites, the strength of the composite consisted of contributions of both indirect and direct strengthening.

composites, the experimental increase in yield strength correlated well with predictions from the modified shear lag model. In composites that were solution treated, quenched, and aged (T6 heat treatment), the strength of the composite consisted of contributions of both indirect and direct strengthening, Fig. 7.10.

7.2 MONOTONIC BEHAVIOR OF CONTINUOUS FIBER REINFORCED MMCS

The monotonic strength and damage evolution in continuous fiber reinforced MMCs are dependent on several factors:

1. *Fiber characteristics*. The volume fraction, strength, and relative orientation of the fibers, with respect to the loading axis.

2. *Strength and nature of the interface*. Interfacial strength has a significant effect on the strengthening and damage tolerance of continuous fiber reinforced MMCs. Interfacial reactions between fiber and matrix, and fiber dissolution, may be detrimental to the strength of the composite. Preferential precipitation at the interface, in age-hardenable systems, can have mixed effects on strength.

3. *Matrix work hardening and strength characteristics*. Of particular importance are the changes in matrix microstructure, during processing, due to the incorporation of the reinforcement (i.e., indirect strengthening, described above).

We now examine each of the three main factors described above. Continuous fiber reinforced MMCs exhibit very high strength in the direction parallel to the fibers, but comparatively low strength perpendicular to the fiber direction. Figure 7.11 shows the anisotropy in strength of Al-2.5Li/$Al_2O_{3,f}$ composites, parallel to the fibers (0°) and perpendicular to the fiber direction (90°), in both tension and compression (Schulte and Minoshima, 1993). Clearly, along the axis of the fibers the degree of strengthening from the fibers will be much higher than that perpendicular to the fiber direction. The mechanical properties of fiber reinforced metal matrix composites are also very much dependent on the strength of the fiber/matrix interface. Figure 7.12 shows a schematic of damage for the case of a relatively weak fiber/matrix interface and that of a relatively strong interface. When a fiber fractures in the composite with a weak interface, fiber debonding and crack deflection take place. These local energy-absorbing mechanisms allow for uniform redistribution of the load, originally borne by the fractured fiber, to the surrounding fibers. This

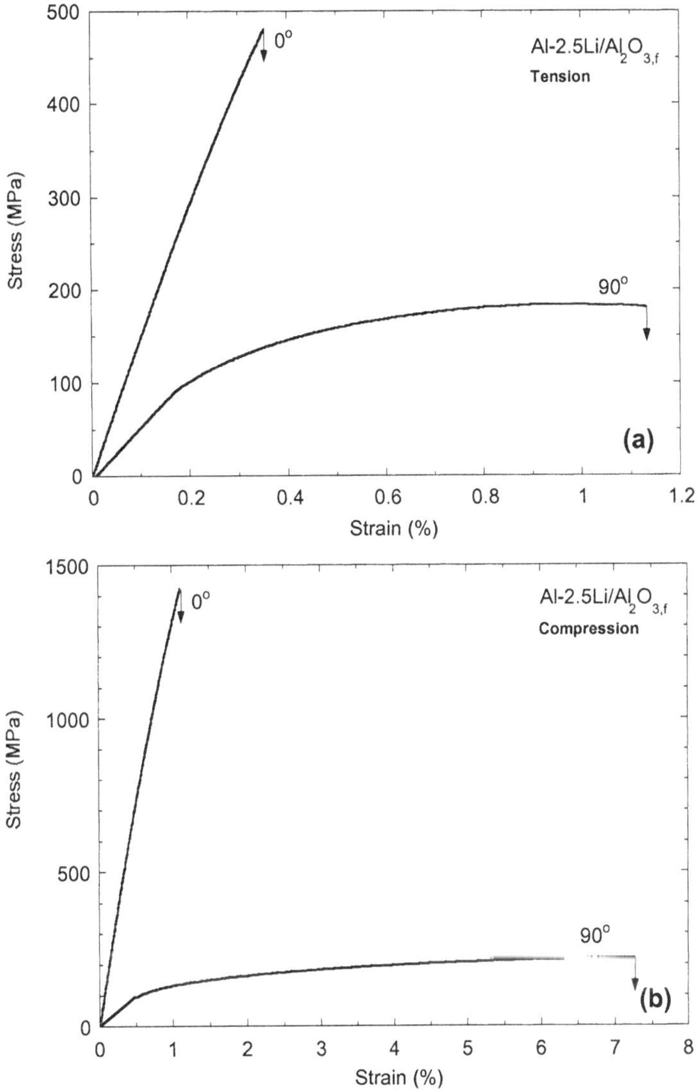

Fig. 7.11 Anisotropy in strength of Al-2.5Li/$Al_2O_{3,f}$ composites, parallel to the fibers (0°) and perpendicular to the fiber direction (90°), in (a) tension and (b) compression (after Schulte and Minoshima, 1993).

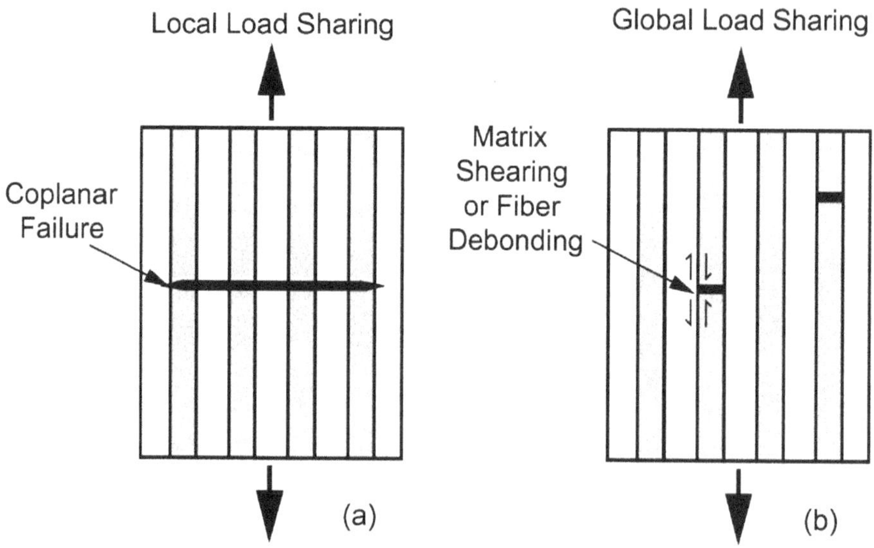

Fig. 7.12 Schematic of damage for the case of: (a) relatively strong fiber/matrix interface, resulting in local load sharing and coplanar failure, and (b) relatively weak interface, where global load sharing is observed, and fiber debonding and matrix shear are predominant.

condition is called *global load sharing* (Curtin, 1993). When the interface strength is very large, fiber fracture will not result in debonding and crack deflection. Rather, the load cannot be redistributed homogeneously, so the single fiber fracture will result in precipitous fracture of adjacent fibers. As the neighboring fibers fracture, more and more fibers will continue to fracture until the composite fails. This series of successive fiber failures due to localization of strain around a single fiber results in *local load sharing* (Gonzalez and LLorca, 2001). It should be noted that a very weak interface is also not desirable in MMCs. Such an interface would not allow efficient load transfer from the matrix to the fiber.

Whether a composite exhibits global or local load sharing behavior may also be influenced by the work hardening characteristics of the matrix. A matrix with high strain hardening rate would be less conducive to plastic relaxation of the stress concentration around the fractured fiber. Thus, more brittle matrix materials are more conducive to local load sharing and a lower ductility. Figure 7.13 shows a comparison of $Al/Al_2O_3/60_f$ and $6061/Al_2O_3/60_f$ fiber reinforced composites (Devé and McCullough, 1995).

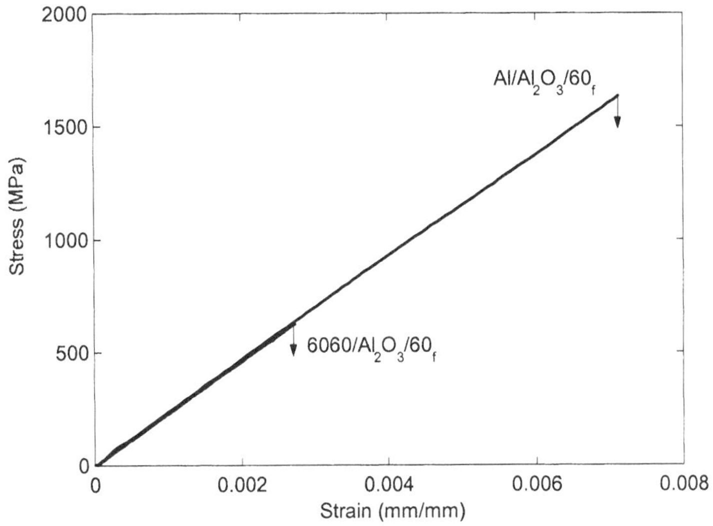

Fig. 7.13 Tensile stress-strain comparison of $Al/Al_2O_3/60_f$ and Al6061/ $Al_2O_3/60_f$ fiber reinforced composites (after Devé and McCullough, 1995). The high interfacial shear strength and higher matrix work hardening rate of the Al6061 matrix, resulted in a much lower ductility and strength than the composite with pure Al matrix.

Because of the combined effects of high interfacial shear strength and higher matrix work hardening rate of the Al 6061 matrix, this composite has a much lower ductility and strength than the composite with the pure Al matrix. Voleti et al. (1998) used a finite element model of a composite consisting of a broken fiber surrounded by the matrix and an intact fiber. They showed that the stress concentration in the intact fiber was influenced by both interfacial debonding and matrix plasticity surrounding the broken fiber. If fiber fracture takes place at very low strains (much below the onset of global plasticity), then the matrix plasticity "propagates" toward the intact fiber resulting in a large degree of stress elevation. Interfacial debonding, on the other hand, reduces the stress concentration around the fractured fiber and the intact fiber.

The degree of global load sharing is also a function of strain rate. Galvez et al. (2001) tested large diameter SiC fiber reinforced Ti-6Al-4V matrix composites at strain rates ranging between $2x10^{-5}$ and 500 s^{-1}. It was shown that at very high strain rates local load sharing conditions were predominant, and a lower composite ductility was observed. Lower strain rates were more conducive to allowing a gradual redistribution of the load, so global load sharing was observed. Microstructural observations showed that at low strain rates the carbon coating on the SiC fiber resulted in crack deflection at the interface, while at very high strain rates the cracks penetrated the coating. These authors did not, however, observe a strain rate dependence on strength of the composite.

Guden and Hall (2000) noticed significant increases in compressive flow stress and strength with increasing strain rate in $Al/Al_2O_{3,f}$ composites. The strain rate sensitivity was observed in both longitudinal and transverse orientations. The strain rate sensitivity of transverse strength was attributed to the matrix strain rate sensitivity, while that in the longitudinal direction was due to strain rate dependent fiber buckling. In static compression, significant kinking and buckling of the fibers also takes place (Deve, 1997). Unfortunately, while relatively weak interfaces are conducive to global load sharing and thus, high tensile strength and ductility, they are quite detrimental to the transverse properties of the composite (Jansson et al., 1991; McCullough et al., 1994; Bushby, 1998). This can be explained by the fact that, under transverse loading, the weak interfaces are perpendicular to the applied load. Similar observations have been made for Al_2O_3 fiber/Mg matrix composites under off-axis loading (Hack et al., 1984).

7.2.1 Criteria for Debonding and Crack Deflection at an Interface

In MMCs, as well as in other types of composites, interfacial bonding affects the fracture behavior of the composite. In very general terms, a strong interfacial bond will allow an oncoming crack to go unimpeded through the interface and the composite will fail, more or less, in one plane and in a brittle manner. The interaction of a crack with a weak interface, on the other hand, is likely to lead to debonding at the interface, followed by crack deflection, crack bridging, fiber fracture, and finally fiber pullout. These are all energy absorbing phenomena that contribute to enhanced fracture toughness and a non-catastrophic failure mode.

A strength-based model for crack deflection or the formation of secondary crack at a weak interface was first proposed by Cook and Gordon (1964). They analyzed the problem of crack deflection at an interface between materials of identical elastic constants, i.e., the same material joined at an interface. Consider a crack advancing perpendicular to the fiber/matrix interface. Cook and Gordon (1964) estimated the strength of the interface necessary to cause a diversion of the crack from its original direction. At the tip of a crack, there exists a triaxial state of stress (plane strain) or a biaxial stress (plane stress), see Fig. 7.14. The principal stress component, σ_y, has a very high value at the crack tip, and decreases sharply with distance from the crack tip. The stress component acting normal to the interface, σ_x, is zero at the crack tip. It rises to a maximum value at a small distance from the crack tip and then falls off in a manner similar to σ_y. If the tensile strength of the interface is less than the maximum value of σ_x, then the interface will fail in front of the crack tip. According to the estimates of Cook and Gordon, an interface with strength equal to or less than 1/5th of σ_x will result in the opening of the interface in front of the crack tip.

One can also analyze the interaction between a crack and an interface in terms of a fracture energy parameter instead of strength (He and Hutchinson, 1989). Two materials meeting at an interface are more than likely to have *different* elastic constants. The modulus mismatch leads to shearing of the crack surfaces. This results in a mixed-mode stress state in the vicinity of an interface crack tip involving both the tensile and shear components. This, in turn, results in mixed-mode fracture, which can occur at the crack tip or in the wake of the crack. Figure 7.15 shows this crack front and crack wake debonding in a fiber reinforced composite. What this means in practical terms is that instead of a simple, one parameter description by the critical stress intensity factor K_{Ic}, one needs a more complex formalism of fracture mechanics to describe the situation in the composite. The parameter K under such a situation becomes scale sensitive, but the critical strain energy release rate, G_{Ic}, is not a scale sensitive parameter. The strain energy release rate, G is a function of the phase angle, ψ, which is a function of normal and shear loading.

One needs to specify G and ψ to analyze the debonding at the interface. G and ψ are related, through the so-called Dundurs parameters, α and β as defined below:

$$\alpha = \frac{G_1(1-\nu_2) - G_2(1-\nu_1)}{G_1(1-\nu_2) + G_2(1-\nu_1)}$$

$$\beta = \frac{1}{2}\frac{G_1(1-2\nu_2) - G_2(1-2\nu_1)}{G_1(1-\nu_2) + G_2(1-\nu_1)}$$

The parameter α can also be defined as:

$$\alpha = \frac{\overline{E}_1 - \overline{E}_2}{\overline{E}_1 + \overline{E}_2}$$

$$\overline{E} = \frac{E}{1-\nu^2}$$

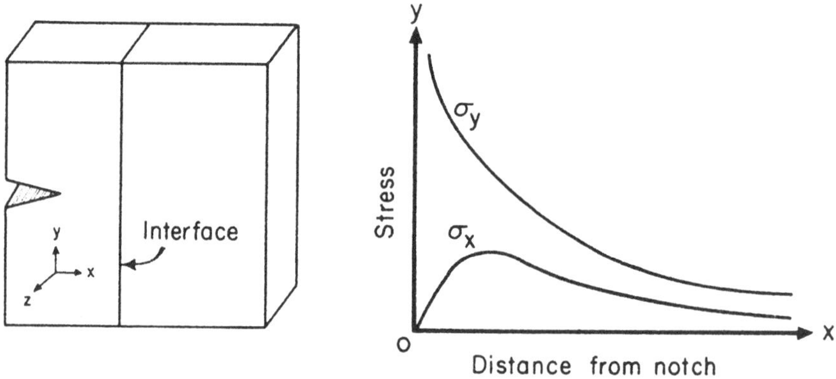

Fig. 7.14 Stress distribution at a crack tip under a uniaxial applied stress in the y-direction (after Cook and Gordon, 1964). σ_y has a very high value at the crack tip, and decreases sharply with distance from the crack tip. The stress component acting normal to the interface, σ_x, is zero at the crack tip; it rises to a maximum value at a small distance from the crack tip and then falls off in a manner similar to σ_y.

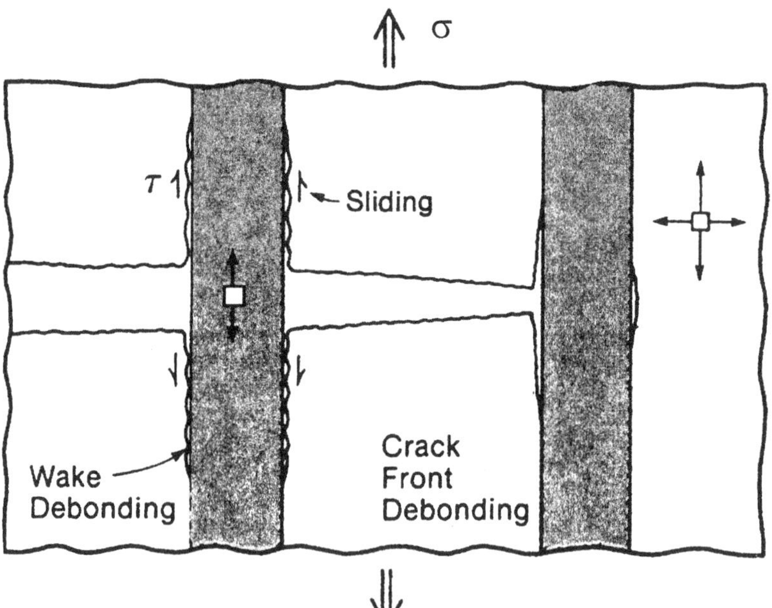

Fig. 7.15 Crack front and crack wake debonding in a fiber reinforced composite under a far field, uniaxial stress, σ. Note the interfacial shear stress, τ causing sliding and triaxial state of stress at the crack tip.

In the above expressions, ν is the Poisson's ratio, E is Young's modulus, and the subscripts 1 and 2 refer to the interface and fiber, respectively. The expression for the phase angle, ψ, in terms of the elastic coefficients of the two media, radius r from the crack tip, and the displacements, u and v, at the crack tip is as follows:

$$\phi = \tan^{-1}\left(\frac{v}{u}\right)$$

$$\psi = \phi - \left(\frac{\ln r}{2\pi}\right)\ln\frac{(1-\beta)}{(1+\beta)}$$

These expressions have been used by several researchers (Ruhle and Evans, 1988; He and Hutchinson, 1989; Evans and Marshall, 1989; Gupta et al., 1993; Chan, 1993) to analyze the conditions for fiber/matrix debonding in terms of the energy requirements. Without going into the details of the model, the main message of such a plot is to display the conditions under which the crack will deflect along the interface or penetrate through the interface into the fiber. The chart of G_i/G_f vs. α is shown in Fig. 7.16. G_i is the mixed- mode interfacial fracture energy of the interface, G_f is the mode I fracture energy of the fiber, and α is a measure of elastic anisotropy as defined above. For opening mode or mode I, ψ = 0° while for mode II, ψ = 90°. For all values of G_i/G_f below the line, interfacial debonding is predicted. For the special condition of α = 0, i.e., zero elastic mismatch, the model

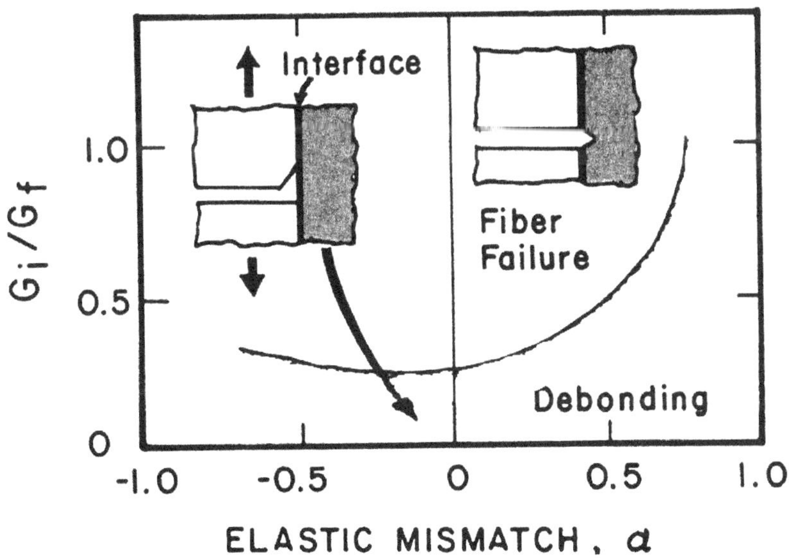

Fig. 7.16 A chart of relative energies, G_i/G_f vs. elastic mismatch, α (after Evans and Marshall, 1989).

predicts debonding at the fiber/matrix interface for G_i/G_f less than about 0.25. Conversely, for G_i/G_f greater than 0.25, the crack will propagate across the fiber. In general, for the elastic mismatch, α greater than zero, the minimum interfacial toughness required for interface debonding increases, i.e., high modulus fiber tends to favor debonding.

Gupta et al. (1993) have also derived strength and energy criteria for crack deflection at a fiber/matrix interface for several composite systems. They took due account of the anisotropic nature of the fiber as well as the fact that a crack can deflect along the interface in one direction (singly) or two directions (doubly). Their experimental technique involved laser spallation of a film from a substrate and the measurement of displacement by a sophisticated laser Doppler displacement interferometer. This technique allows the measurement of the tensile strength of the planar interface between the film and the substrate. The strength determined in the laser spallation experiment is thought to be independent of any inelastic processes because the interface separation takes place at a very high strain rate, about 10^6 s^{-1}. The tensile strength determined in this fashion is then related to the *intrinsic* interface toughness. According to the analysis of Gupta et al., for most pairs of materials, the energy release rate is higher for the doubly deflected crack than that for the singly deflected case. In this formulation, a generalized interface delamination chart based on the energy criterion cannot be made. However, the authors did provide the ratio of energies of crack deflection and crack penetration for a few chosen interface systems (Gupta, 1991; Gupta et al., 1993).

7.2.2 Work Done in Fiber Pullout

Fiber pullout can be an important feature of the failure process in fiber reinforced composites. We now derive an expression for the work done in the pullout process. Consider the situation depicted in Fig. 7.17. Let us say that the fiber, of diameter d, fractures at some distance k below the main crack plane, such that $0 < k < \ell_c/2$, where ℓ_c is the critical length for load transfer. Locally, debonding of the fiber/matrix interface occurs. When the fiber is pulled out of the matrix, an interfacial frictional shear stress, τ_i, will be generated. In our simple analysis here, we assume that this shear stress, τ_i, resisting fiber sliding is a constant. More complex treatments involving a Coulomb friction law governing fiber sliding resistance (Shetty, 1988; Gao et al., 1988) and treatments taking into account residual stresses are available in the literature (Cox, 1990; Hutchinson and Jensen, 1990; Kerans and Parthasarathy, 1991).

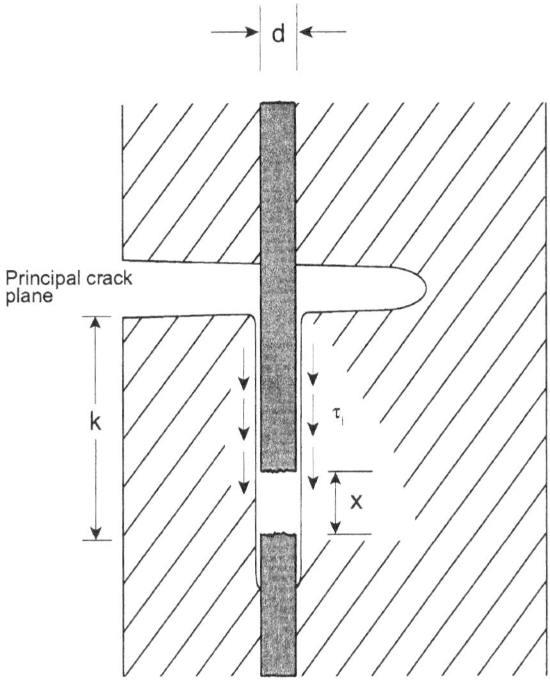

Fig. 7.17 Fiber pulled out through a distance x after fiber fracture. A shear stress, τ_i, comes into play at the interface during pullout.

Let the fiber be pulled out a distance x. The interfacial shear force opposing this motion of fiber is $\tau_i \pi d\,(k - x)$, where $\pi d\,(k - x)$ is the cylindrical surface area over which the shear stress is acting. Let the fiber be pulled out through a small distance dx. Then the work done by the interfacial shear force is $\tau_i \pi d\,(k - x)\,dx$. The total work done in the fiber pullout process over the length k is obtained by integration:

$$\int_0^k \tau_i \pi d(k - x)\,dx = \frac{\tau_i \pi d k^2}{2}$$

The length of the fiber pulled out will vary between 0 and $\ell_c/2$, where ℓ_c is the critical length for load transfer. Therefore, the average work done in the fiber pullout process is:

$$W_{fp} = \frac{1}{\ell_c/2} \int_0^{\ell_c/2} \frac{\tau_i \pi d k^2\,dk}{2} = \frac{\tau_i \pi d \ell_c^{\,2}}{24}$$

This expression assumes that all of the broken fibers are pulled out. Experimental observations, however, show that only fibers with broken ends within a distance $\ell_c/2$ from the principal fracture plane undergo pullout. Thus, we should expect a fraction (ℓ_c/ℓ) of fibers to pullout, and the average work done per fiber in fiber pullout can be written as

$$W_{fp} = \frac{\ell_c}{\ell}\frac{\tau_i \pi d \ell_c^{\,2}}{24}$$

7.2.3 Effect of Interfacial Reactions on Monotonic Behavior

In chapter 4, we described the general features of the interface and provided some examples of interfacial reactions in MMCs. Here we explore the effects of interfacial reactions on the monotonic properties of MMCs. Interfacial reactions can play an important role in damage of continuous fiber reinforced MMCs (Page et al., 1984). In Ti matrix composites reinforced with SCS-6 fibers, the C-rich fiber coating reacts with the Ti matrix to form layers of brittle TiC and Ti_5Si_3 (Konitzer and Loretto, 1989; Leyens et al., 2003). In B fiber reinforced Al, AlB_2 is formed at the interface at temperatures as low as 500°C (Grimes et al., 1977). Tensile loading in the longitudinal axis results in circumferential cracks in the reaction layer, which severely impair the strength of the composite (Grimes et al., 1977; Mikata and Taya, 1985; Kyono et al., 1986). Figure 7.18(a) shows that the longitudinal strength decreases with increasing exposure time at 500°C (Kyono et al., 1986). In the transverse orientation, however, Figure 7.18(b), there was a slight increase in strength. This is because the damage mechanisms in transversely loaded composites were quite different. Here, the fracture surfaces of as-fabricated samples showed significant interfacial debonding. With increasing exposure time and an increase in the reaction layer thickness, the interface strength increased, so microcracks formed in the reaction layer and propagated through the boron fiber, resulting in fiber splitting.

Transverse strength may also be affected by binders used to densify the matrix of the composite. Eldridge et al. (1997) fabricated sapphire fiber reinforced NiAl matrix composites with and without a polymethyl methacrylate (PMMA) organic binder. Transverse fracture surfaces showed that the composites with the binder exhibited a large amount of carbon residue at the fiber/matrix interface, which prevented a strong interfacial mechanical bond and led to low interfacial strength, Fig. 7.19(a). Composites processed without binder had a much “cleaner” fracture surface, Figure 7.19(b), and stronger bond strength.

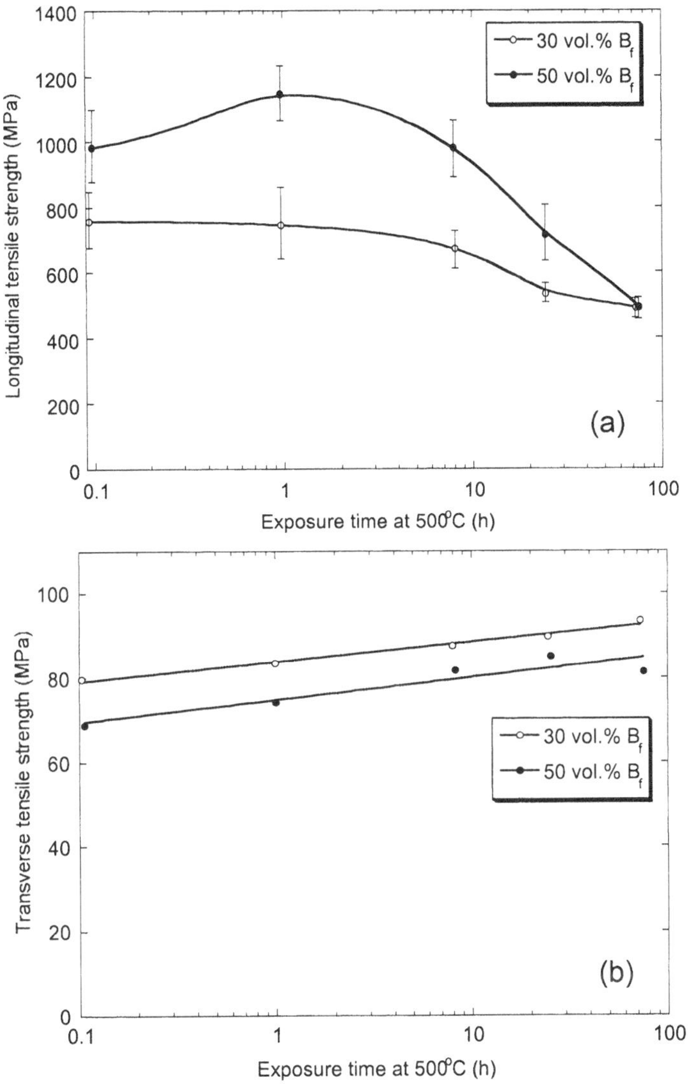

Fig. 7.18 Effect of exposure time at 500°C on the tensile behavior of Al/B_f composites in (a) longitudinal orientation and (b) transverse orientation (after Kyono et al., 1986). AlB_2 is formed at the interface and its thickness increases with exposure time. Circumferential cracks in the brittle interface form during tensile loading along the longitudinal direction. In the transverse orientation, the interface strength increases slightly with exposure time.

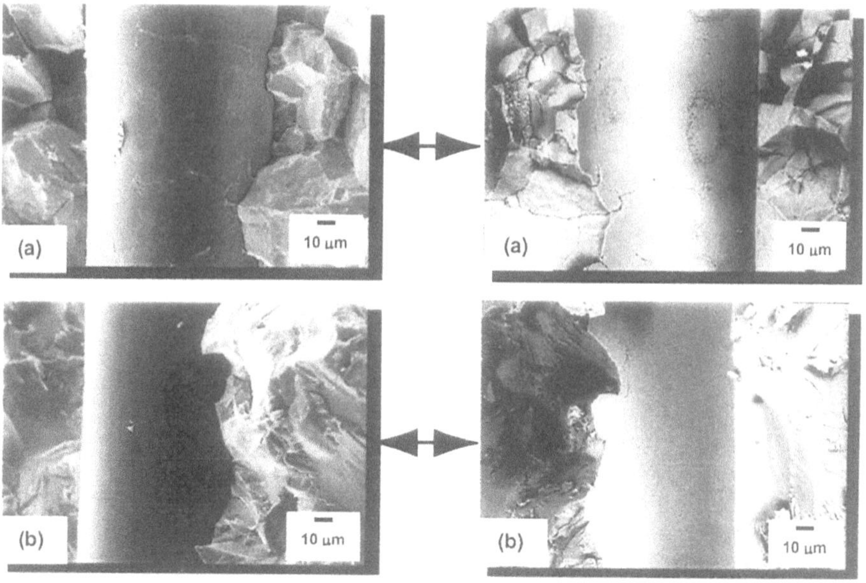

Fig. 7.19 Matching transverse fracture surfaces of sapphire fiber reinforced NiAl matrix composites (after Eldridge et al., 1997; courtesy of J.I. Eldridge): (a) with polymethyl methacrylate (PMMA) binder and (b) without binder (after Eldridge et al., 1997). Composites with the binder exhibited a large amount of carbon residue at the fiber/matrix interface, which prevented a strong interfacial mechanical bond and led to lower interfacial strength.

The extent of interfacial reaction can be reduced by the application of a thin fiber coating. In SiC fiber reinforced W composites, for example, a reaction zone of tungsten silicide is formed, which embrittles the interface and reduces the strength of the composite. Deposition of a TiC coating significantly limited the reduction in strength (Faucon et al., 2001). It should be noted that while the incorporation of a fiber coating may hinder interfacial reaction, it may also contribute to a lower degree of wetting by the matrix and poorer densification of the composite.

As mentioned above, in composites with a precipitation hardenable matrix, the interface may be affected by precipitates in the matrix which typically nucleate heterogeneously at the fiber/matrix interface (see Chapter 3). Cornie et al. (1993) tailored heat treatments to control the precipitate size and spacing at the interface. They found that a minimum in precipitate spacing (which corresponds to a minimum in precipitate size) resulted in a maximum in longitudinal strength and a minimum in transverse strength. This was caused by the decrease in interfacial strength due to precipitation at the interface. With increasing annealing time, coarsening of the precipitates took place, increasing the precipitate spacing, and increasing the interfacial bond strength. In this case, the transverse strength was maximized, although the longitudinal strength degraded by a small amount. The increase in transverse strength was explained by the increase in the interfacial area fraction of fiber/pure matrix (precipitate-free) bond.

The discussion above points to the difficulty in obtaining a combination of longitudinal *and* transverse strengthening in continuous fiber reinforced MMCs. The longitudinal properties are controlled primarily by the strength and volume fraction of the fibers, while the transverse properties are dictated primarily by the matrix (Rao et al., 1993) and fiber/matrix interface strength (Warrier and Majumdar, 1997). An increase in fiber volume fraction, however, increases the residual stresses during cooling from processing, while also decreases the transverse strength (Rosenberger et al., 1999). Thus, increasing the fiber strength would be a logical step toward increasing the longitudinal strength and simultaneously retaining the transverse properties of the composite. Rosenberger et al. (1999) compared the strength of composites with a high strength ultra-SCS fiber in a Ti alloy matrix, to those with conventional SCS fibers, and showed an increase in longitudinal strength with no corresponding decrease in transverse strength.

Slight changes in composite modulus may also be achieved by controlling the matrix microstructure through heat treatments (Miller and Lagoudas, 2000). This is confined to matrix materials, such as Ti alloys, where a change in crystal structure morphology of second phases takes place during heat treatment. Such manipulation of the matrix microstructure and crystal structure can also be used to tailor composite strength and ductility (Boehlert et al., 1997). In addition, other processing parameters may be used to tailor the matrix microstructure. Blucher et al. (2001) for example, studied the tensile behavior of composite wires of $Al/Al_2O_{3,f}$ (Nextel 610), $6061/Al_2O_{3,f}$, and Al/C_f. With increasing wire drawing velocity, an increase in strength was observed, Figure 7.20. This was attributed to the refinement in the microstructure, due to higher solidification rate, at higher drawing velocity. In the Al/C_f composites, a reduction in interfacial reaction with increasing drawing velocity may have also contributed to the observed behavior.

7.2.4 Modeling of monotonic behavior of continuous fiber reinforced MMCs

The tensile behavior of continuous fiber reinforced MMCs has been modeled extensively by finite element modeling (Brockenbrough et al., 1991; Gonzalez and LLorca, 2001; Rossoll et al., 2005). Gonzalez and LLorca

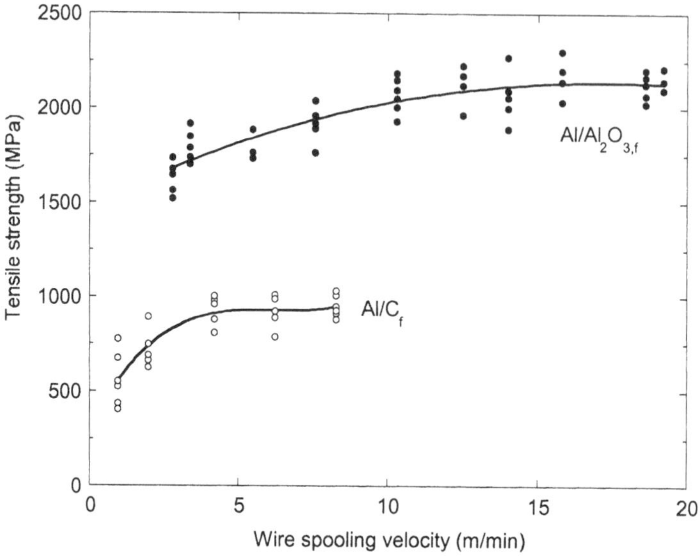

Fig. 7.20 Effect of wire spooling velocity on the tensile strength of $Al/Al_2O_{3,f}$ and Al/C_f composites (after Blucher et al., 2001). Increasing drawing velocity results in faster solidification rate and finer microstructure. In the case of Al/C_f it also reduces the time for interfacial reaction.

(2001) studied the tensile behavior of SCS-6 fiber reinforced Ti-6Al-4V matrix composites at ambient and elevated temperatures. The fibers were modeled using an axisymmetric model shown in Fig. 7.21(a). The fibers are embedded in a "homogeneous composite" with average composite properties. A comparison of the stress-strain behavior predicted by the model with the experiment is shown in Fig. 7.21(b). The experimental behavior is shown as the gray shaded region (due to slight scatter in the data), while the model prediction is shown in the solid black line. The composite exhibited an initial linear portion followed by a pronounced deviation from linearity and fracture of the composite. Note that due to the processing-induced residual stress, at zero applied stress the matrix is in a state of residual tension while the fiber is in compression. With increasing applied stress, the load on both fiber and matrix increase, although the rate of loading in the fiber is higher because of load transfer from the matrix. When the matrix yields, the stress in the matrix reaches a plateau. This stress corresponds to the deviation from linear stress-strain behavior of the composite. The stress in the fiber, on the other hand, continuously increases until fracture of the composite takes place.

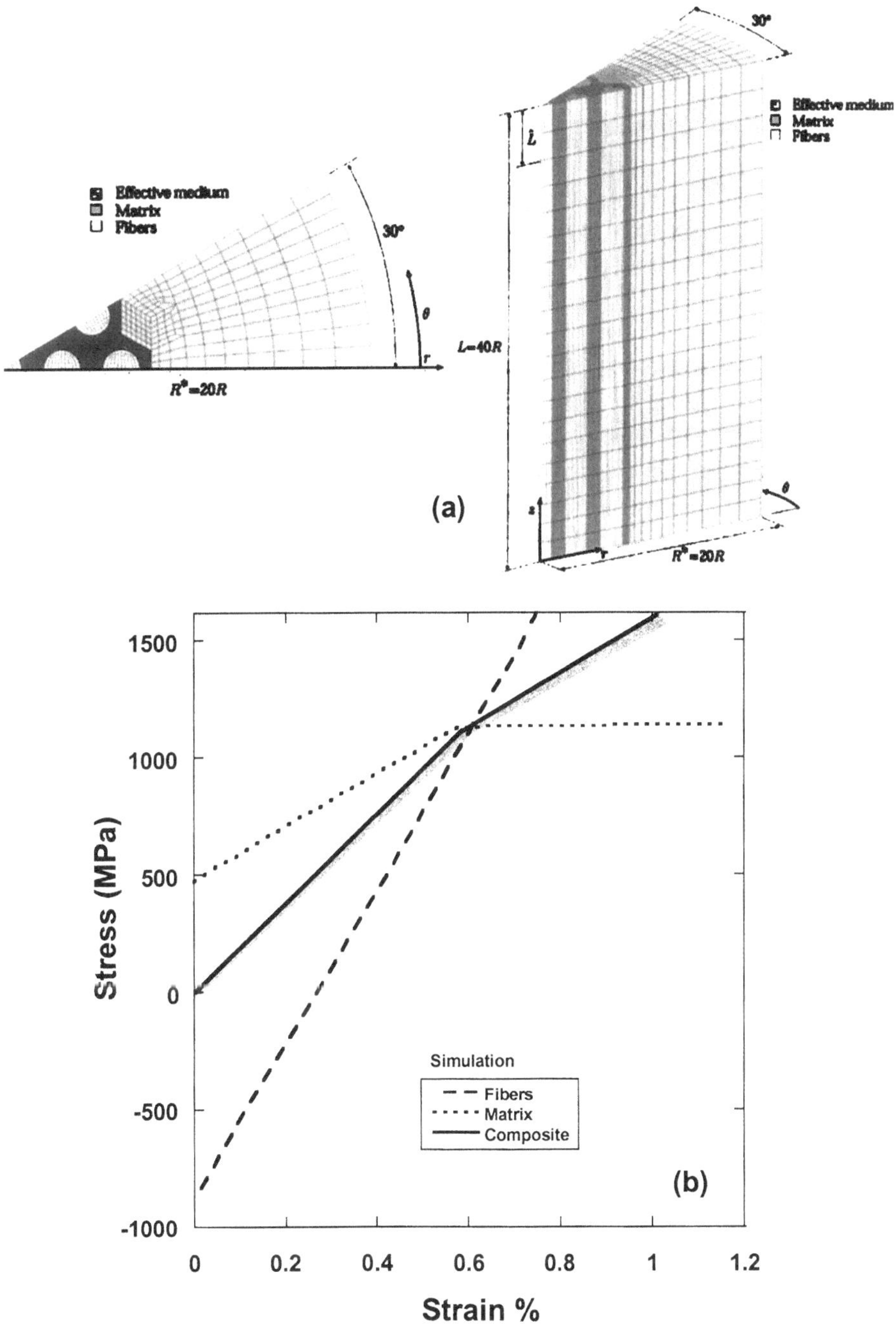

Fig. 7.21 Finite element modeling of tensile behavior of fiber reinforced Ti-6Al-4V matrix composites (after Gonzalez and LLorca, 2001; courtesy of J. LLorca): (a) finite element model and (b) simulated composite, fiber, and matrix response. The simulated response correlates well with the experiment.

When a fiber fractures, the stress on neighboring fibers is also affected, since more load is carried by these fibers. The stress state in neighboring fibers is shown in Fig. 7.22. The stress in the neighboring fibers is highest in the plane of fracture of the original fiber. The first-nearest neighbor fiber has the highest stress, followed by the second and third-nearest neighbor, respectively.

As shown above, during longitudinal loading the load is carried primarily by the fibers, although plastic deformation of the matrix between the fibers also takes place. Under transverse loading, however, significant plastic deformation between the fibers takes place. Thus, under transverse loading, the distribution of the fibers plays an important role in the composite response. Brockenbrough et al. (1991) modeled the longitudinal and transverse response of a 6061/B/46$_f$ composite with varying fiber distributions. As one would expect, under longitudinal loading, fiber distribution did not have a significant influence on the modeled behavior, Fig. 7.23(a), since the composite behavior was dominated by the response of the fibers. The experimental behavior compared well with the model

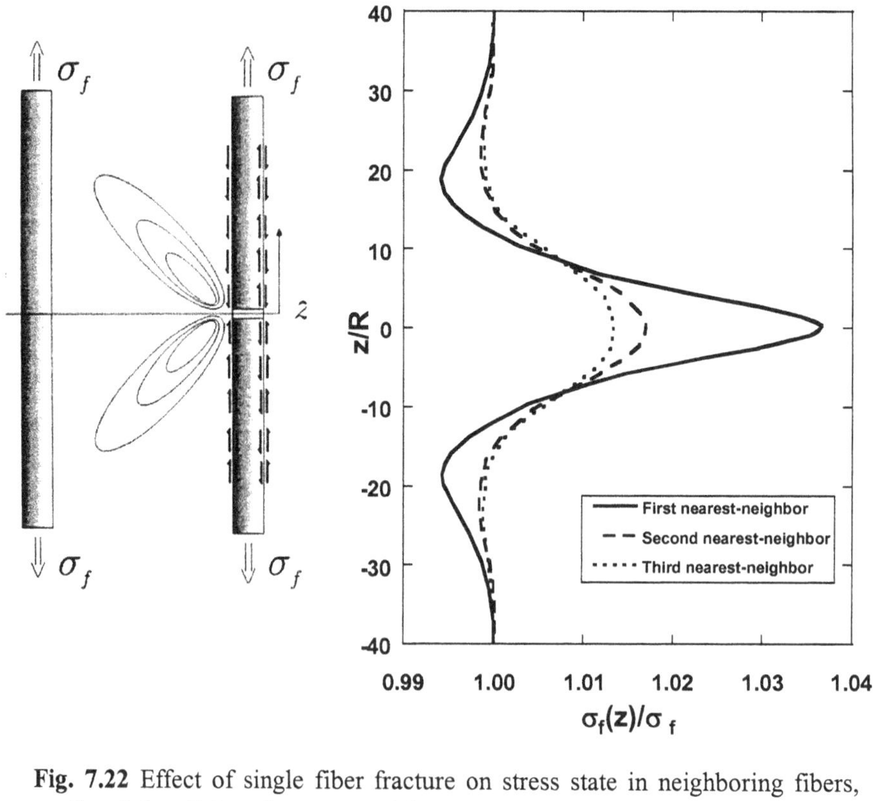

Fig. 7.22 Effect of single fiber fracture on stress state in neighboring fibers, predicted by finite element modeling (after Gonzalez and LLorca, 2001; courtesy of J. LLorca). The stress is highest in the first nearest-neighbor.

prediction. Under transverse loading, the fiber distribution was modeled as square, square-diagonal, and triangle-packed, Fig. 7.23(b). The elastic regime was somewhat unaffected by the fiber distribution. Once the matrix yielded, the effect of fiber distribution became more apparent. The distribution of square packing exhibited the highest work-hardening rate, followed by triangle and square-diagonal packing, respectively.

Rossoll et al. (2005) used finite element analyses (FEA) for varying fiber distributions, ranging from single fiber unit cells to complex cells to study the damage evolution during a tensile test. The *in situ* flow stress of the matrix in the composite was found to be different from that of the unreinforced alloy. This is the result of constrained deformation of the metal matrix in the composite, especially in the presence of clustering of fibers. This results in a deviation from the rule-of-mixtures, based on isostrain, because of a stiffening effect of matrix confinement when surrounded by touching fibers.

7.3 MONOTONIC BEHAVIOR OF DISCONTINUOUSLY REINFORCED MMCS

The monotonic behavior of discontinuously reinforced MMCs is dependent on several factors, such as reinforcement volume fraction, particle size, shape, and matrix microstructure. Chawla et al. (1998a) examined the effect of particle volume fraction (at a constant SiC particle size) on monotonic tensile behavior. Figure 7.24 shows the tensile behavior of an Al-Cu-Mg (2080)/SiC_p-T8 composite with varying volume fraction (at a constant particle size of 5 μm). With an increase in volume fraction, higher elastic modulus, macroscopic yield and tensile strengths, and lower ductility were observed. A comparison of the measured increase in yield strength, with increasing reinforcement volume fraction, compared very well with predictions from a simple modified shear lag analysis by Nardone and Prewo (1989) (see sec. 7.1.1 in this chapter).

As mentioned in section 7.1, microsplasticity takes place at a fairly low stress, which corresponds to the proportional limit stress in the stress-strain curve. This microplasticity originates from stress concentrations at the poles and sharp corners of the particle, Fig. 7.25 (Chawla et al., 1998b). The initial microyielding stress decreases with increasing volume fraction of reinforcement, as the number of stress concentration points increases. The work hardening rate increases with increasing volume fraction of reinforcement (and decreasing matrix volume). The lower ductility can be attributed to the earlier onset of void nucleation with increasing reinforcement. The high stress concentration at the tips of the cracked

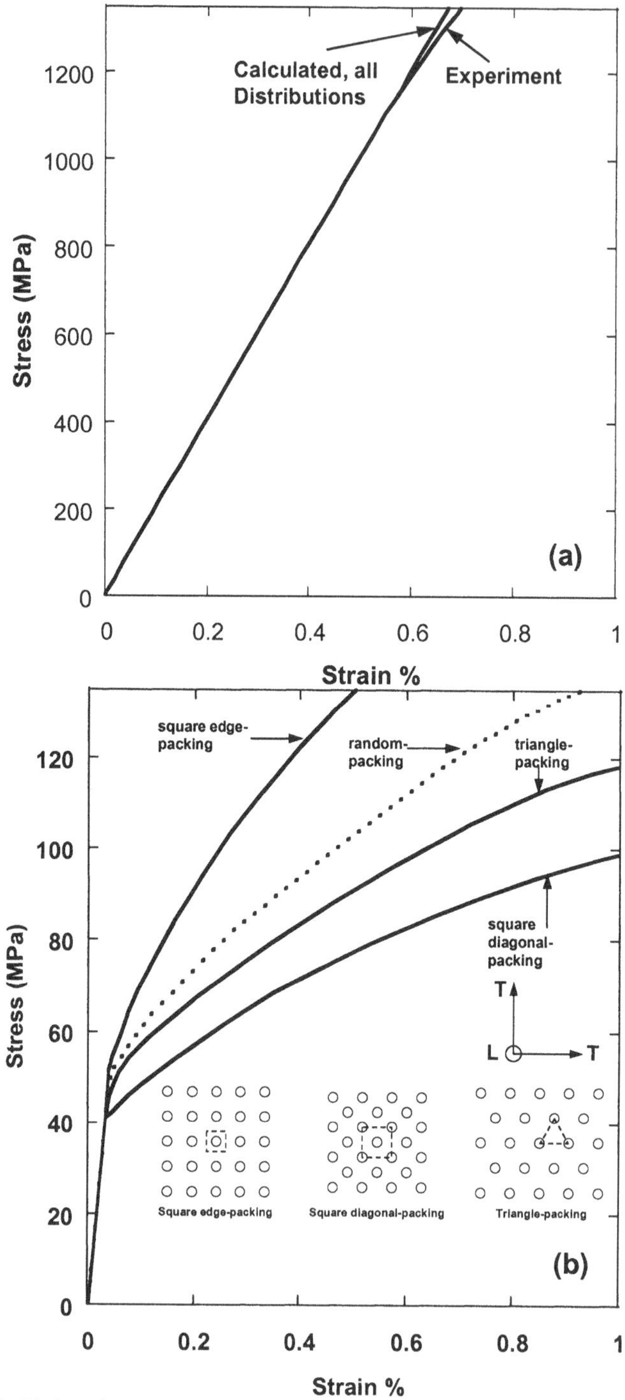

Fig. 7.23 Finite element model predictions of tensile behavior of continuous fiber reinforced composites with varying fiber arrangement (after Brockenbrough et al., 1991): (a) longitudinal and (b) transverse.

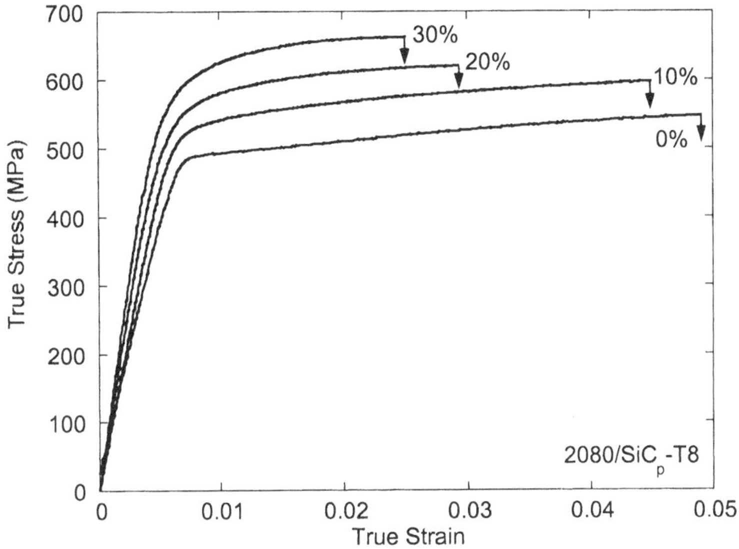

Fig. 7.24 Tensile behavior of an Al-Cu-Mg (2080)/SiC$_p$-T8 composite with varying volume fraction, at a constant particle size of 5 μm (after Chawla et al., 1998a). With an increase in volume fraction, higher elastic modulus, macroscopic yield and tensile strengths result, coupled with lower ductility.

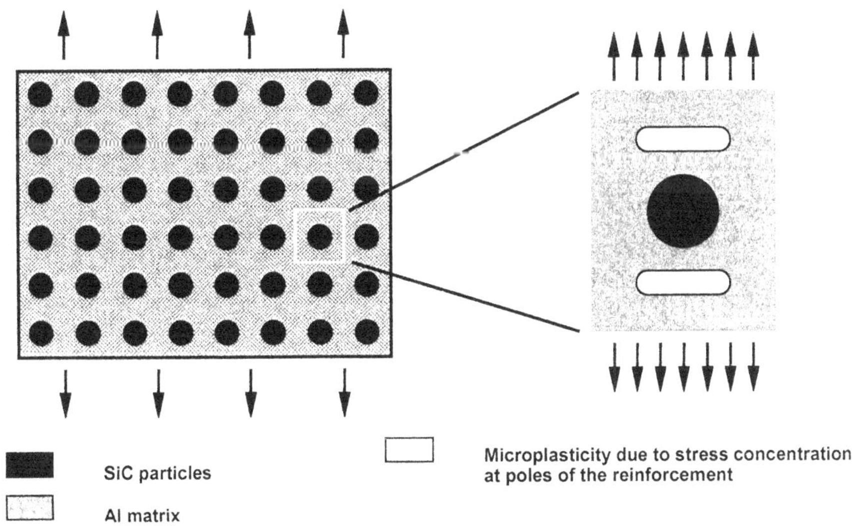

Fig. 7.25 Microplasticity in the matrix of particle reinforced MMCs due to stress concentrations at the poles of the reinforcement and/or at sharp corners of the reinforcing particles (after Chawla et al., 1998b).

particles could also contribute to a lower ductility in the composite, compared to the unreinforced alloy.

The effect of particle size on tensile behavior has been documented by several investigators (Mummery et al., 1991; Manoharan and Lewandowski, 1992; Chawla et al., 1998a). Figure 7.26 shows the general trend of increase in strength and ductility with a decrease in particle size. This may be attributed to an increase in the SiC particle strength with a decrease in particle size. This inverse relationship between particle size and particle strength can be explained as follows. As the volume of the particle increases, the probability of a strength-limiting flaw existing in the volume of the material also increases. At relatively large particle sizes of this material, a significant amount of particle cracking takes place during extrusion prior to testing. Chawla et al. (1998a), showed that in Al 2080/SiC_p composites, with volume fractions ranging from 10 to 30 vol. %, particle cracking was observed above an average particle size of 20 μm. Cracked particles don't carry any load, so the strength of the composite is lower than that of the unreinforced material, Fig. 7.26.

A lower particle size also means a lower interparticle spacing (for a given

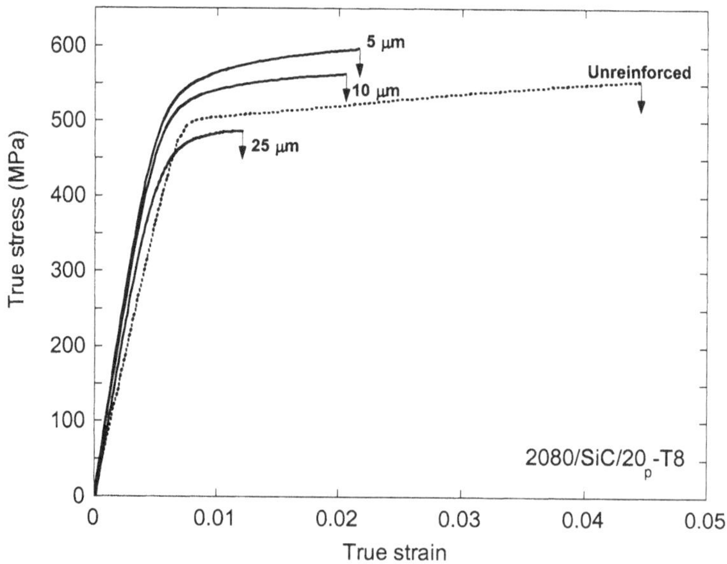

Fig. 7.26 Effect of reinforcement particle size on tensile behavior of an Al-Cu-Mg (2080)/SiC_p-T8 composite, at a constant volume fraction of 20% (after Chawla et al., 1998a). Very large particles are detrimental to strength because of processing-induced fracture, prior to testing. Above about 20 μm or so, strength and ductility both increased with decreasing particle size.

volume fraction of particles) so that nucleated voids in the matrix are unable to coalesce as easily (Mummery et al., 1991). A higher work hardening rate has also been observed with decreasing particle size (Lewandowski et al., 1991; Manoharan and Lewandowski, 1992). This is attributed to the formation of dislocation tangles around the particles and the formation of a dislocation cell structure with a cell size directly proportional to the interparticle spacing (Kamat et al., 1989).

The fracture of particle reinforced MMCs is very much dependent on particle strength and particle/matrix interface strength. A schematic of the two possible types of damage evolution is shown in Fig. 7.27. If the strength of the interface is greater than the particle strength (often observed in peak-aged composites) then the particles fracture before the interface. Matrix void growth takes place, and shear localization between fractured particles results in failure of the composite. In order to quantify the extent of particle fracture and particle pullout during tensile loading, both mating fracture surfaces need to be examined. An example of mating fracture surfaces is shown in Figure 7.28. A particle found on both mating fracture surfaces has fractured. The nature of fracture of the brittle SiC particles is quite interesting. Figure 7.29 shows the fractured surface of a SiC particle after tensile loading. Note the spherical flaws or voids on the fracture surface, which presumably were responsible for crack initiation in the particles. The flaws arise during processing of the particles. The second scenario for damage is when the interface strength is much lower than the particle strength. Here, void

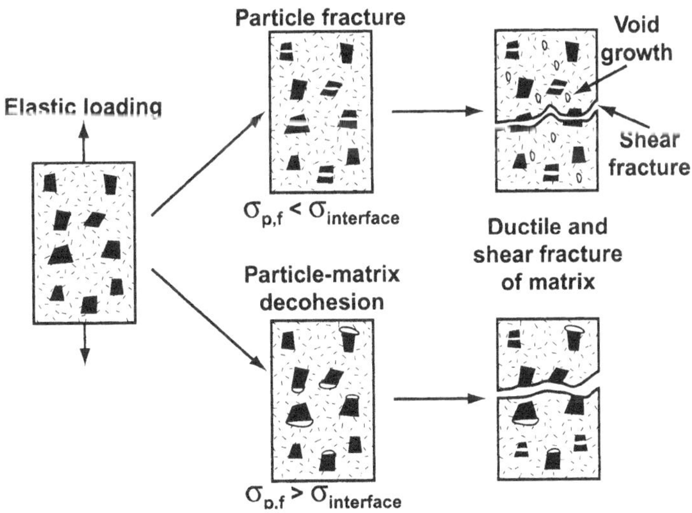

Fig. 7.27 Schematic of the two possible types of tensile damage evolution in particle reinforced MMCs: (a) interface strength greater than particle strength and (b) interface strength less than particle strength.

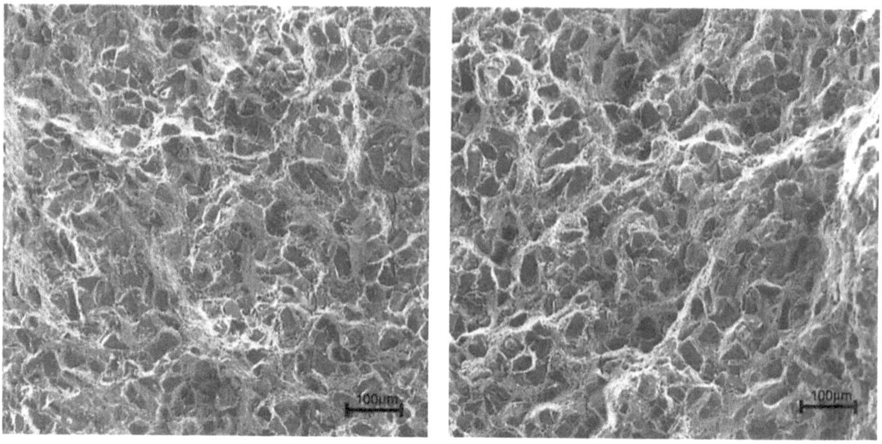

Fig. 7.28 Mating tensile fracture surfaces of a 2080/SiC/20p composite showing significant amount of particle fracture (after Chawla et al., 2002a).

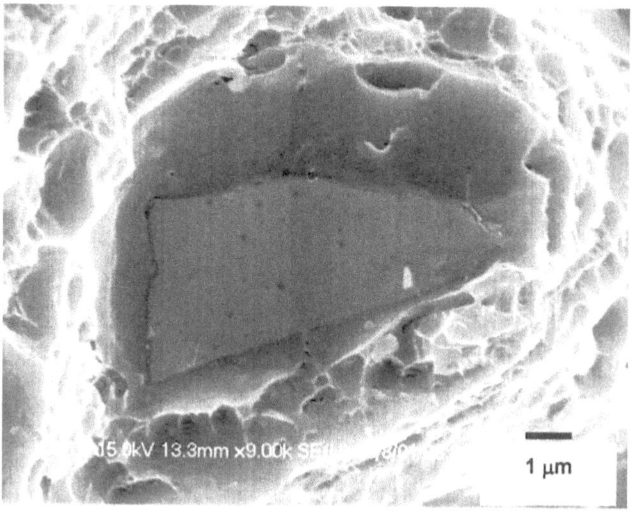

Fig. 7.29 Fracture surface of a SiC particle after tensile loading (after Chawla et al., 2002b). Note the spherical flaws or voids on the fractured particle surface, which presumably are responsible for crack initiation. The flaws arise during processing of the particles.

nucleation and growth will take place at the interface, due to decohesion of the matrix from the particle. This will be followed by ductile shear fracture through the matrix alone. An example of this is shown in Figure 7.30, for an overaged composite, with relatively weak interface strength. Note that a thin matrix layer has been left on the SiC particle surface.

Particle reinforced composites, especially those subjected to extrusion,

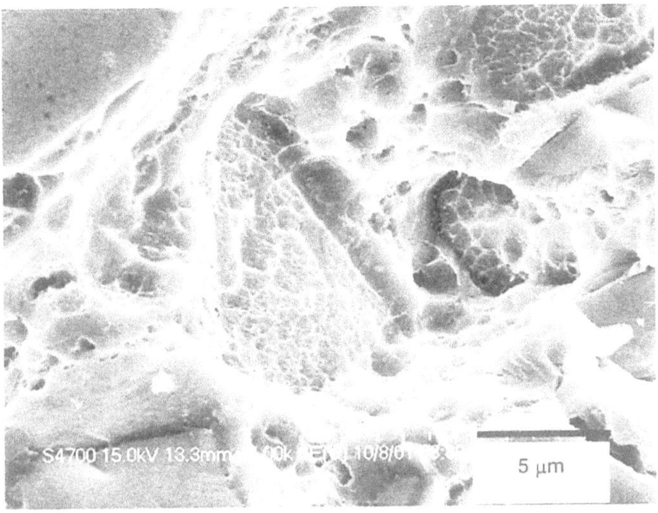

Fig. 7.30 Void nucleation and growth at the particle/matrix interface, due to relatively weak interface strength (overaged condition). Ductile shear fracture is also observed in the matrix of the composite (after Chawla et al., 2002b).

exhibit a fair amount of anisotropy due to preferential particle orientation along the extrusion axis. Logsdon and Liaw (1986) studied the tensile strength anisotropy behavior in SiC particle and whisker reinforced aluminum alloys, and noted that the strength was higher parallel to the extrusion axis, than perpendicular to the extrusion axis. Jeong et al. (1994) also noted a higher Young's modulus of the composite along the extrusion axis. Ganesh and Chawla (2004, 2005) noted that the extent of orientation was highest for lowest volume fraction composites, since with increasing volume fraction the mean free path for rotation and alignment of a given particle decreased. The Young's modulus and tensile strength of the composites, independent of orientation, increased with increasing volume fraction of reinforcement, Fig. 7.31. Thus, although the degree of microstructural anisotropy was greatest for 2080/SiC/10_p, the greatest anisotropy in mechanical behavior was observed in 2080/SiC/30_p.

The mechanical properties of lower cost processing techniques, such as sinter-forging, have been compared to existing hot-pressed and extruded materials. Chawla et al. (2002a) examined the strength of composites fabricated by a low-cost sinter-forging approach. The SiC particle size was relatively coarse, e.g., 25 μm. Materials processed by the sinter-forged approach exhibited similar tensile properties to those of the extruded material of similar composition, reinforcement volume fraction, and particle size, Fig. 7.32. The microstructure of the sinter-forged composites exhibited relatively uniform distribution of SiC particles, which appeared to be

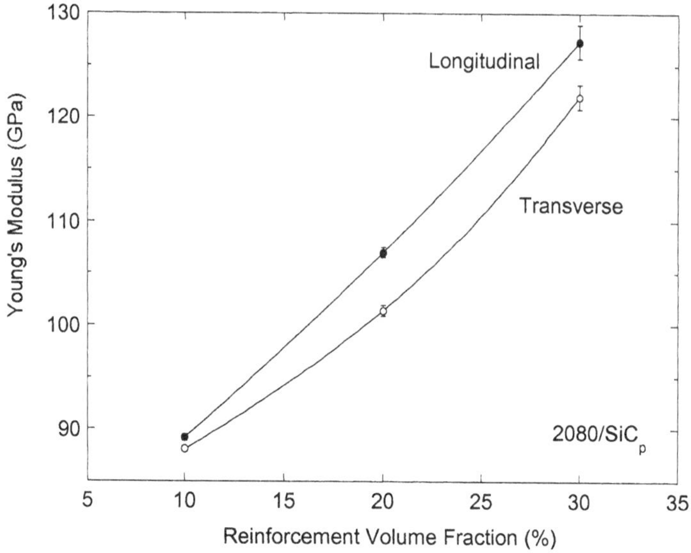

Fig. 7.31 Anisotropy in Young's modulus in a 2080/SiC$_p$ composite (after Ganesh and Chawla, 2005). The longitudinal orientation is parallel to the extrusion axis, while transverse is perpendicular.

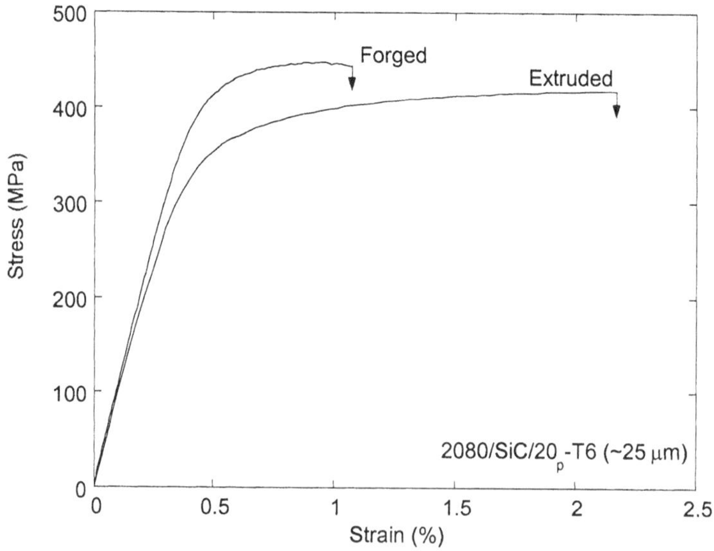

Fig. 7.32 Tensile strength of 2080/SiC/20$_p$-T6 fabricated by low-cost sinter-forging and extrusion approach (after Chawla et al., 2002a). The sinter-forged material exhibited similar tensile properties, with slightly lower ductility, than the extruded material.

somewhat aligned perpendicular to the forging direction. The degree of alignment and interparticle bond strength was not as high as that observed for the extruded composite. The sinter-forged composite exhibited higher Young's modulus and ultimate tensile strength than the extruded material, but lower strain-to-failure. The higher modulus and strength were attributed to the absence of any significant processing-induced particle fracture, while the lower strain-to-failure was caused by poorer matrix interparticle bonding compared to the extruded material. Indeed, secondary working of the composite, such as extrusion after initial casting, can significantly improve the ductility of the composite, Fig. 7.33 (Lloyd, 1997). The ductility of the composite is also very much a function of the degree of particle clustering. Murphy et al. (1998) controlled the degree of particle clustering by controlling the cooling rate of the composite. The degree of particle clustering was measured by tesselation techniques (see inset), to determine a clustering severity parameter, P. The ductility of the composite was found to significantly decrease with increasing P.

Matrix microstructure also plays an important role on strength of the composite. Overaging heat treatments modify the matrix microstructure, resulting in coarsening of the precipitate structure, while retaining a homogeneous precipitate distribution, which directly influences the strength

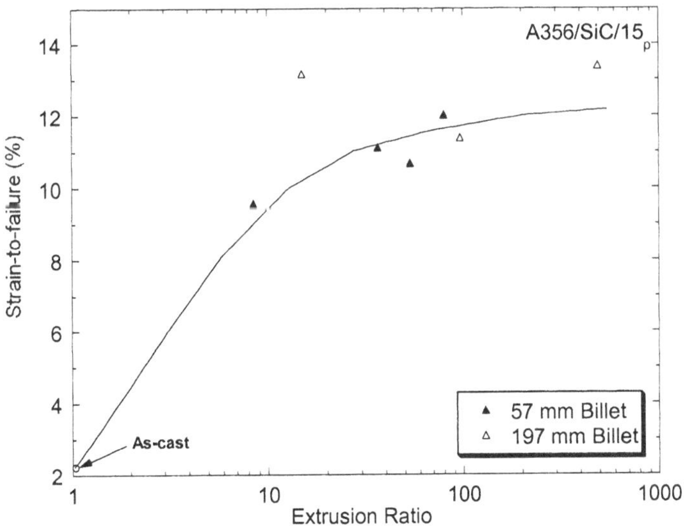

Fig. 7.33 Effect of degree of secondary working (i.e., extrusion ratio) on ductility of Al/SiC/15_p (after Lloyd, 1997). Extrusion significantly improved the ductility of the composite.

Quantifying Reinforcement Clustering

During processing of continuous fiber or particle reinforced MMCs, clustering of the reinforcement is likely to take place (see chapter 4). While individual fibers can be distributed relatively homogenously (e.g., when woven in a fiber fabric), controlling particle distribution can be more challenging. As shown in this chapter, reinforcement clusters have a profound effect on mechanical properties, because the clusters act as stress concentrations. Many techniques have been used to quantify the degree of reinforcement clustering. Here we describe two techniques that can be used to obtain some measure of reinforcement clustering: (a) Dirichlet and finite-body tessellation and (b) digital image dilation.

The technique is name after the mathematician Dirichlet (1850) who proposed a tessellation scheme to quantify the arrangement of geometric objects in space. Consider a hypothetical two-dimensional microstructure consisting of elliptical particles in space (see figure below). Let us mark the centroid of each ellipse. We can now construct cells around the particles, such that each cell wall is equidistant between two centroids (part (b) in the figure below). This construction is called a tessellation. The tessellation is very useful because it yields information about the cell size, nearest-neighbor spacing distributions between particles, etc. Inspection of the Dirichlet tessellation, however, shows that when the objects are not perfect spheres (in this case ellipses), the particles are not always fully contained within the cell boundaries. This is because the centroid of the ellipse is used to construct the tessellation. Thus, for the case of non-spherical reinforcement, a conventional tessellation is not quite adequate for quantifying the microstructure.

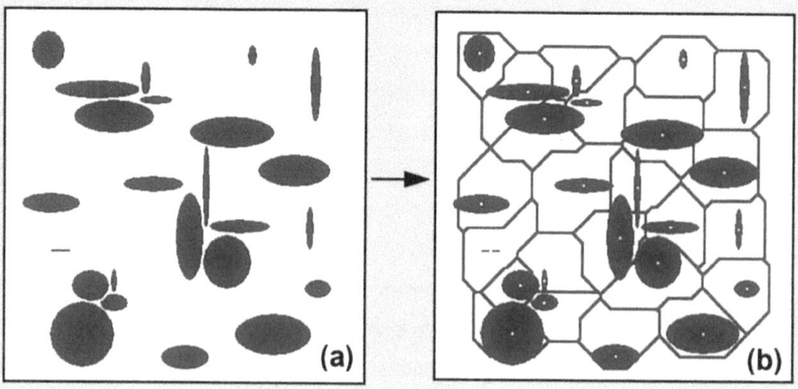

The conventional tessellation scheme can be enhanced by employing a finite body tessellation (Chawla et al., 2002a). The steps involved in this process are shown in the panel below. An optical or scanning electron micrograph, (a), is segmented into a black and white image, (b). A watershed image operation is then conducted, (c), which analyzes distances between particles based on the centroid and perimeter of each particle. A tessellation is then constructed, (d), whereby each of the irregular particles resides within the boundaries of the cell. Similar statistics can be obtained from the finite body tessellation as those from the conventional tessellation.

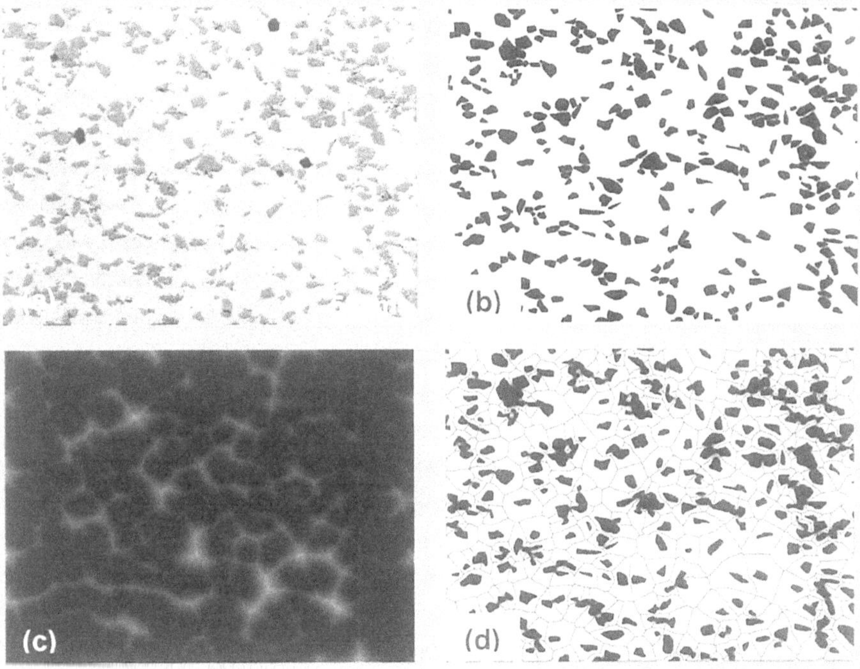

An alternative method for quantifying the degree of clustering in a composite is to use a digital image dilation technique (Chawla et al., 2002a). This technique consists of, again, analyzing a segmented microstructure (part (a) below). The perimeter of each particle is "dilated" by a certain increment, part (b). The dilation increment can be calculated by analytical expressions for the average interparticle spacing for a random distribution of particles of single (Meyers and Chawla, 1999) or multiple sizes (Torquato, 2002). If the particles are touching after the dilation, then this set of particles is identified as a cluster (shaded region in part (b)). The cluster may be quantified by the number of particles in the cluster or by the area fraction of the cluster, relative to the whole microstructure. Ayyar et al. (2005) used this technique to show the relative degree of clustering between two microstructures (also shown below). The microstructure on the left is relatively homogeneous and the one on the right is highly clustered. Part (c) shows a histogram of the cluster size distribution for the two microstructures. Note that the clustered microstructure has a much higher fraction of larger clusters. This information is particularly useful from a cyclic fatigue design perspective, since the fatigue life is controlled by the largest defect (cluster) size.

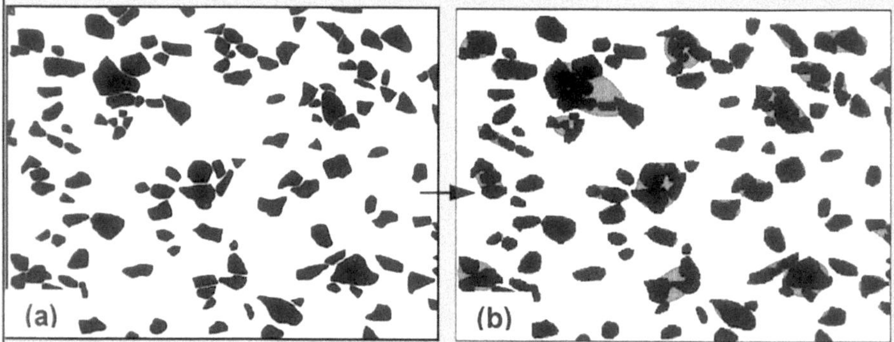

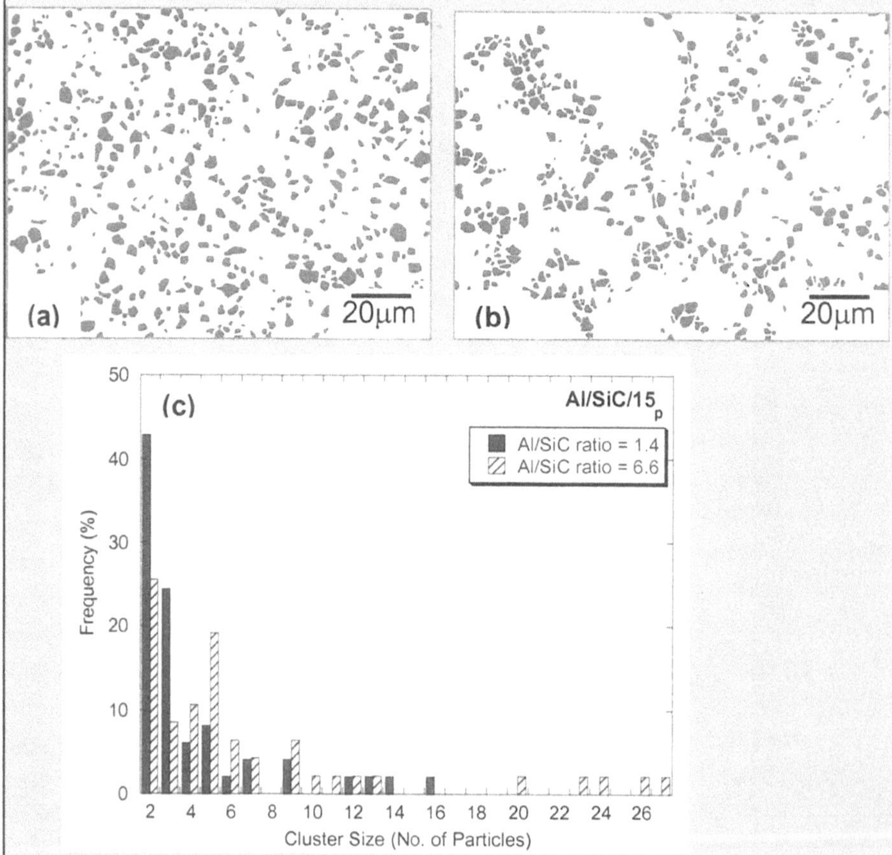

of the composite (Chawla et al., 2000). Figure 7.34 shows the coarsening and increase in precipitate spacing in the matrix of MMCs overaged at various temperatures for 24 h. Increasing precipitate spacing decreases tensile strength, Fig. 7.35. This is to be expected since coarser precipitates result in a larger interprecipitate spacing and easier bypass of dislocations. For the composites subjected to higher overaging temperatures, the yield strength also decreased with an increase in precipitate spacing.

7.3.1 Modeling of Monotonic Behavior of Particle Reinforced MMCs

Similar to modeling of continuous fiber reinforced MMCs, FEM modeling has also been used to model the behavior of discontinuously reinforced MMCs. Figure 7.36(a) shows several single particle unit-cell models, in which the left vertical boundary represents the axially symmetric axis and mirror symmetry exists about the horizontal boundary. A periodic arrangement of particles with the shape of a "unit cylinder," "truncated cylinder," "double-cone," and "sphere" may be simulated using the appropriate boundary conditions (Shen et al., 1994). The matrix is modeled as an isotropically hardening elastic-plastic solid (following the experimental stress-strain curve of a peak-aged Al-3.5Cu alloy) and the SiC particles as an elastic solid. The calculated tensile stress-strain response of Al/SiC/20_p composites, having the four particle shapes described above, is shown in Fig. 7.36(b). Clearly, particle shape has a significant influence on the overall tensile behavior of the composite. The unit-cylinder particles clearly strengthen the composite more than the other three shapes for a given reinforcement fraction. This, however, does not imply that particles with sharp corners have a more pronounced strengthening effect, as shown by the case of "double-cone" particles, possessing the "sharpest" type of corners. A detailed analysis (Shen et al., 1995) showed that the unit-cylinder and double-cone particles result in the highest and lowest degrees of "disturbance" of the local plastic flow paths in the matrix, respectively. This directly reflects the different extents of constrained plastic flow and hence the strengthening behavior in the composite. The reader is referred to other simple unit-cell approaches focusing on various aspects of elasto-plastic behavior of short fiber and particle reinforced composites, such as reinforcement particle fracture (LLorca, 1995; Steglich et al., 1999), reinforcement clustering (Christman et al., 1989; Toda et al., 1998), matrix void growth (LLorca et al., 1991), and residual stress from thermal treatments and processing (Levy and Papazian, 1991; Dutta et al., 1993).

The thermal expansion mismatch between the reinforcement and the matrix results in thermal stresses within the composite upon cooling from the

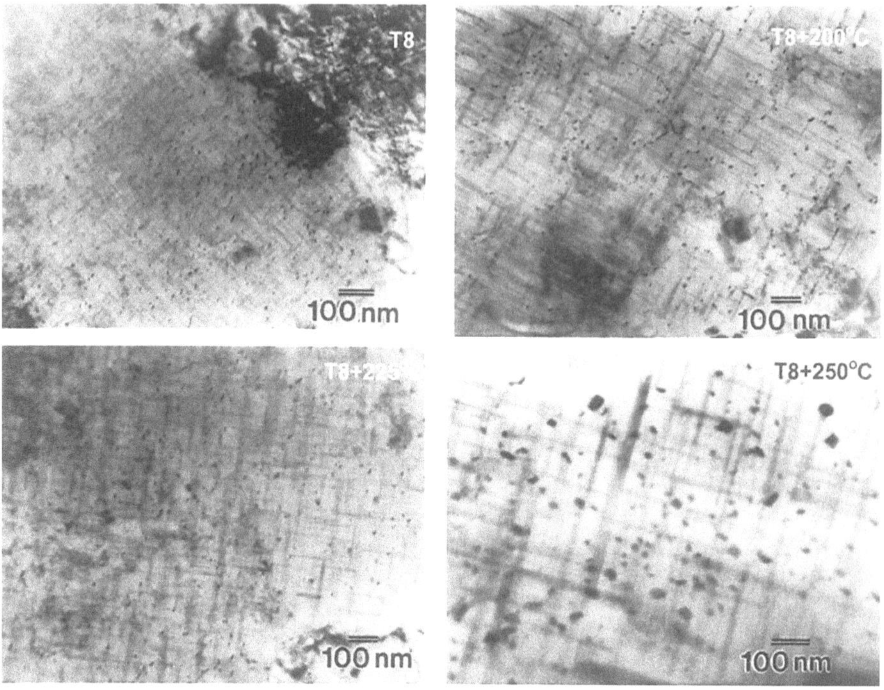

Fig. 7.34 Coarsening and increase in precipitate spacing in the matrix of 2080/SiC/20_p-T8 overaged at various temperatures for 24 h (after Chawla et al., 2000).

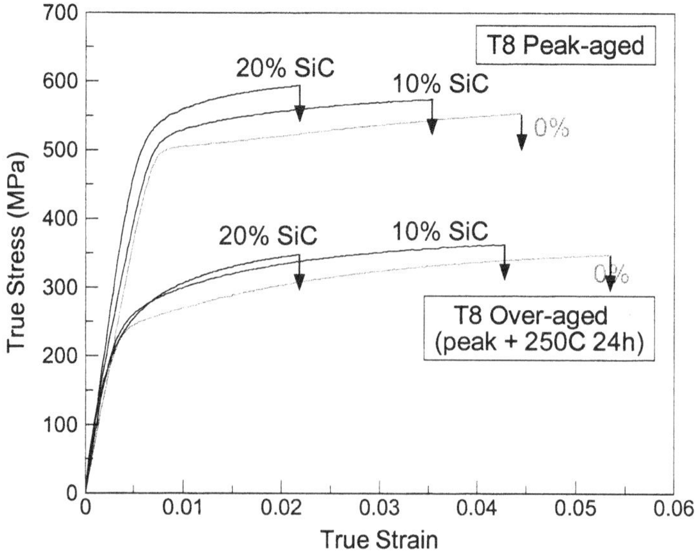

Fig. 7.35 Decrease in strength due to overaging of 2080-T8 and 2080/SiC_p-T8 composites (after Shen and Chawla, 2001). The reduction in strength is similar for all materials. Coarser precipitates result in a larger interprecipitate spacing and easier bypass by dislocations.

processing temperature to ambient temperature. As mentioned above, in actual composites thermal residual stresses are relieved by plastic deformation in the matrix, resulting in indirect strengthening. Figure 7.37 shows the calculated tensile stress-strain curves for 20% SiC reinforced Al alloy utilizing the unit-cylinder particles (Fig. 7.36(a)), with and without the presence of thermal residual stress. As before, the matrix was taken to be an isotropically hardening elastic-plastic material in the model. The thermal residual stresses were calculated for a composite cooled from the solutionizing temperature of 500°C, where the composite is in a relatively stress-free state, to room temperature, 20°C. Also included in the figure is the stress-strain response used for the pure matrix. During cooling the matrix near the particle/matrix interface undergoes yielding. This has direct bearing on subsequent loading of the material. It can be seen in Fig. 7.37 that in the presence of thermal stresses a smaller slope is observed at the early stage of deformation, due to the slightly smaller apparent modulus arising from prior plastic deformation. When compared with the material free of residual stresses, higher values of the average axial stress were observed. This means that the existence of residual stresses enhanced the initial strain hardening rate in the material. Comparing the curves for the pure matrix and for the composite without thermal residual stresses, direct strengthening effects are observed. The higher flow stress for the composite is a direct consequence of load transfer from the matrix to the reinforcement, which is also related to the constrained plastic flow within the matrix. A comparison of the two curves of the composite reveals the indirect strengthening effect. In the model strain hardening caused by cooling-induced plasticity leads to subsequent higher strength for the composite with thermal stresses incorporated (after the crossover point). In actual materials thermal mismatch-induced dislocation punching renders higher matrix strength due to strain hardening. Thus, when appropriate constitutive models are chosen (e.g., hardening plasticity rather than perfect plasticity in the present case), continuum-based numerical modeling can provide insights into the deformation mechanisms.

Simple shaped single particle and multiparticle models can provide useful insight into deformation. Microstructure-based models, however, more accurately predict the deformation behavior of the composites (Chawla et al., 2003; Chawla et al., 2004; Ganesh and Chawla, 2004). This is because in actual composites the particles are highly irregular in shape and commonly contain sharp corners, so spherical particles are not necessarily a realistic choice for simulation. Thus, while simplifications in unit cell models may aid in computation, they fail to capture the complex morphology, size, and

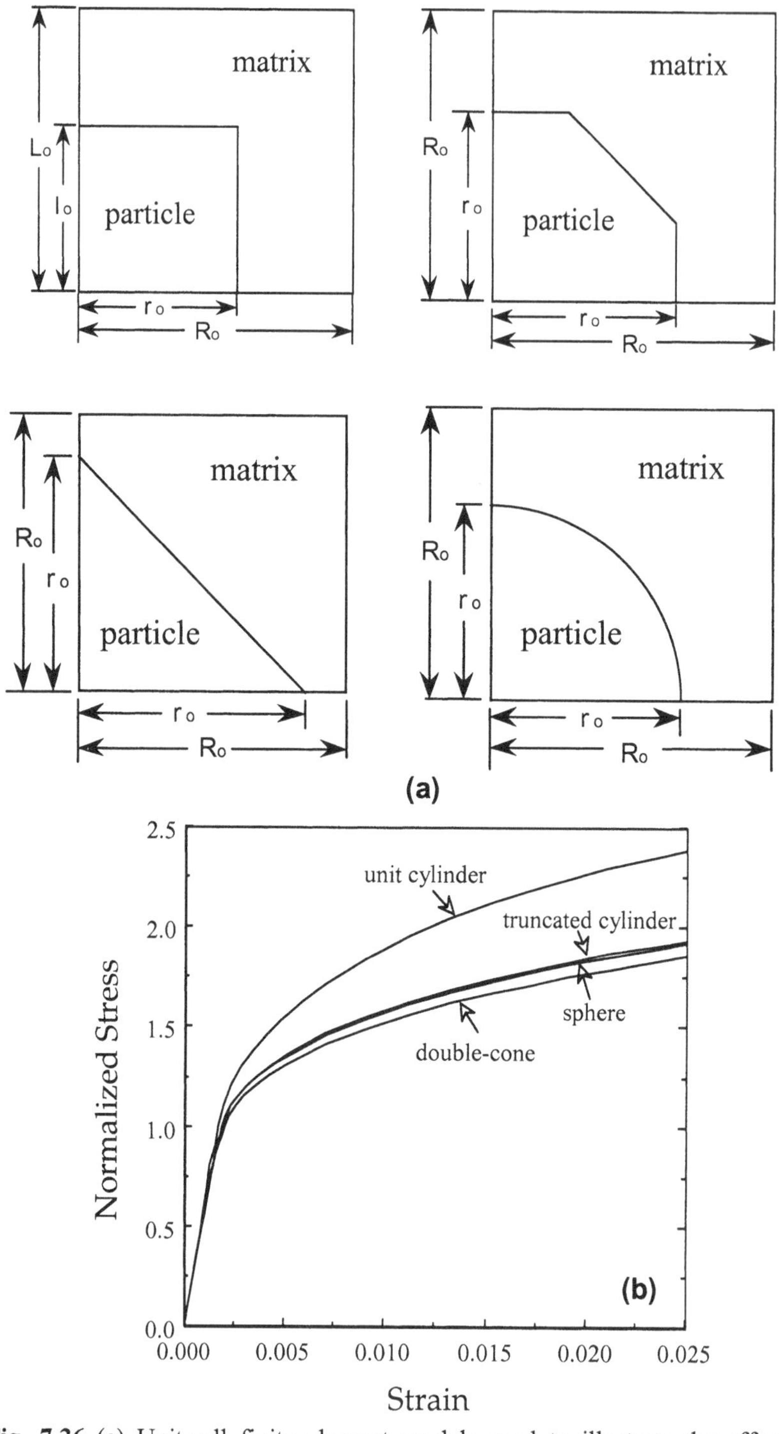

Fig. 7.36 (a) Unit-cell finite element models used to illustrate the effect of particle shape and (b) predicted stress-strain behavior (after Shen et al., 1994).

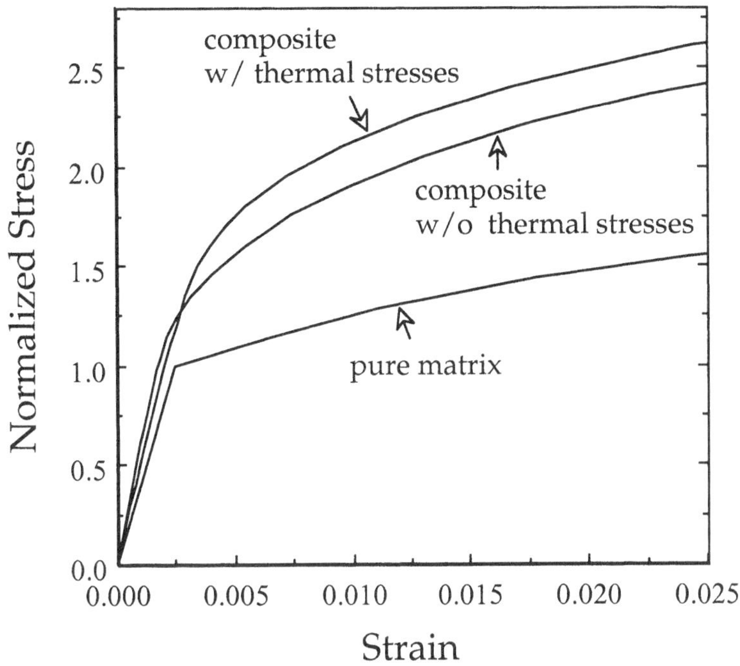

Fig. 7.37 Tensile stress-strain curves for Al/SiC/20_p utilizing the unit-cylinder particles with and without thermal residual stress (after Shen et al., 1994). Hardening caused by cooling-induced plasticity leads to higher strength for the composite.

spatial distribution of the reinforcement. It follows that an accurate simulation of the mechanical behavior of material can only be obtained by incorporating actual three-dimensional (3D) microstructure morphologies as a basis for the model.

A comparison of the modeled 3D response using the actual microstructure versus a simplified representation of spherical particles is shown in Fig. 7.38. The spatial distribution of the particles in both models is about the same. Note that the angular particles are under a much larger stress than the spherical particles, indicating more load transfer to the angular particles. The stress in the spherical particles is quite uniform, while that in the angular particles is not. The plastic strain contours in the matrix are also quite different. More localization of strain is observed in the model with angular particles. This simple comparison shows that, indeed, the microstructure-based model predictions are quite different from those of simplified spherical particles. Thus, modeling of the material using the actual microstructure is extremely important.

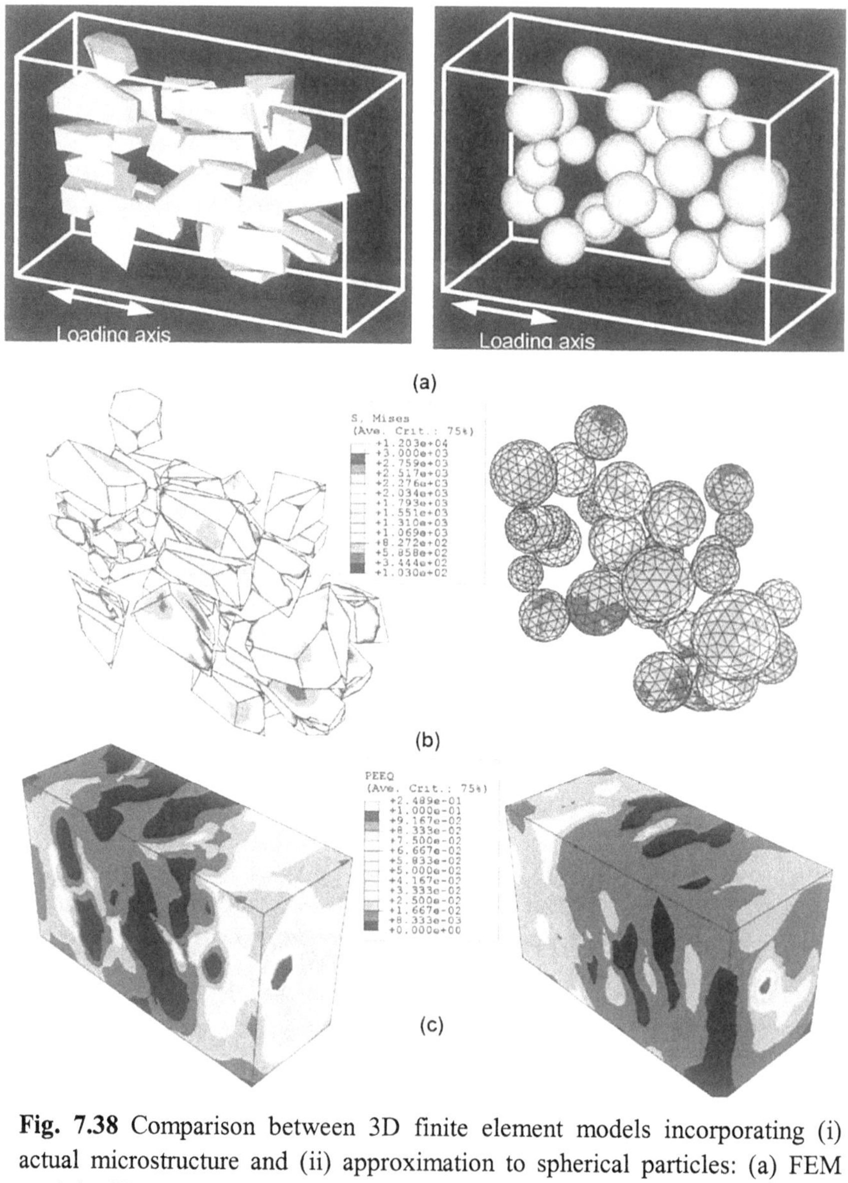

Fig. 7.38 Comparison between 3D finite element models incorporating (i) actual microstructure and (ii) approximation to spherical particles: (a) FEM models, (b) stress distribution in particles, and (c) plastic strain in matrix. Note that the microstructure model exhibits much higher stress in the particles and larger and more inhomogeneous plastic strain than the simplified spherical particle model.

A comparison of Young's modulus predicted by unit cell models versus the microstructure-based model was shown in chapter 6. The microstructure-based model was closest to the experiment (Chawla et al., 2004). A comparison of the overall stress-strain curve (elastic and plastic parts) of the 3D microstructure simulation to the experiment is shown in Fig. 7.39. These simulations incorporate a cooling step in the model, from the solution treating temperature of 493°C to 25°C. The prismatic rectangle and microstructure-based models both predict the experimental behavior quite well. Nevertheless, the microstructure-based models more faithfully represent the experimental behavior. More importantly, the localized plasticity that results from the sharp and angular nature of SiC particles, can only be captured in the microstructure-based model. Thus, other models that approximate the shape of the particles to ellipsoids underestimate the extent of strengthening.

The effect of particle clustering has also been modeled using FEM. Segurado et al. (2003) conducted 3D finite element simulations of clustered composites. They showed that within a particle cluster, the stresses in the particles are much higher than the average particle stress, Fig. 7.40(a). With an increase in particle clustering, the standard deviation of the particle stress increased significantly, Fig. 7.40(b). In a practical sense, this would cause

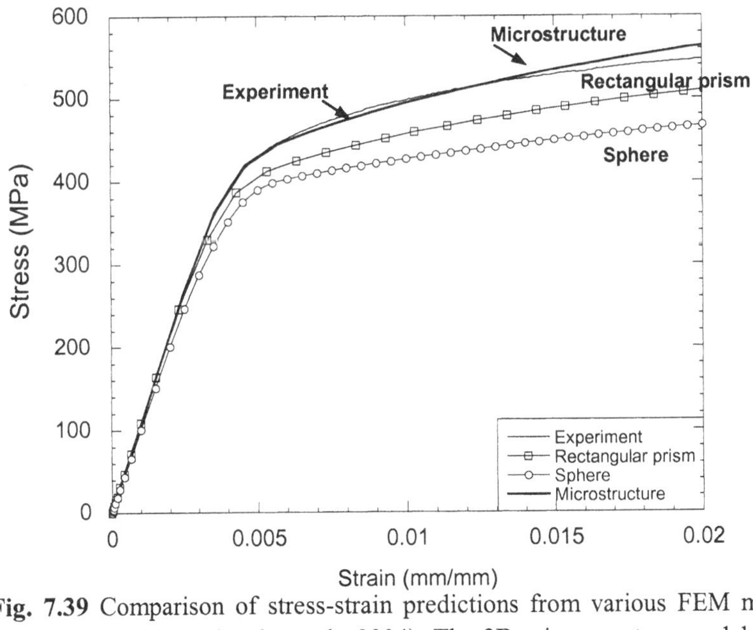

Fig. 7.39 Comparison of stress-strain predictions from various FEM models after cooling (after Chawla et al., 2004). The 3D microstructure model (from two random regions in the microstructure) is most accurate in predicting the experimentally observed behavior.

particle fracture at a much lower far-field applied stress, Fig. 7.40(c) compared to a composite with homogeneous distribution.

Deng and Chawla (2005) developed model microstructures, consisting of circular reinforcement particles in a metallic matrix. The microstructures had varying degrees of particle clustering, as quantified by the coefficient of variance in particle spacing (see inset). The tensile behavior of the microstructure was modeled using two-dimensional (2D) finite element analysis. Matrix plasticity and particle fracture were explicitly modeled. All the particles had a homogeneous strength of 1 GPa. The plastic strain distribution in the matrix is much more homogeneous in the homogeneous particle arrangement, Fig. 7.41(a), although the plastic strain is intensified in the region of particle fracture. In the clustered particle microstructure, larger stresses develop which results in particle fracture within the cluster, Fig. 7.41(b). A lower degree of plastic strain is observed. The modeled stress-strain response shows that the clustered microstructure has a lower "ductility," although matrix fracture was not modeled. This model verifies the experimental results of Murphy et al. (1998), described above.

Shen and Chawla (2001) and Shen et al., (2001) explored the correlation between macro-hardness and tensile properties of particle-reinforced metal matrix composites. It was shown that, unlike most monolithic metals and alloys, hardness does not necessarily scale with the overall strength of particle-reinforced metal matrix composites, Fig. 7.42. The hardness test may significantly overestimate the overall tensile and yield strengths of composites containing large reinforcement particles, which are prone to fracture during deformation processing and/or tensile loading. The predominant local compressive stress state in a hardness test prevents the pre-existing fractured particles from weakening the material during indentation. For composites having relatively small reinforcement particles, a unique relationship between hardness and tensile/yield strength did not exist, even when the material was essentially free of pre-existing fractured particles. This was especially true in cases where the strength of the Al matrix was relatively low. The tendency of higher hardness for particle-reinforced composites can be attributed to the localized increase in particle concentration directly underneath the indenter during hardness testing, Fig. 7.43. This was illustrated by micromechanical modeling using the finite element method. Under indentation, the material system with discrete particles exhibits a higher resistance to deformation than the homogenized system having exactly the same overall stress–strain behavior, Fig. 7.44 (Shen et al., 2001). These findings are true even when the particle size is

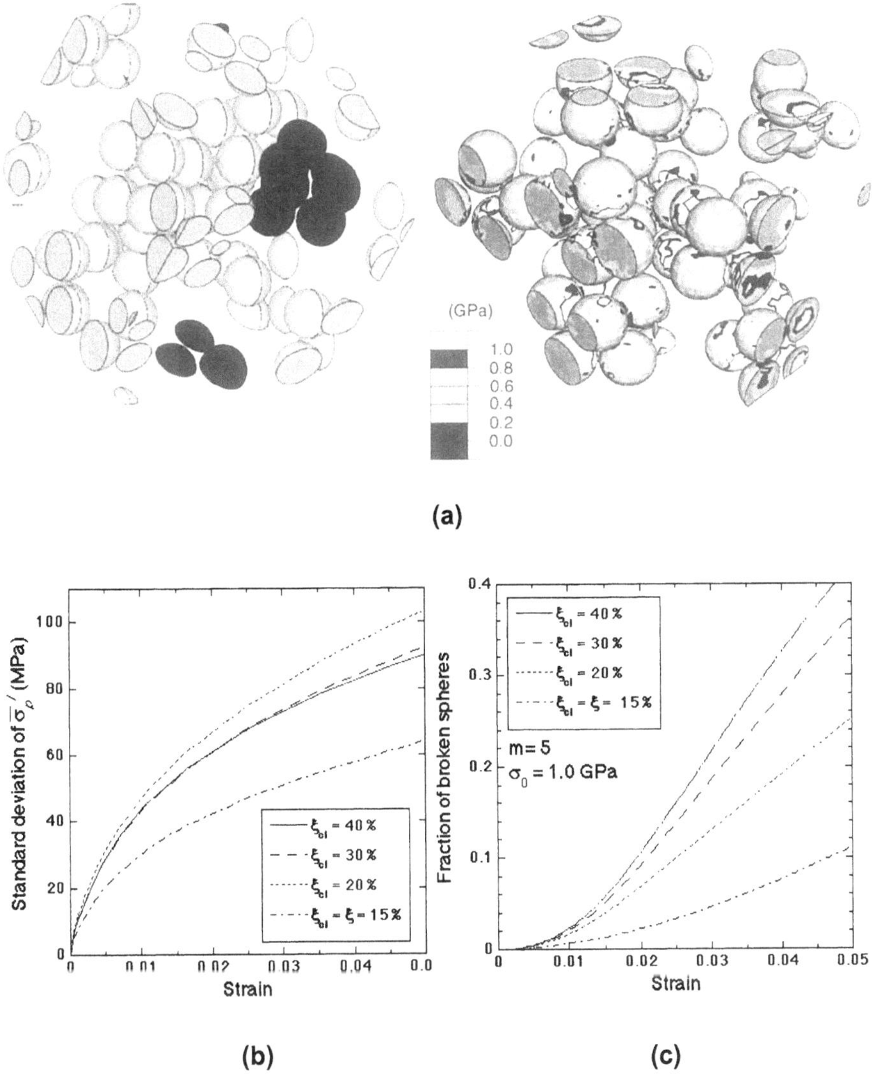

Fig 7.40 (a) 3D finite element model consisting of perfectly spherical SiC particles in an Al matrix (after Segurado et al., 2003; courtesy of J. LLorca). The model consists of 49 particles and 7 "clusters." The stress within the cluster is higher than the average stress. (b) Standard deviation of stress in particles with increasing clustering, for a given strain, and (c) fraction of broken particles predicted. ξ and ξ_{cl} are the volume fraction of particles (15%) in the composite and within the cluster, respectively.

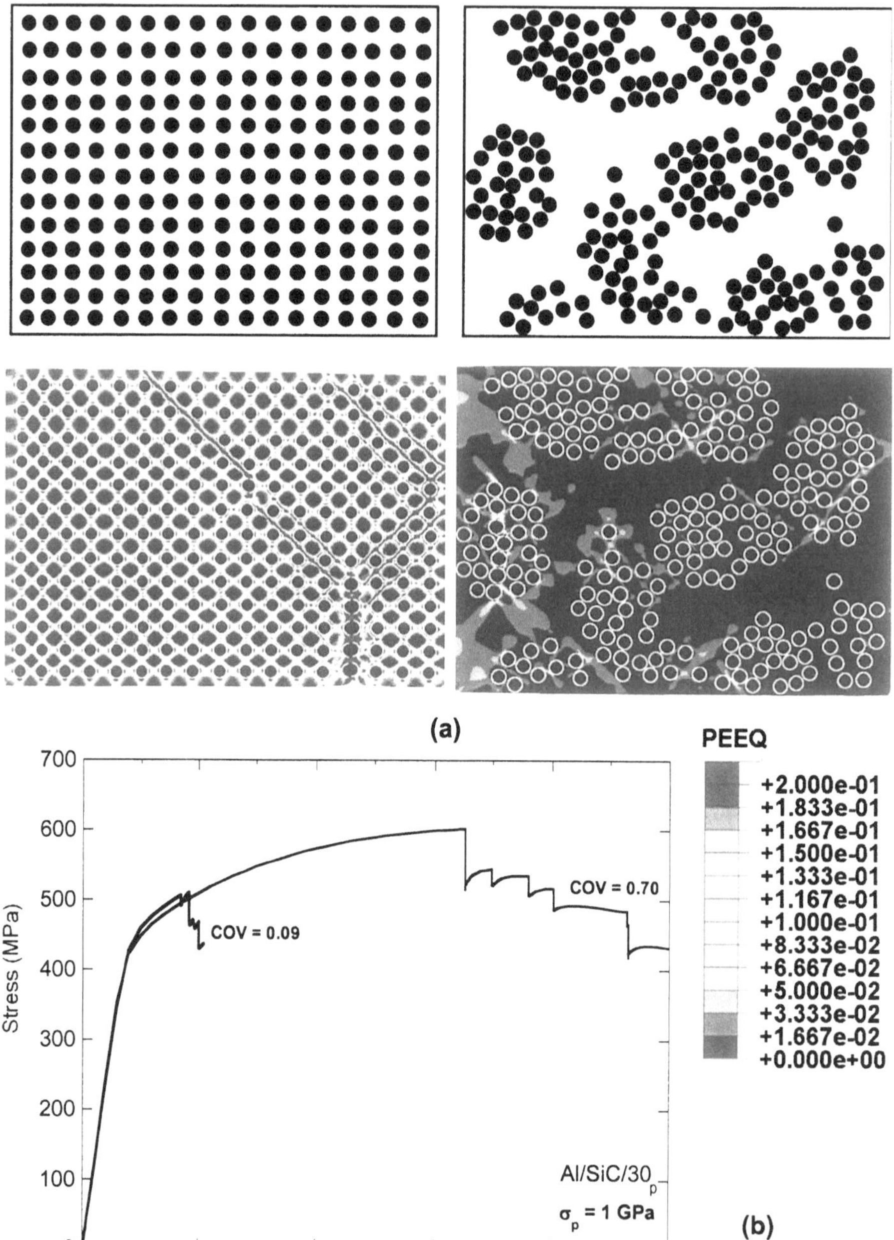

Fig 7.41 Two-dimensional (2D) finite element analysis of the effect of particle clustering (after Deng and Chawla, 2005): (a) model microstructures consisting of circular SiC particles in Al, showing particle fracture within the cluster and (b) Predicted tensile stress-strain behavior; all particles were assumed to have a strength of 1 GPa. The clustered microstructure has a lower "ductility."

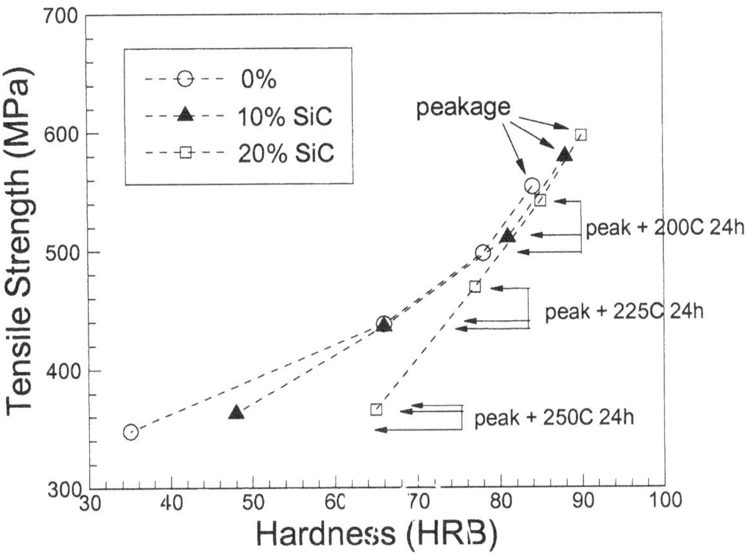

Fig. 7.42 Correlation between macro-hardness and tensile properties of particle-reinforced MMCs (after Shen et al., 2001). Unlike most monolithic metals and alloys, hardness does not necessarily scale with strength of MMCs.

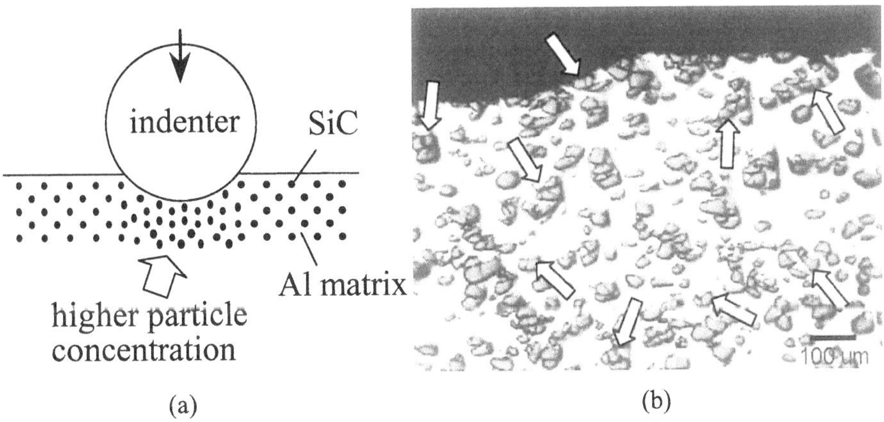

Fig. 7.43 Deformation behavior of particle reinforced MMCs during microindentation (after Shen and Chawla, 2001): (a) localized increase in particle concentration due to plastic flow of the matrix and (b) localized particle fracture, indicated by the arrows.

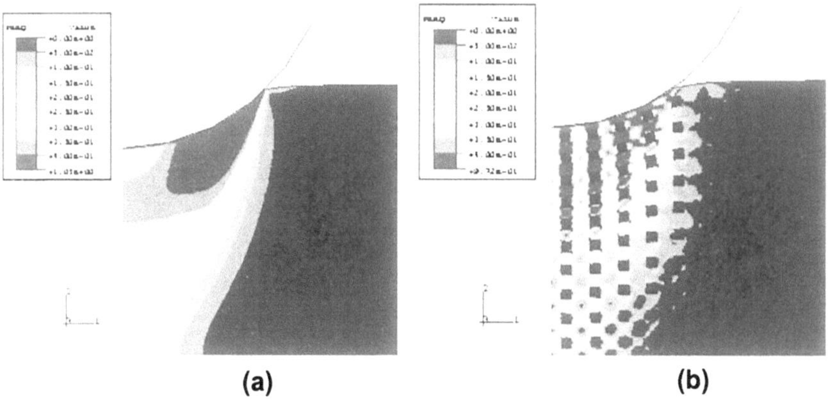

Fig. 7.44 Finite element model of indentation in (a) homogeneous material and (b) particle reinforced composite. In the model, both materials had the same macroscopic tensile constitutive behavior (after Shen et al., 2001). Under indentation, the material system with discrete particles exhibited a higher resistance to deformation.

7.4 FRACTURE TOUGHNESS

The fracture toughness of particle reinforced MMCs is controlled by several factors. These include: (a) volume fraction of reinforcement, (b) interparticle spacing and strength of particles, (c) spatial distribution of particles (i.e., particle clustering), and (d) microstructure of matrix and interface region (as controlled by heat-treatment in age-hardenable alloys). Figure 7.45 is a compilation of toughness in several composite systems, as a function of reinforcement volume fraction (Manoharan et al., 1993, Hunt et al., 1993, and Beck Tan et al., 1994). Note that all the composites show a decrease in toughness with increasing volume fraction of reinforcement. The toughness appears to reach a "plateau" at volume fractions of 20% and above. The decrease in toughness, shown in above, can be explained by an increase in stress triaxiality with increasing volume fraction of particles (similar to what takes place during tensile loading of the composite).

The effect of particle size is less clear-cut. This is because decreasing particle size results in a lower interparticle spacing. In addition, it also results in an increase in strength of ceramic particle, because of a lower probability of encountering a strength-limiting flaw. It has been shown that the degree of clustering increases with decreasing particle size (Hunt et al., 1993), and that the degree of indirect strengthening increases with

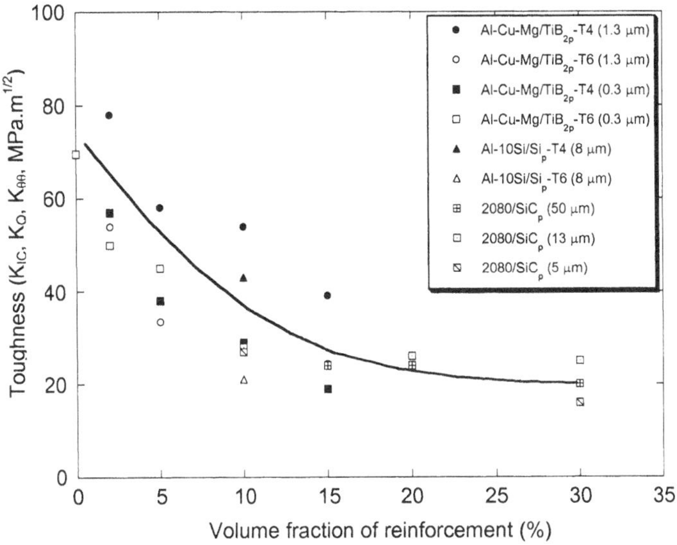

Figure 7.45 Toughness of several particle reinforced MMCs as a function of reinforcement volume fraction (after Manoharan et al., 1993, Hunt et al., 1993, and Beck Tan et al., 1994). All the composites show a decrease in toughness with increasing volume fraction, reaching a "plateau" at volume fraction of about 20% and above.

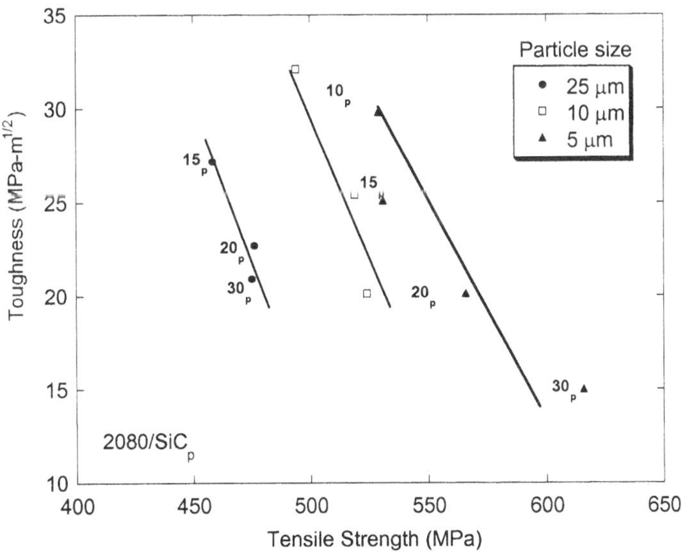

Figure 7.46 Effect of particle size on toughness of a 2080/SiC$_p$ composite, at three volume fractions (after Hunt et al., 1993). For a given volume fraction, the toughness of the composite decreases slightly with decreasing particle size.

decreasing particle size (Arsenault and Shi, 1986). Figure 7.46 shows the decrease in toughness of a 2080/SiC_p composite with increasing strength, at three volume fractions and as a function of particle size. The curve for the largest particle size is farthest left, because of the decrease in the strength of the composite with increasing particle size. It is interesting to note that, for a given volume fraction, the toughness of the composite decreases slightly with decreasing particle size. The effect of particle size on toughness was also investigated by Kamat et al. (1989). They studied the fracture toughness behavior of Al_2O_3 particle reinforced Al matrix composites at volume fractions of 10 and 20% Al_2O_3, and several particle sizes ranging between 5 and 50 μm. At large particle sizes (> 15 μm), particle fracture was hypothesized to cause unstable crack growth and much lower crack growth toughness. At smaller particle diameter, interfacial decohesion took place. In this lower particle size (and lower interparticle spacing) regime, the behavior of the composites was found to follow the model of Rice and Johnson, where particle/matrix decohesion is assumed to take place at the interface ahead of the main crack. The coalescence of voids arising from particle/matrix decohesion is facilitated by a decrease in interparticle spacing. Their model showed the ratio of toughness-to-fracture strength, J_{Ic}/σ_f, to be directly proportional to the interparticle diameter, λ:

$$\frac{J_{Ic}}{\sigma_f} = \lambda$$

The fracture toughness is also influenced by the degree of clustering of particles. Lloyd (1995) obtained varying degrees of clustering by varying the cooling rate of a cast A356/SiC/15_p composite, Fig. 7.47. As mentioned in Chapter 4, faster cooling rates result in less time for particle pushing from dendrites, resulting in a more homogeneous distribution of particles. This work clearly shows that with increasing clustering (quantified by an increase in minimum edge-to-edge spacing of particles), the toughness decreases. This stems from an increase in stress triaxiality caused by particle clusters. The toughness is also very much affected by matrix microstructure. Manoharan and Lewandowski (1990) studied the fracture toughness behavior of SiC particle reinforced Al matrix composites, Fig. 7.48. The materials were heat treated to the underaged and overaged condition and had the same nominal tensile strength. The fracture toughness in the two heat treated conditions was quite different. In the UA condition the toughness was twice as large as that of the OA material. This was attributed to a transition in fracture mode from particle-fracture controlled (UA) to interfacial decohesion (OA). The ease of interfacial decohesion was explained by coarsening of precipitates and weakening at the particle/matrix

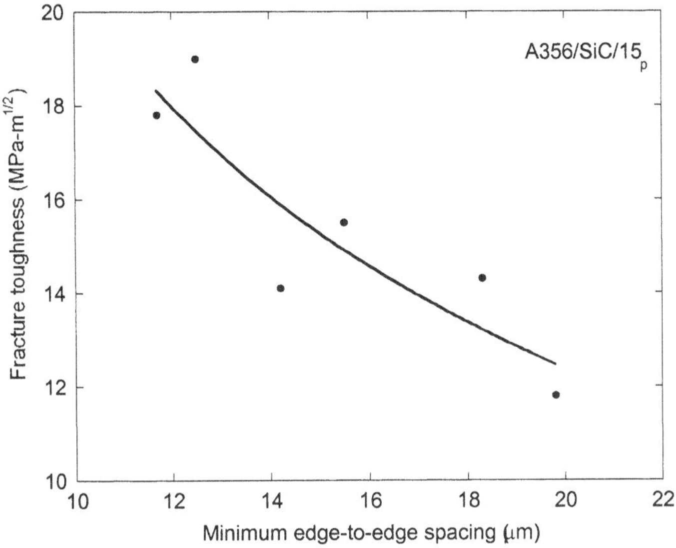

Fig. 7.47 Effect of particle clustering on toughness of a cast A356/SiC/15_p composite (after Lloyd, 1995). With increasing clustering (quantified by minimum edge-to-edge spacing of the particles), the toughness decreases.

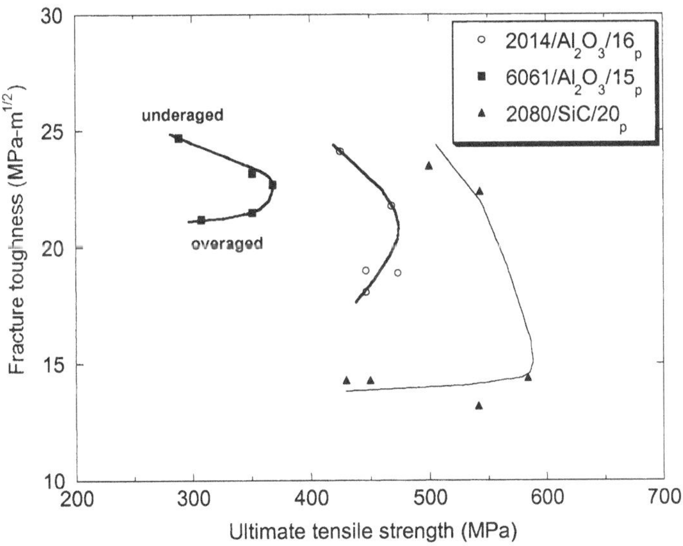

Fig. 7.48 Effect of matrix microstructure on the fracture toughness-tensile strength relationship in particle reinforced MMCs (after Lewandowski, 2000; Hunt et al.; 1993, Kamat et al., 1989). For a given tensile strength, the toughness is higher in the underaged condition vis-à-vis the overaged condition.

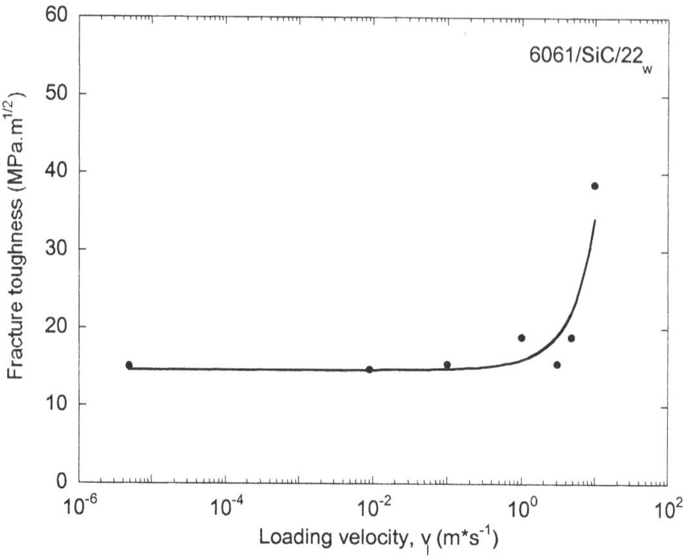

Fig. 7.49 Effect of loading velocity on the fracture toughness of a $6061/SiC/22_w$ (after Wang and Kobayashi, 1997). Above a loading velocity of 10 m/s, the fracture toughness increased significantly.

interface. In the unreinforced alloy the toughness in both UA and OA conditions was similar.

The toughness of particle reinforced MMCs has also been shown to be a function of loading velocity. Wang and Kobayashi (1997) examined the effect of loading velocity on the fracture toughness of a $6061/SiC/22_w$, Fig. 7.49. At a loading velocity above 10 m/s, the fracture toughness increased significantly. Fractographic analysis showed that matrix dimples were deeper and the damage zone extended to a longer distance at the faster loading rate. It was postulated that the very fast loading rate did not allow enough time for interaction and coalescence between cracks and voids.

References

Arsenault, R.J., and N. Shi, (1986) *Mater. Sci. Eng.*, **81**, 175.

Ayyar, A., J.J. Williams, and N. Chawla, (2005) *Comp. Sci. Tech.*, submitted.

Beck Tan, N.C., R.M. Aikin, Jr., and R.M. Briber, (1994) *Metall. Mater. Trans.*, **25A**, 2461-2468.

Blucher, J.T., U. Narusawa, M. Katsumata, and A. Nemeth, (2001) *Composites*, **32**, 1759-1766.

Boehlert, C.J., B.S. Majumdar, S. Krishnamurty, and D.B. Miracle, (1997) *Metall. Mater. Trans.*, **28A**, 309-323.

Brockenbrough, J.R., S. Suresh, and H.A. Wienecke, (1991) *Acta Metall. Mater.*, **39**, 735-752.

Bushby, R.S., (1998) *Mater. Sci. Tech.*, **14**, 877-886.

Chan, K.S., (1993) *Metall. Trans.*, **24A**, 1531-1542.

Chawla, K.K., (1973a) *Phil. Mag.*, **28**, 401.

Chawla, K.K., (1973b) *Metallography*, **6**, 155-169.

Chawla, K.K., (1998) *Composite Materials - Science and Engineering*, 2nd Ed., Springer-Verlag, New York, p. 102.

Chawla, K.K., and M. Metzger, (1972) *J. Mater. Sci.*, **7**, 34.

Chawla, N., C. Andres, J.W. Jones, and J.E. Allison, (1998a) *Metall. Mater. Trans.*, **29A**, 2843.

Chawla, N., C. Andres, J.W. Jones, and J.E. Allison, (1998b) *Scripta Mater.*, **38**, 1596.

Chawla, N., U. Habel, Y.-L. Shen, C. Andres, J.W. Jones, and J.E. Allison, (2000) *Metall. Mater. Trans.*, **31A**, 531-540.

Chawla, N., and Y.-L. Shen, (2001) *Adv. Eng. Mater.*, **3**, 357-370.

Chawla, N., J.J. Williams, and R. Saha, (2002a) *J. Light Metals*, **2**, 215-227.

Chawla, N., J.J. Williams, and R. Saha, (2002b) *Metall. Trans.*, **33A**, 3861-3869.

Chawla, N., B.V. Patel, M. Koopman, K.K. Chawla, E.R. Fuller, B.R. Patterson, and S. Langer, (2003) *Mater. Charac.*, **49**, 395-407.

Chawla, N., V.V. Ganesh, and B. Wunsch, (2004) *Scripta Mater.*, **51,** 161-165.

Cheskis, H.P., and R.W. Heckel, (1970) *Metall. Trans.*, **1**, 1931-1942.

Christman, T., A. Needleman and S. Suresh, (1989) *Acta Metall. Mater.*, **37**, 3029.

Cook, J., and J.E. Gordon, (1964) *Proc. Roy. Soc. London*, **A228**, 508

Corbin, S.F. and D.S. Wilkinson, (1994) *Acta Metall. Mater.*, **42**, 1319.

Cornie, J.A., M.L. Seleznev, M. Ralph, and F.A. Armatis, (1993) *Mater. Sci. Eng.*, **A162**, 135-142.

Cox, B.N., (1990) *Acta Metall. Mater.*, **38**, 2411.

Cox, H.L., (1952) *Brit. J. App. Phys.*, **3**, 122.

Curtin, W.A., (1993) *Composites*, **24** 98-102.

Davis, L.C., and J.E. Allison, (1993) *Metall. Trans.*, **24A**, 2487.

Deng, X., and N. Chawla, (2005) *J. Mater. Sci.*, submitted.

Deve′, H.E., (1997) *Acta Mater.*, **45**, 5041-5046.

Deve′, H.E., and C. McCullough, (1995) *JOM*, **7,** 33-37.

Dirichlet, G.L., (1850) *J. Reine Angew. Math.*, **40**, 209-227.

Dunand, D.C., and A. Mortensen, (1991) *Acta Metall. Mater.*, **39**, 1417-1429.

Dutta, I., J.D. Sims and D.M. Seigenthaler, (1993) *Acta Metall. Mater.*, **41**, 885.

Eldridge, J.I., D.R. Wheeler, R.R. Bowman, and A. Korenyi-Both, (1997) *J. Mater. Res.*, **12**, 2191-2197.

Evans, A.G., and D.B. Marshall, (1989) *Acta Metall.*, **37**, 2567.

Faucon, A., T. Lorriot, E. Martin, S. Auvray, Y. Lepetitcorps, K. Dyos, and R.A. Shatwell, (2001) *Comp. Sci. Tech.*, **61**, 347-354.

Galvez, F., C. Gonzalez, P. Poza, and J. LLorca, (2001) *Scripta Mater.*, **44**, 2667-2671.

Ganesh, V.V., and N. Chawla, (2004) *Metall. Mater. Trans.*, **35A**, 53-62.

Ganesh, V.V., and N. Chawla, (2005) *Mater. Sci. Eng.*, **A391**, 342-353.

Gao, Y.C., Y.W. Mai, and B. Cotterell, (1988) *SAMPE*, **39**, 550.

Gonzalez, C., and J. LLorca, (2001) *Acta Mater.*, **49**, 3505-3519.

Goodier, J.N., (1933) *J. Appl. Mech.*, **55-7**, 39.

Grimes, H.H., R.A. Lad, and J.E. Maisel, (1977) *Metall. Trans.*, **8**, 1999-2005.

Guden, M., and I.W. Hall, (2000) *Computers and Struc.*, **76**, 139-144.

Gupta, V., (1991) *MRS Bull.*, **16** (No. 4), 39.

Gupta, V., J. Yuan, and D. Martinez, (1993) *J. Amer. Ceram. Soc.*, **76**, 305.

Hack, J.E., R.A. Page, and G.R. Leverant, (1984) *Metall. Trans.*, **15A**, 1389-1396.

He, M.Y., and J.W. Hutchinson, (1989) *J. App. Mech.*, **56**, 270.

Humphreys, F.J., (1977) *Acta Metall.*, **25** 1323-1344.

Humphreys, F.J., (1991) *Mater. Sci. Eng.*, **135A,** 267-273.

Hunt, W.H., T.M. Osman, and J.J. Lewandowski, (1993) *JOM*, **45**, 30-35.

Hutchinson, J.W., and H.M. Jensen, (1990) *Mech. Matls.*, **9**, 139-163.

Isaacs, J.A., and A. Mortensen, (1992) *Metall. Trans.*, **23A**, 1207-1219.

Jansson, S., H.E. Deve, and A.G. Evans, (1991) *Metall. Mater. Trans.*, **22A**, 2975.

Jeong, H., D.K. Hsu, R.E. Shannon, P.K. Liaw, (1994) *Metall. Mater. Trans.*, **25A**, 799-809.

Kamat, S., J.P. Hirth, and R. Mehrabian, (1989) *Acta Metall.*, **37**, 2395.

Kelly, A., (1973) *Strong Solids*, Clarendon Press, Oxford, p. 157.

Kelly, A., and H. Lilholt, (1969) *Phil. Mag.*, **20**, 311-328.

Kerans, R.J., and T.A. Parthasarathy, (1991) *J. Am. Ceram. Soc.*, **74**, 1585-1596.

Krajewski, P.E., J.E. Allison, and J.W. Jones, (1993) *Metall. Mater. Trans.*, **24**, 2731.

Konitzer, D.G., and M.H. Loretto, (1989) *Mater. Sci. Eng.*, **A107**, 217-223.

Kyono, T., I.W. Hall, and M. Taya, (1986) *J. Mater. Sci.*, **21**, 4269-4280.

Levy, A., and J. M. Papazian, (1991) *Acta Metall. Mater.*, **39**, 2255.

Lewandowski, J.J., (2000) in *Comprehensive Composite Materials*, T.W. Clyne, A. Kelly, and C. Zweben (eds.), Elsevier, vol. 3, pp. 151-187.

Lewandowski, J.J., D.S. Liu, and C. Liu, (1991) *Scripta Metall.*, **25**, 21.

Leyens, C., J. Hausmann, and J. Kumpfert, (2003) *Adv. Eng. Mater.*, **5**, 399-410.

LLorca, J., (1995) *Acta Metall. Mater.*, **43**, 181-192.

LLorca, J., A. Needleman and S. Suresh, (1991) *Acta Metall. Mater*, **39**, 2317.

Lloyd, D.J., (1995) in *Intrinsic and Extrinsic Fracture Mechanisms in Inorganic Composite Systems*, J.J. Lewandowski and W.H. Hunt (eds.), TMS-AIME, Warrendale, PA, pp. 39-47.

Lloyd, D.J., (1997) in *Composites Engineering Handbook* (P.K. Mallick, ed.), Marcel Dekker, New York, pp. 631-669.

Logsdon, W.A., and P.K. Liaw, (1986) *Eng. Frac. Mech.*, **24**, 737-751.

Manoharan, M., and J.J. Lewandowski, (1990) *Acta Metall. Mater.*, **38**, 489-496.

Manoharan, M., and J.J. Lewandowski, (1992) *Mater. Sci. Eng.*, **A150**, 179-186.

Manoharan, M., J.J. Lewandowski, and W.H. Hunt, (1993) *Mater. Sci. Eng.*, **A172**, 63-69.

Meyers, M.A., and K.K. Chawla, (1999) *Mechanical Behavior of Materials*, Prentice-Hall, Upper Saddle River, NJ, p. 493.

Miller, D.A., and D.C. Lagoudas, (2000) *J. Eng. Mater. Tech.*, **122**, 74-79.

Mikata, Y., and M. Taya, (1985) *J. Compos. Mater.*, **19**, 554.

McCullough, C., H.E. Devé, and T.E. Channel, (1994) *Mater. Sci. Eng.*, **A189**, 147-154.

Mummery, P.M., B. Derby, D.J. Buttle, and C.B. Scruby, (1991) in *Proc. of Euromat 91*, T.W. Clyne and P.J. Withers (eds.), vol. 2, Cambridge, UK, pp. 441-447.

Murphy, A.M., S.J. Howard, and T.W. Clyne, (1998) *Mater. Sci. Tech.*, **14**, 959-968.

Nardone, V.C., and K.M. Prewo, (1986) *Scripta Metall.*, **23**, 291.

Page, R.A., J.E. Hack, R. Sherman, and G.R. Leverant, (1984) *Metall. Trans.*, **15A**, 1397-1405.

Rao, K.T. Venkateshwara, S.C. Siu, and R.O. Ritchie, (1993) *Metall. Trans.*, **24A**, 721-734

Rosenberger, A.H., P.R. Smith, and S.M. Russ, (1999) *J. Comp. Tech. Res.*, **21**, 164-172.

Rossoll, A., B. Moser, and A. Mortensen, (2005) *Mech. Mater.*, **37**, 1.

Ruhle, M., and A.G. Evans, (1988) *Mater. Sci. Eng.*, **A107**, 187.

Segurado, J., C. Gonzalez, and J. LLorca, (2003) *Acta Mater.*, **51**, 2355-2369.

Shen, Y.-L., and N. Chawla, (2001) *Mater. Sci. Eng.*, **A297**, 44-47.

Shen, Y.-L., M. Finot, A. Needleman, and S. Suresh, (1994) *Acta Metall. Mater.*, **42**, 77.

Shen, Y.-L., M. Finot, A. Needleman, and S. Suresh, (1995) *Acta Metall. Mater.*, **43**, 1701.

Shen, Y.-L., J.J. Williams, G. Piotrowski, N. Chawla, and Y.L. Guo, (2001) *Acta Mater.*, **49**, 3219-3229.

Schulte, K., and K. Minoshima, (1993) *Composites*, **24**, 197-208.

Shetty, D.K., (1988) *J. Amer. Ceram. Soc.*, **71**, C-107.

Steglich, D., T. Siegmund, and W. Brocks, (1999) *Comput. Mater. Sci.*, **16**, 404-413.

Suresh, S., and K.K. Chawla, (1993) in *Fundamentals of Metal Matrix Composites*, Suresh, S., A. Mortensen, and A. Needleman, eds., Butterworth-Heinemann, Stoneham, MA, p. 119.

Termonia, Y., (1987) *J. Mater. Sci.*, **22**, 504-508.

Toda, H., T. Gouda, and T. Kobayashi, (1998) *Mater. Sci. Tech.*, **14**, 925-932.

Torquato, S., (2002) *Random Heterogeneous Materials: Microstructure and Macroscopic Properties*, Springer-Verlag, New York, pp. 160–176.

Vogelsang, M., R.J. Arsenault, and R.M. Fisher, (1986) *Metall. Trans.*, **17A**, 379.

Voleti, S.R., C.R. Ananth, and N. Chandra, (1998) *J. Comp. Tech. Res.*, **29**, 203-209.

Wang, L., and T. Kobayashi, (1997) *Mater. Trans. JIM.*, **38**, 615-621.

Warrier, S.G., and B.S. Majumdar, (1997) *Mater. Sci. Eng.*, **237**, 256-257.

Young, R.J., (1994) in *High Performance Composites: Commonalty of Phenomena*, K.K. Chawla, P.K. Liaw, and S.G. Fishman, eds., TMS, Warrendale, PA, p. 263.

Chapter 8

CYCLIC FATIGUE

Fatigue is the phenomenon of mechanical property degradation under cyclic loading. The cyclic loads may be mechanical, thermal, or a combination of the two. Many high-volume applications of composite materials involve cyclic-loading situations, e.g., automobile components, aircraft structures, etc. Below we provide a brief description of the two main approaches that have been used to quantify fatigue behavior of materials (for a more complete description, the reader may consult the texts by Meyers and Chawla (1999) and Suresh (1998).

Stress versus cycles (S-N): This approach involves cyclic fatigue testing to develop *S-N* curves, where S is the stress amplitude and *N* is the number of cycles to failure. In general, for ferrous metals, a clear fatigue limit or endurance limit is observed. For stress levels below this endurance limit, theoretically, the material can be cycled infinitely without failure. In nonferrous materials, such as aluminum, a true endurance limit does not exist. Here one can arbitrarily define a certain number of cycles, say 10^7, as a point of "fatigue runout," at which point the experiment is stopped. The fatigue behavior of structural materials can be conveniently divided into two stages: crack initiation and propagation. In high cycle fatigue (HCF) of metals, most of the fatigue life is spent initiating a fatigue crack and a very small fraction of the life is spent in propagation. In low cycle fatigue (LCF), a much larger fraction of the fatigue life is spent in crack propagation. The phenomena of work hardening and work softening also play a more important role in LCF. One major drawback of the *S-N* approach is that no distinction can be made between the crack initiation phase and the crack propagation phase.

Fatigue crack growth: This approach is based on fracture mechanics and it enables the study of fatigue crack propagation. Fatigue crack propagation experiments are generally conducted on notched samples. The crack length, a, is measured as a function of fatigue cycles, N, over a range of cyclic stress intensity factors, ΔK. The applied cyclic stress intensity range is given by:

$$\Delta K = Y \Delta\sigma \sqrt{\pi a}$$

where Y is a geometric parameter that depends on the nature of the notch and specimen configuration, $\Delta\sigma$ is the cyclic stress range, and a is the crack length. The results are then presented as log da/dN (crack growth per cycle) versus log ΔK. Crack growth rate, da/dN, is related to ΔK according to the power-law relationship first formulated by Paris et al. (1961, 1963):

$$\frac{da}{dN} = C(\Delta K)^m$$

where C and m are constants that depend on the material and test conditions.

The application of a fracture mechanics based approach to composites is not straightforward. The main reason for this is the inherent heterogeneity and anisotropic nature of the composites. These attributes of composites result in damage mechanisms that are very different from those encountered in conventional, homogeneous materials. Despite these limitations, conventional approaches have been used and modified to quantify the fatigue behavior of composites. As in previous chapters, we divide our discussion of fatigue into two major types of MMCs: continuous fiber reinforced and discontinuously reinforced (primarily particle reinforced) MMCs.

8.1 STRESS VERSUS CYCLES (S-N) FATIGUE

Several material variables play an important role in fatigue of MMCs. These include modulus, strength, ductility, and work hardening characteristics of the constituents (fiber and matrix) as well as the characteristics of the interface. As mentioned above, conventional approaches to fatigue involve *S-N* curves and fatigue crack propagation studies. Another approach involves the monitoring of damage accumulation. Specifically, measurement of modulus loss as a function of cycles has been used extensively in polymer matrix and ceramic matrix composites (Chawla, K.K., 1997). In the following, we describe the fatigue behavior of continuous fiber reinforced MMCs.

8.1.1 Continuous Fiber Reinforced MMCs

Incorporation of fibers generally improves the fatigue resistance in the fiber direction. In general, in composites containing fibers aligned along the stress axis and in large volume fractions, high monotonic strength and modulus translates into high fatigue strength. This can be explained by the fact that with increasing stiffness and strength, an increasing fraction of the load is borne by the fibers. High strength, brittle fibers such as carbon or boron do not fatigue as readily as metals, although ceramic fibers have been shown to be susceptible to cyclic fatigue (Chawla et al., 2005; Kerr et al., 2005).

Typically, the *S-N* curves of fiber reinforced MMCs are close to horizontal. Figure 8.1 shows the *S-N* curves for unidirectionally reinforced 6061/B/40_f, Al/Al_2O_3(FP)/50_f, and Mg/Al_2O_3(FP)/50_f under tension-tension fatigue (Champion et al., 1978). The cyclic stress is normalized with respect to the ultimate tensile strength. The unidirectional MMCs show better fatigue properties than the matrix, when loaded parallel to the fibers. Note the rather flat *S-N* curves of all the composites. In unidirectional composites, the fatigue strength will be a maximum along the fiber direction, and the greatest efficiency will be achieved if the fibers have uniform properties, are as defect-free as possible, and are much stronger and stiffer than the matrix. McGuire and Harris (1974) observed that increasing the fiber volume fraction from 0 to 24% resulted in increased fatigue resistance of tungsten

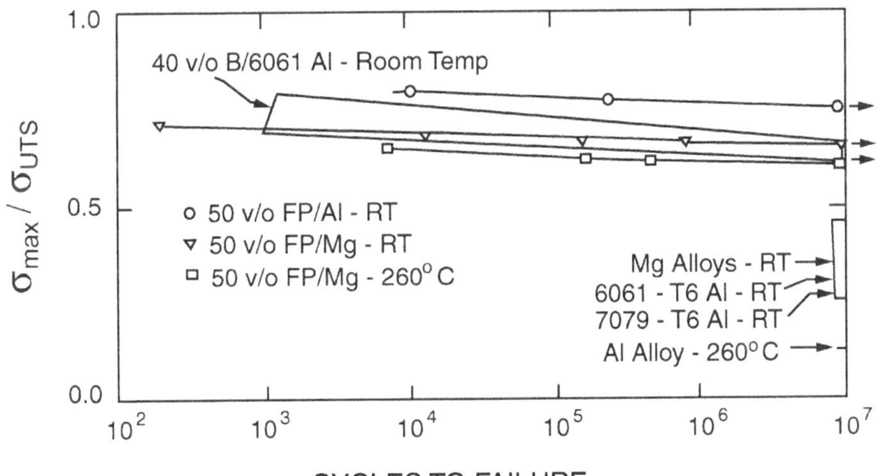

Fig. 8.1 Stress versus cycles (*S-N*) curves for unidirectionally reinforced 6061/B/40_f, Al/Al_2O_3(FP)/50_f, and Mg/Al_2O_3(FP)/50_f under tension-tension fatigue (after Champion et al., 1978). The fatigue strength of the composites is significantly enhanced over that of the monolithic alloys. FP is the Du Pont tradename for alumina fibers.

fiber reinforced Al-4Cu alloy under tension-compression cycling ($R = \sigma_{min}/\sigma_{max} = -1$). This was a direct result of increased monotonic strength and stiffness of the composite as a function of the fiber volume fraction. The enhancements in fatigue life due to increased stiffness of the composite are most pronounced in the HCF regime. Thus, in stress-controlled fatigue, for a given applied stress, the matrix in the composite will undergo a much lower strain than that of the unreinforced alloy.

Some work has been done to understand the fatigue damage mechanisms in continuous fiber reinforced MMCs. Baker et al. (1972) conducted one of the earliest studies on fatigue of continuous fiber reinforced MMCs. The *S-N* fatigue behavior of Al/B_f (125 μm fiber diameter) and Al/C_f (8 μm fiber diameter) composites was studied. The carbon fiber reinforced composites exhibited poor fatigue resistance, because of processing-induced fracture of the fibers and poor interfacial bonding. The following mechanisms were observed during fatigue: (a) fiber fracture dominated damage, where the rate of fiber fracture was much higher than cyclic plasticity in the matrix, and (b) matrix shear and plasticity induced localized and progressive fiber fracture. Mechanism (a) was shown to be predominant in low cycle fatigue, where the applied stresses were much closer to the ultimate strength of the fibers (and the composite). In high cycle fatigue, the fatigue damage in the composite is more likely to be controlled by matrix plasticity, followed by fiber fracture, or mechanism (b). Gouda et al. (1981) also observed crack initiation early in the fatigue life at defects in boron fibers in unidirectionally reinforced Al/B_f composites. These cracks then grew along the fiber/matrix interface and accounted for a major portion of the fatigue life, as would be the case in a composite with high fiber-to-matrix strength ratio. In composites with low fiber-to-matrix strength ratio, crack propagation may be a major portion of fatigue life, but the fatigue cracks grow through the fibers, resulting in poor fatigue resistance. Uniform spacing of fibers is also crucial, because clustered or touching fibers will result in enhanced stress concentration and easier crack nucleation.

The reader should note that because of the highly anisotropic nature of the fiber reinforced composites in general, the fatigue strength of off-axis MMCs, as in any fibrous composite, will be expected to decrease with increasing angle between the fiber axis and the applied stress axis. This has been confirmed by studies involving *S-N* behavior of alumina fiber reinforced magnesium composites (Hack et al., 1987; Page et al., 1987). It was found that the fatigue strength mirrored the tensile strength of the composites. Increased fiber volume fractions resulted in enhanced fatigue lifetimes in the axial direction, but little or no improvement was observed in the off-axis directions. Fatigue crack initiation and propagation occurred primarily through the magnesium matrix. Thus, alloy additions to increase the strength of the matrix and fiber/matrix interface were incorporated. The alloy additions did improve the off-axis properties but decreased the axial properties. The reason for this was that while the alloy additions resulted in strengthening of the matrix and interface, they decreased the fiber strength.

Few direct microstructural observations of fatigue damage in continuous fiber reinforced MMCs have been conducted. One such study involved a single-crystal copper matrix containing tungsten fibers (Chawla, 1975). Tungsten/copper is an unusual metal matrix composite system. The two metals are mutually insoluble, yet molten copper wets tungsten. This allows for a strong mechanical bond between the tungsten fiber and the copper matrix, without any attendant chemical complications at the interface. The composites were made by liquid metal infiltration of fibers in vacuum. It was observed that the process of fabrication involving cooling from a temperature >1080°C (melting point of copper) to room temperature resulted in thermal stresses large enough to deform the copper matrix plastically (Chawla and Metzger, 1972). Dislocation etch-pitting technique was used to characterize the dislocation density in the single-crystal copper matrix. The dislocation density in the matrix was higher ($>10^8$ cm^{-2}) near the fiber/matrix interface than away from the interface. The dislocation distribution had a cellular structure, with the cell structure being better defined and smaller near the fiber than away from the fiber. This very heterogeneous distribution became homogeneous on cycling stress. The copper matrix structure after cycling had a high dislocation density distributed in a more or less uniform cellular structure.

Zhang et al. (2003) studied the *S-N* fatigue behavior of a 45 vol.% Nextel 610 alumina fiber reinforced pure Al matrix composite. The composite exhibited a tensile stress-strain behavior that was linear to failure, with about 0.7% strain-to-failure. In fatigue, the *S-N* curve was linear, with a fatigue strength (fatigue runout taken at 10^7 cycles) around 700 MPa. A clear and definitive fatigue strength was not apparent. The damage mechanisms varied considerably with applied stress. At very low stress (HCF regime), longitudinal cracking between fibers was the dominant failure mechanism, Fig. 8.2(a). It was suggested that these cracks initiated at fractured fibers and then grew parallel to the fiber direction. With increasing cyclic stress amplitude, in the intermediate regime between HCF and LCF, longitudinal matrix cracking was accompanied by transverse matrix cracks between fibers. Finally, at very high stresses (LCF regime), a single crack propagated in a catastrophic fashion perpendicular to the fibers, with no significant toughening, Fig. 8.2(b). Persistent slip band formation in the matrix has also

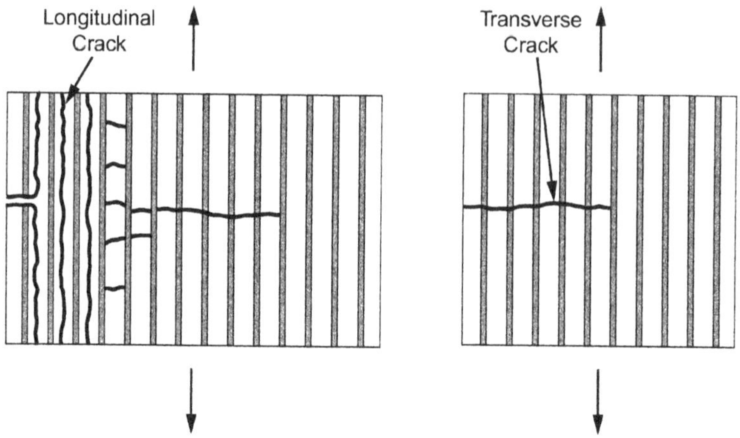

Fig. 8.2 Fatigue damage mechanisms in $Al/Al_2O_{3,f}$ composites: (a) low stress (high cycle fatigue), longitudinal splitting between fibers is the dominant failure mechanism and (b) high stress (low cycle fatigue), a single dominant crack propagates (after Zhang et al., 2003).

been shown to contribute to the development of fatigue cracks in the fibers, leading to fiber fracture (Majumdar and Newaz, 1995).

The fracture morphology is directly related to the fatigue regime of the material, i.e., crack velocity. Figure 8.3 shows the fracture surface of a SiC fiber reinforced Ti alloy matrix composite in the power law regime (also called the Paris regime) and in the fast fracture regime. In the Paris law regime, the localized fatigue striations are observed in the matrix of the composite. In the fast fracture region, the fracture is similar to that observed under monotonic tension, because of the very high crack velocities. Here, void nucleation and growth in the matrix are predominant, along with debonding at the fiber/matrix interface.

Finally, in composite systems where the interfacial bond strength is relatively weak, interfacial wear may contribute to fiber fracture. Interestingly, this behavior has been shown in ceramic matrix composites and can result in significant temperature rise of the composite at higher frequencies (>100 Hz), due to frictional heating (Chawla et al., 1996, Chawla, N., 1997). Walls and Zok (1994) quantified the extent of interfacial wear during fatigue in a fiber pushout test (see Chapter 4). The interfacial strength in the composite with increasing fatigue cycles was significantly lower than the as-received material, with a continuously increasing fiber displacement Fig. 8.4. This indicates that significant wear processes were taking place during fatigue (Walls et al., 1993). Figure 8.5 shows

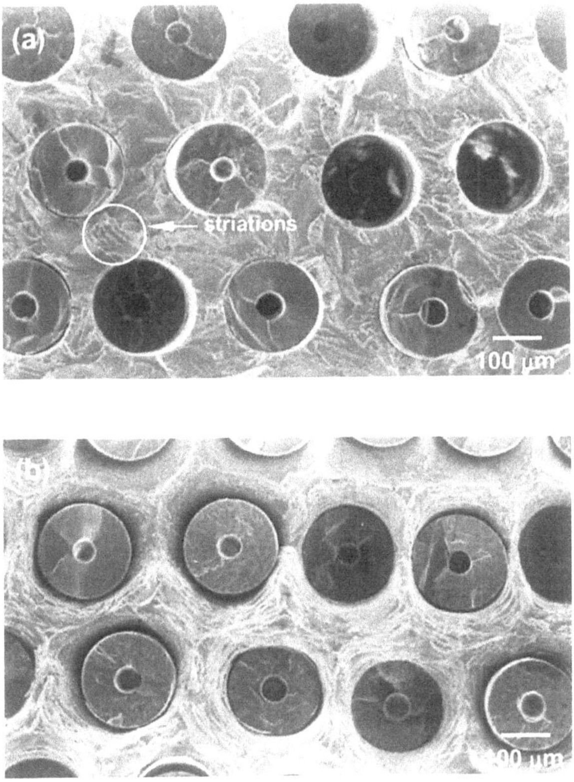

Fig. 8.3 Fatigue fracture in a Ti-β-21s/SiC$_f$(SCS-6) composite: (a) Paris law regime, showing localized striations in the matrix, and (b) fast fracture regime, where the matrix morphology is characteristic of void nucleation and growth (courtesy of J. Liu).

quantitative microstructural evidence of the interfacial wear process during fatigue of a Ti-β-21s (Ti-15Mo-2.7Nb-3Al-0.2Si)/SiC$_f$(SCS-6) composite obtained by *in situ* SEM (Liu and Bowen , 2003).

In addition to the mechanisms described above, elevated temperature also exacerbates fatigue damage. The mechanisms of fatigue damage are very much dependent on the composite system. In large-diameter SiC fibers (such as SCS-6), the surface of the fiber is C-rich. The fibers also contain a central carbon core. Thus, at elevated temperatures, the carbon will react with the matrix, such as Ti, to form hard and brittle carbides at the interface. The brittle carbides can act as fatigue crack initiation sites and catastrophic crack propagation can result. Foulk et al. (1998) also noted that oxygen dissolution of the interface combined with mechanical wear at the interface also degraded fatigue life. It was found that the stress on the fiber increased with

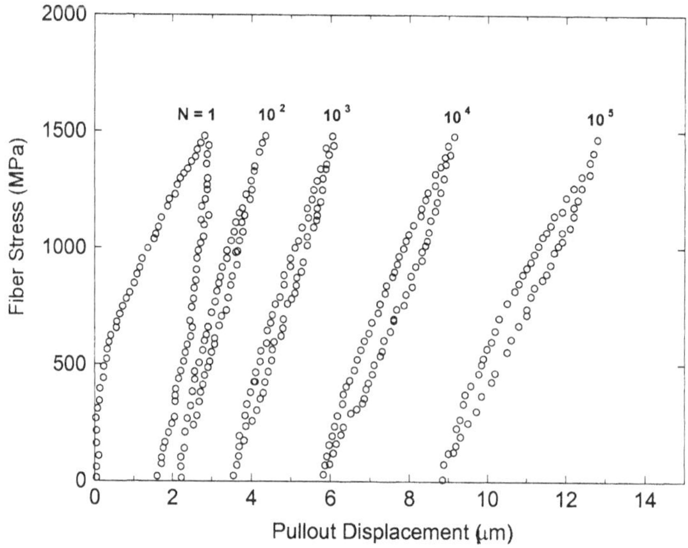

Fig. 8.4 Stress-strain hysteresis of a single fiber SiC fiber (SCS-6) in a Ti alloy matrix during fatigue. The hysteresis loops become wider with increasing cycles, indicating inelastic deformation in the form of wear at the fiber/matrix interface (after Walls and Zok, 1994).

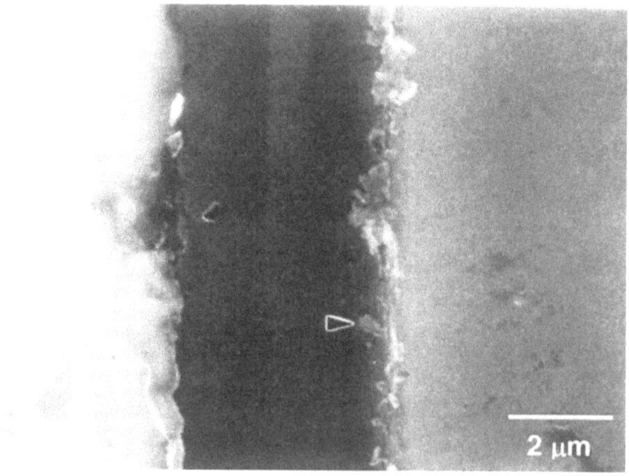

Fig. 8.5 In situ SEM micrograph of interfacial wear process during fatigue (indicated by arrow) of a Ti-β-21s/SiC_f(SCS-6) composite (courtesy of J. Liu).

increasing degradation of the interface. Significant degradation in interfacial shear strength, as measured by fiber pushout, due to interfacial wear and degradation due to elevated temperature fatigue has also been observed (Blatt et al., 1995). The fatigue damage is also controlled by cycling frequency. Lower frequencies allow more time for interfacial reaction to take place, leading to lower fatigue lives (in cycles) in LCF, although in the HCF regime, the fatigue strength is about the same (Mall and Portner, 1992).

Sanders and Mall (1996) investigated the elevated temperature fatigue response (in strain control) of a Ti-15V-3Cr/SiC(SCS-6)/36_f in the transverse orientation. At relatively high strains, termed Region I, the composite had lower fatigue resistance than the matrix material, Fig. 8.6. This was attributed to matrix cracks, originating at the fiber/matrix interface, which resulted in premature failure of the composite, relative to the unreinforced alloy. With decreasing applied strain, Region II, the composite had equivalent transverse fatigue strength as the unreinforced alloy, although the damage mechanisms were quite different. In the monolithic alloy, creep-fatigue mechanisms took place. In the composite however, the interfacial damage took place which gradually progressed into matrix cracking. The matrix cracks, assisted by matrix creep, resulted in failure of the composite.

In non-structural composites, such as filamentary superconductor reinforced MMCs, the degradation of other functional properties is important. Salazar et al. (2004) studied the degradation in critical current density of a $Bi_2Sr_2Ca_2Cu_3O_x$(BSCCO)/Ag-Mg composite superconductor as a function of fatigue cycles at 77 K. Figure 8.7 shows the microstructure of the composite,

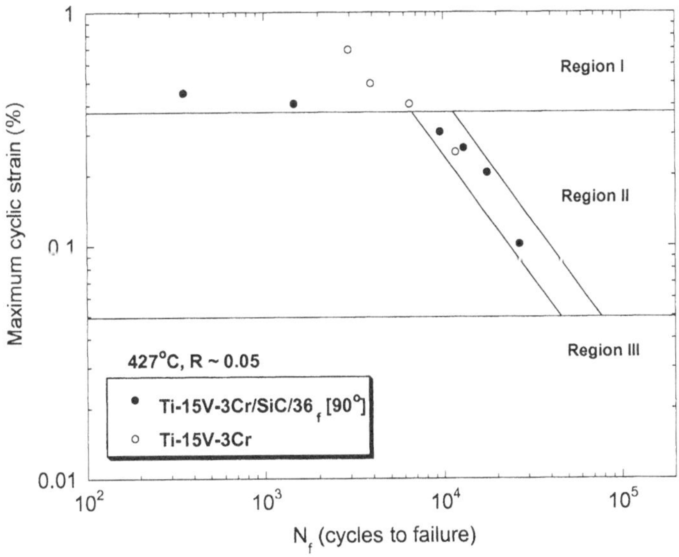

Fig. 8.6 Elevated temperature fatigue response (in strain control) of a Ti-15V-3Cr/SiC(SCS-6)/36_f in the transverse orientation. At relatively high strains, termed Region I, the composite has lower fatigue resistance than the matrix material. With decreasing applied strain, Region II, the composite has equivalent transverse fatigue strength as the unreinforced alloy (after Sanders and Mall, 1996).

which contains some processing-induced longitudinal cracks in the BSSCO filaments. An electric fatigue limit was defined, analogous to the mechanical fatigue limit, which corresponds to the stress below which degradation in electrical properties is negligible. In the BSSCO/Ag-Mg system, this corresponded to between 80-90% of the yield strength of the composites. Fracture of the ceramic superconductor filaments also took place during fatigue, eventually resulting in fatigue failure of the composite. Figure 8.8 shows fatigue microcracks that nucleated in the superconductor sheaths, which are not fully-dense. These cracks propagated until the superconductor/matrix interface was reached. Necking of the matrix between filaments occurred followed by fracture of the composite.

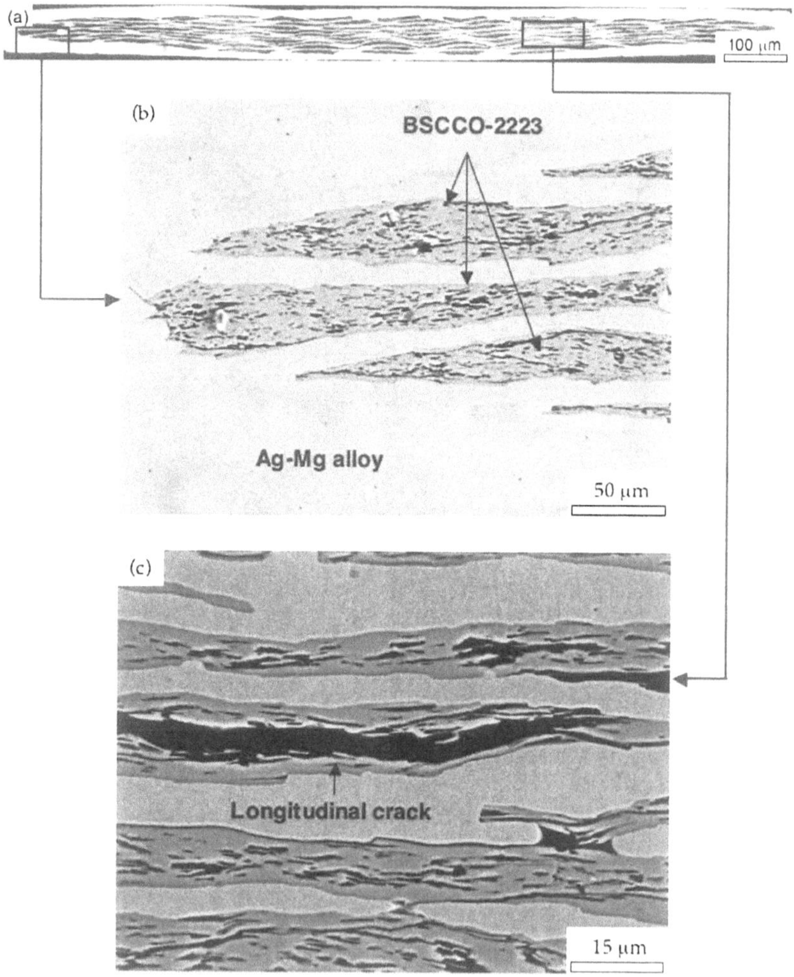

Fig. 8.7 Microstructure of $Bi_2Sr_2Ca_2Cu_3O_x$(BSCCO)/Ag-Mg composite superconductor (courtesy of J. LLorca). The BSSCO filaments contain some processing-induced longitudinal cracks.

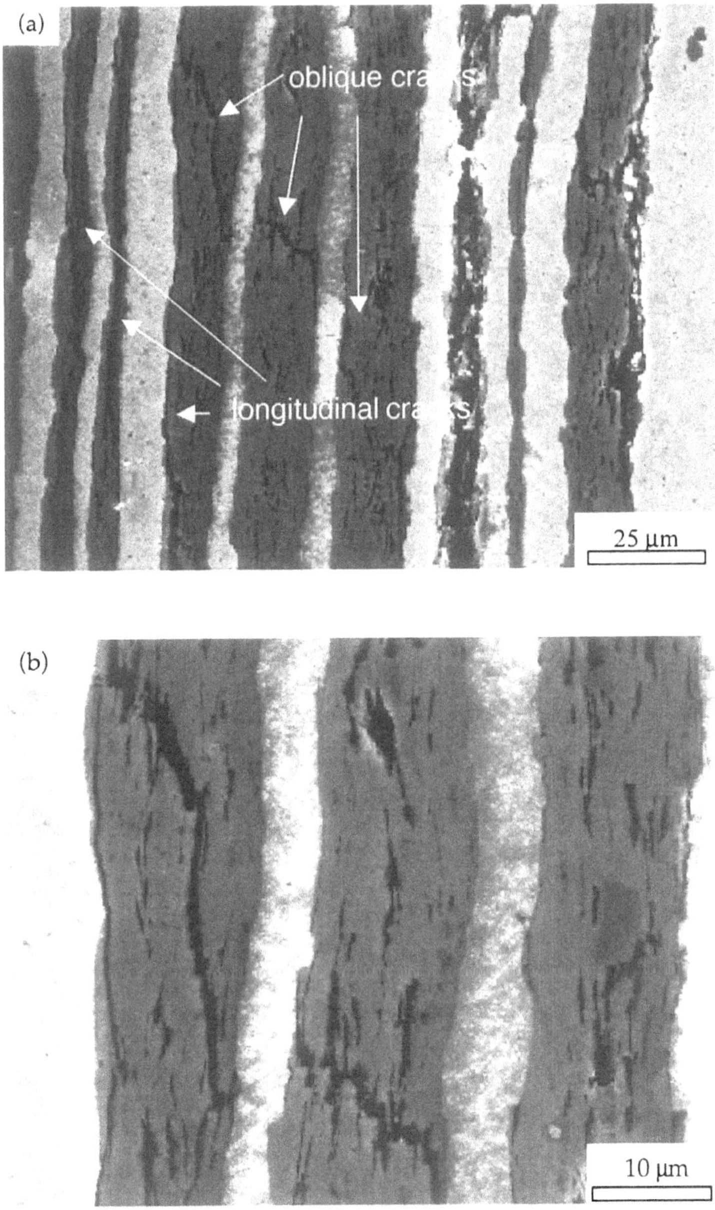

Fig. 8.8 Fatigue microcracks in $Bi_2Sr_2Ca_2Cu_3O_x$(BSCCO)/Ag-Mg composite superconductor: (a) lower magnification and (b) higher magnification (courtesy of J. LLorca). The cracks nucleate in the superconductor sheaths, which are not fully-dense. These cracks propagate until the superconductor/matrix interface is reached.

8.1.2 Stiffness Loss

The major problem in quantifying the stress intensity range during fatigue crack growth of fiber reinforced MMCs is the absence of *one and only* one dominant crack that is propagating. The lack of *self-similar* crack growth, i.e., when the crack propagates in the same plane and direction as the initial crack, is caused by the many modes of damage in MMCs, such as matrix cracking, fiber fracture, interfacial delamination and debonding, void growth, multidirectional cracking, and so on. These modes appear rather early in the fatigue life of composites. One manifestation of such damage is the stiffness loss as a function of cycles. In general, one would expect the scatter in fatigue data of composites to be much greater than that in fatigue of monolithic, homogeneous materials. This is because of the existence of a variety of damage mechanisms in composites. Thus, with continued cycling, there occurs an accumulation of damage. This accumulated damage results in a reduction of the overall stiffness of the composite. Measurement of stiffness loss as a function of cycling has been shown to be quite a useful technique of assessing the fatigue damage in polymer matrix composites (Chawla, 1997). In MMCs, the fatigue behavior of boron fiber and silicon carbide fiber reinforced aluminum and titanium alloy matrix composite laminates having different stacking sequences has been examined using the stiffness-loss measurement technique (Johnson, 1982; Johnson and Wallis, 1986; Johnson, 1988). It was observed that on cycling below the fatigue limit but above a distinct stress range, $\Delta\sigma$, the plastic deformation and cracking (internal damage) in the matrix led to a reduced modulus. Johnson (1988) has proposed a model that envisioned the specimen reaching a "saturation damage state" (SDS) during constant amplitude fatigue testing.

Gomez and Wawner (1988) also observed stiffness loss in silicon carbide/aluminum composites subjected to tension-tension fatigue ($R = 0.1$) at 10 Hz. The Young's modulus was measured at periodic intervals between cycles. The SCS coating fractured at high cycles, and the fracture surface showed the coating clinging to the matrix. Sanders and Mall (1996) also observed a distinct decrease in Young's modulus and maximum stress under strain-controlled fatigue of Ti-15V-3Cr/SiC(SCS-6)36_f, under transverse loading, Fig. 8.9. Two distinct stages of damage were observed. In stage I, the damage in modulus and stress was relatively stable. In stage II, however, significant damage, likely due to interfacial fracture, resulted in a significant decrease in the Young's modulus and the stress carried by the composite.

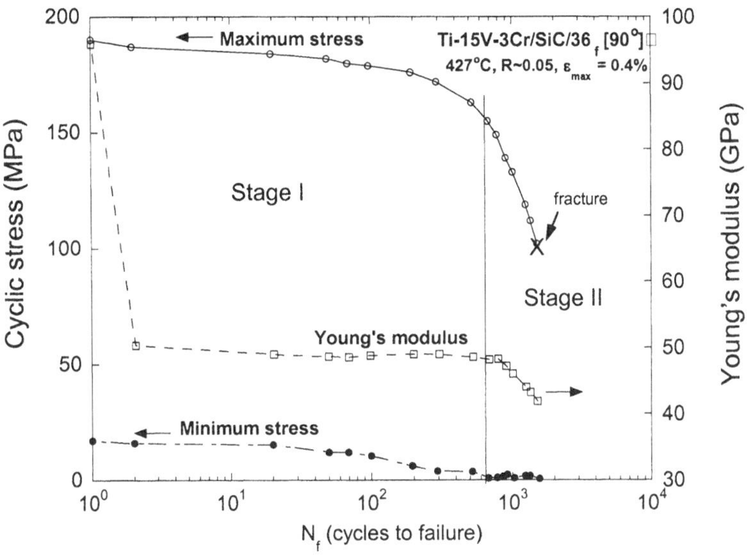

Fig. 8.9 Decrease in Young's modulus and maximum stress in strain-controlled fatigue of Ti-15V-3Cr/SiC(SCS-6)36_f, under transverse loading (after Sanders and Mall, 1996). In stage I, the damage in modulus and stress is relatively stable. In stage II, however, significant damage, likely due to interfacial fracture, resulted in a significant decrease in the stress carried by the composite and the Young's modulus.

8.1.3 Particle Reinforced MMCs

The use of a high stiffness ceramic reinforcement in particulate form can result in a substantial increase in fatigue resistance while maintaining cost at an acceptable level. The fatigue resistance of particulate MMCs depends on a variety of factors, including reinforcement particle volume fraction, particle size, matrix and interfacial microstructure, the presence of inclusions or defects that arise from processing, and testing environment (Chawla and Shen, 2001; Chawla and Allison, 2001; Lewandowski, 2000; LLorca, 2002; Ganesh and Chawla, 2004). The effect of these factors on the fatigue behavior of particle reinforced MMCs is summarized in this section.

Several studies have shown that increasing volume fraction and decreasing particle size both result in enhanced fatigue resistance (Hall et al., 1994; Poza and LLorca, 1999; Chawla et al., 1998a; Chawla et al., 2000a; Chawla et al., 2000b). In the composite, most of the load is carried by the high modulus, high strength reinforcement, so for a given stress, the composite undergoes a lower average strain than the unreinforced alloy. Thus, the fatigue lives of particle reinforced metal matrix composites are generally superior to those of unreinforced metals, Fig. 8.10. These improvements are

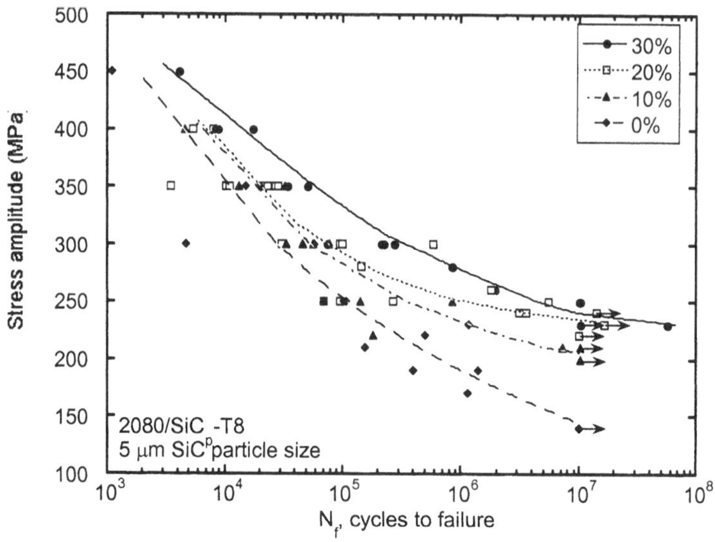

Fig. 8.10 Effect of SiC volume fraction on the stress versus cycles fatigue behavior of 2080 Al/SiC$_p$ composites. Increasing volume fraction results in higher fatigue life (after Chawla et al., 1998).

most pronounced at low cyclic stresses, i.e., in the high cycle fatigue regime, while at higher cyclic stresses the differences between reinforced and unreinforced materials are reduced. This is called "ductility exhaustion" of the composites, which takes place in the low cycle fatigue regime. The higher ductility of the unreinforced alloy in this regime contributes to higher fatigue life than that of the composite. With decreasing particle size, for a given reinforcement volume fraction, the reinforcement interparticle spacing decreases, resulting in more barriers to the reversible slip motion that takes place during fatigue, and a decrease in strain localization by cyclic slip refinement. Above a critical particle size, reinforcement fracture is predominant and contributes to a low fatigue life, because of the increased propensity for particle cracking as the particle size increases (Chawla et al., 1998a). Narrowing of the particle size range distribution also results in higher fatigue life, particularly when eliminating larger particles that are more prone to cracking (Couper and Xia, 1991).

The cyclic stress-strain behavior of the composite is also affected by the onset of microplasticity at relatively low stress (see chapter 7). Chawla et al., (1998b) compared the cyclic stress-strain behavior of a 2080/SiC/30$_p$ composite with that of the unreinforced alloy. The composite exhibited microplasticity at relatively low stresses, presumably due to stress concentrations at SiC particles and localized plasticity at the poles of the reinforcement, 8.11(a). The cyclic stress-strain curve of the 2080 Al alloy at the same stress was elastic in nature, although it was more compliant than the composite. At an applied stress higher than the yield strength of the unreinforced alloy, the hysteresis loop for the unreinforced alloy was much larger than that of the composite, 8.11(b). It should be noted that the onset of cyclic microplasticity in the composite at low stress did not appear to affect the fatigue life of the composite in a detrimental fashion. Rather, the composites had higher fatigue life than the unreinforced alloy. Chawla and Shen (2001) used a unit cell finite element model, consisting of spherical particles arranged in a staggered arrangement, shown in the inset of Fig. 8.12. The cyclic stress-strain hysteresis obtained from this simple model yielded surprisingly good agreement with the experimental behavior, see Fig. 8.12. The effects of reinforcement aspect ratio, shape, and matrix hardening characteristics on cyclic fatigue have also been elucidated (LLorca et al., 1992; LLorca, 1994).

The cyclic stress-strain behavior of MMCs is also affected by a phenomenon called the Bauschinger effect. The Bauschinger effect is defined as a decrease in flow stress that occurs when we reverse the loading direction, e.g., from a tensile to compressive stress or vice versa (Bauschinger, 1886). Thus, an understanding of the Bauschinger effect is essential to understanding of work hardening and cyclic loading.

One of the origins of the Bauschinger effect in monolithic materials has been attributed to dislocation pileup. Pileup of dislocations, in the forward loading direction, creates a "back-stress" which aids deformation upon reversal in loading, resulting in a lower flow stress (Mott, 1952; Seeger et al., 1958). In precipitation-hardened materials, Orowan loops associated with the precipitate particles also induce a back-stress (Orowan, 1959). The phenomenon has also been observed in particle reinforced MMCs (Arsenault and Wu, 1987; LLorca et al., 1990). Here, the back-stress is enhanced by tensile residual stresses that are generated upon cooling from the processing temperature, as well as constrained deformation of the matrix by the rigid particles. This results in a loading asymmetry (Arsenault and Wu, 1987; Arsenault and Pillai, 1996) where a more significant Bauschinger effect was observed under compression-tension sequence than for a tension-compression sequence.

Indeed, finite element modeling shows that even when the matrix is assumed to exhibit isotropic hardening behavior (i.e., no Bauschinger effect exists), the composite shows a distinct Bauschinger effect upon reversed loading (LLorca et al., 1990), as was observed experimentally. An examination of the evolution of local stress field showed an apparent early reversed yielding for the composite, due to non-uniformity of deformation in the matrix caused

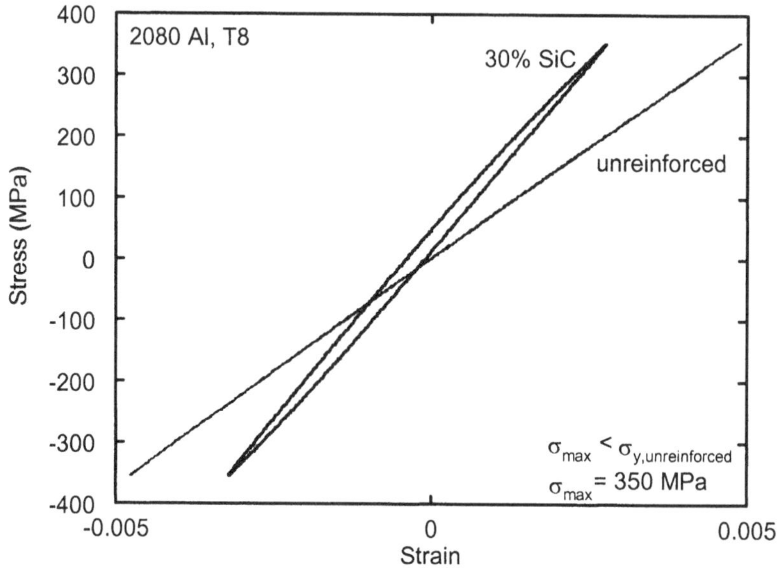

(a)

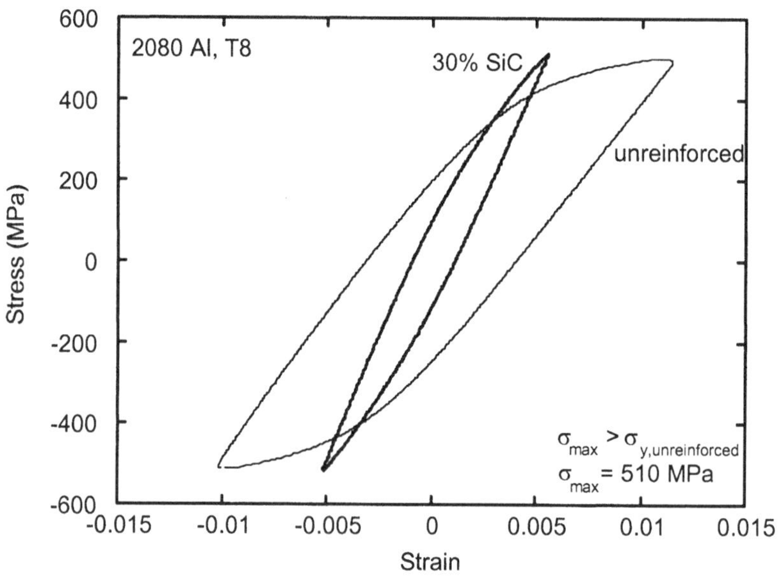

(b)

Fig. 8.11 Comparison of cyclic stress strain behavior of a 2080/SiC/30p composite and 2080 Al unreinforced alloy (Chawla et al., 1998b): (a) low cyclic stress ($\sigma_{max} < \sigma_{y,unreinforced}$), and (b) high cyclic stress ($\sigma_{max} > \sigma_{y,unreinforced}$). The composite exhibits microplasticity at very low stress, due to localized plasticity at the poles of the SiC reinforcement and at sharp particle corners.

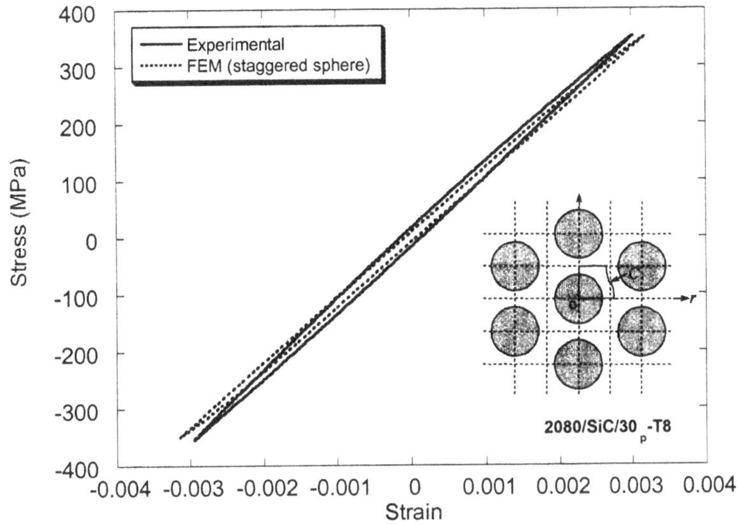

Fig. 8.12 Cyclic stress-strain hysteresis behavior of a 2080/SiC/30_p composite (after Chawla and Shen, 2001). A finite element model consisting of spherical, staggered SiC particles in Al predicts the experimental behavior very well.

by the constraint imposed by the brittle reinforcement (Shen et al., 1995). Thus, high *local* effective stresses trigger early *local* yielding after the load is reversed, which is reflected in the macroscopic stress-strain behavior.

In addition to particle reinforcement, the matrix microstructure also significantly influences the fatigue behavior of the composite. Factors affecting the matrix microstructure include size, shape, and spacing of precipitates, grain size, and non-reinforcement dispersoids or inclusions (such as Fe-rich inclusions in aluminum that are commonly formed during processing). It would appear that with regard to grain size, composites follow the same trend as monolithic materials, i.e., for a given matrix alloy composition and volume fraction of reinforcement, finer grain sizes generally result in improved properties. Contrary to conventional wisdom for monolithic materials, in MMCs high matrix yield and ultimate tensile strength do not necessarily reflect high fatigue strength (defined here as fatigue runout at 10^7 cycles). Vyletel et al. (1991) showed that there was no significant difference in fatigue behavior between naturally aged and artificially aged MMC, even though the naturally aged material had a much lower yield and ultimate strength. Chawla et al. (2000b) compared two materials with constant reinforcement volume fraction and particle size, but very different microstructures. A thermomechanical treatment (T8) of the Al-Cu-Mg alloy produced a fine and homogeneous distribution of S' precipitates, while a thermal treatment only (T6) resulted in coarser and inhomogeneously distributed S' precipitates. Because of finer and more closely spaced precipitates, the composite that underwent a T8 treatment exhibited higher yield strength than the T6 material. Despite its lower yield strength, however, the T6 matrix composites exhibited higher fatigue resistance than the T8 matrix composites. This contrasting behavior between monotonic and cyclic loading can be attributed to the strong influence of the presence, stability, and morphology of the S' precipitates in the matrix of the composite (Calabrese and Laird, 1974a; Calabrese and Laird, 1974b; Starke and Luetjering, 1979). In fatigue, failure processes are affected by a variety of microstructural factors, which include resistance to dislocation motion and possible dislocation pileup at precipitates and/or reinforcement particles and cracking along slip bands. In aluminum alloys, small precipitates with a coherent interface are easily sheared (Calabrese and Laird, 1974a; Calabrese and Laird, 1974b). Coarser precipitates, which have a semi-coherent or incoherent interface with the matrix, cause the dislocation to loop around the precipitate. In the T8 materials, the precipitates are fine enough that it is believed that the precipitates are cut by dislocations and precipitate slip bands (PSBs) are formed during fatigue, which reduces the strengthening effect of precipitates, thereby impairing the fatigue strength. In the T6 materials, the larger precipitate size allows them to retain their precipitates structure and strength during fatigue.

Overaging heat treatments also modify the matrix microstructure, resulting in coarsening of the precipitate structure, while retaining a homogeneous precipitate distribution, which directly influences fatigue life (Chawla et al., 2000b). Figure 8.13 shows the coarsening and increase in precipitate spacing in the matrix of MMCs overaged at various temperatures for 24 hours. Increasing precipitate spacing decreases both fatigue strength and fatigue lifetime, Fig. 8.14. This is to be expected since coarser precipitates result in a larger interprecipitate spacing and easier bypass of dislocations. For the composites subjected to higher overaging temperatures, the yield strength and fatigue strength also decrease with increasing precipitate spacing. It is important to realize that precipitate size alone should not be taken as the determining factor for fatigue resistance. Rather, the precipitates should be of sufficient size not to be susceptible to precipitate shearing, but should be semi- or completely coherent with the matrix to impede dislocation motion.

Processing-related defects in the form of intermetallic inclusions or particle clusters are also part of the matrix microstructure, and can play a role in fatigue strength, particularly in powder metallurgy processed materials (Chawla et al., 1998a; Chawla et al., 2000a; Li and Ellyin, 1996). Figure 8.15 shows a fracture surface of a 2080/SiC/20_p composite (Chawla et al., 1998a). The crack initiates at an Fe-rich inclusion. This is followed by a

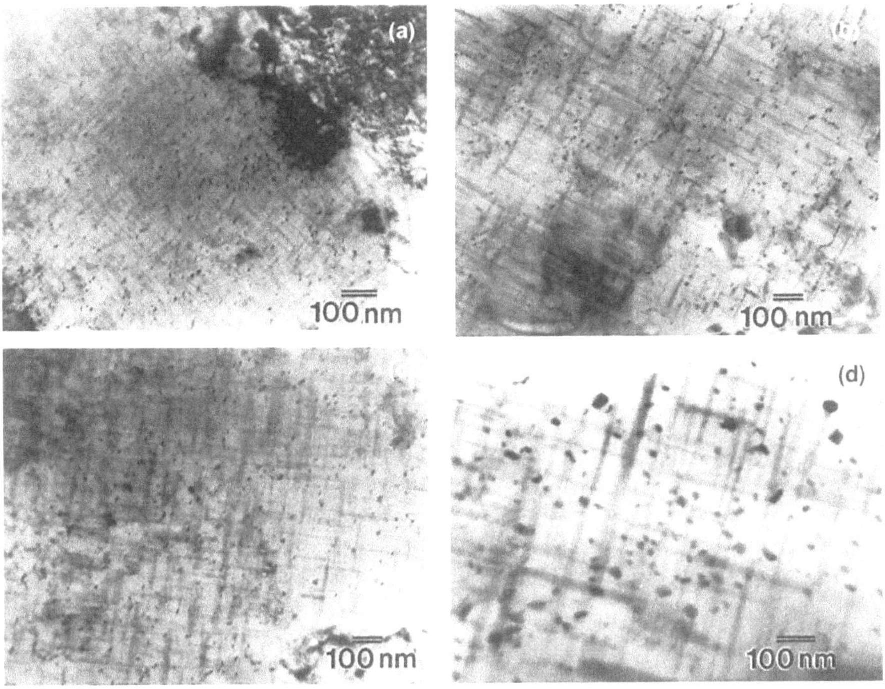

Fig. 8.13 Coarsening of S' precipitates due to overaging: (a) solution treated, rolled, and aged (T8), (b) T8+24 h at 200°C, (c) T8+24 h at 225°C, and (d) T8+24 h at 250°C (Chawla et al., 2000a).

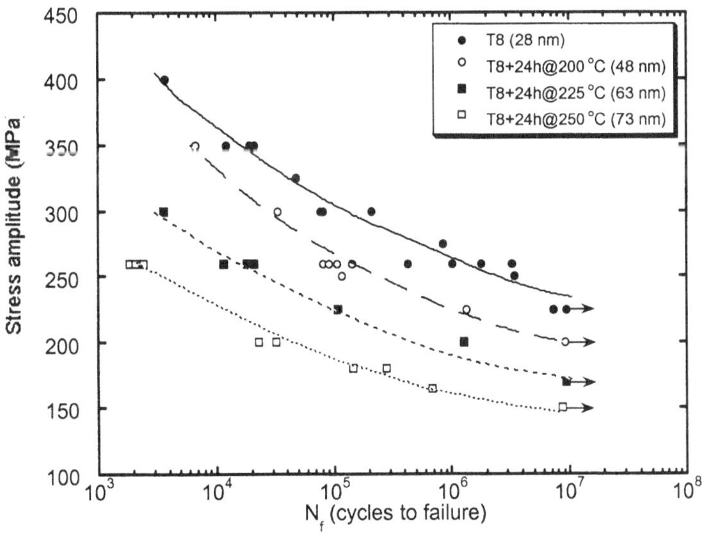

Fig. 8.14 Influence of matrix precipitate spacing on fatigue life of 2080/SiC/20_p-T6 composite. Increase in precipitate spacing results in a reduction in fatigue life (after Chawla et al., 2000a).

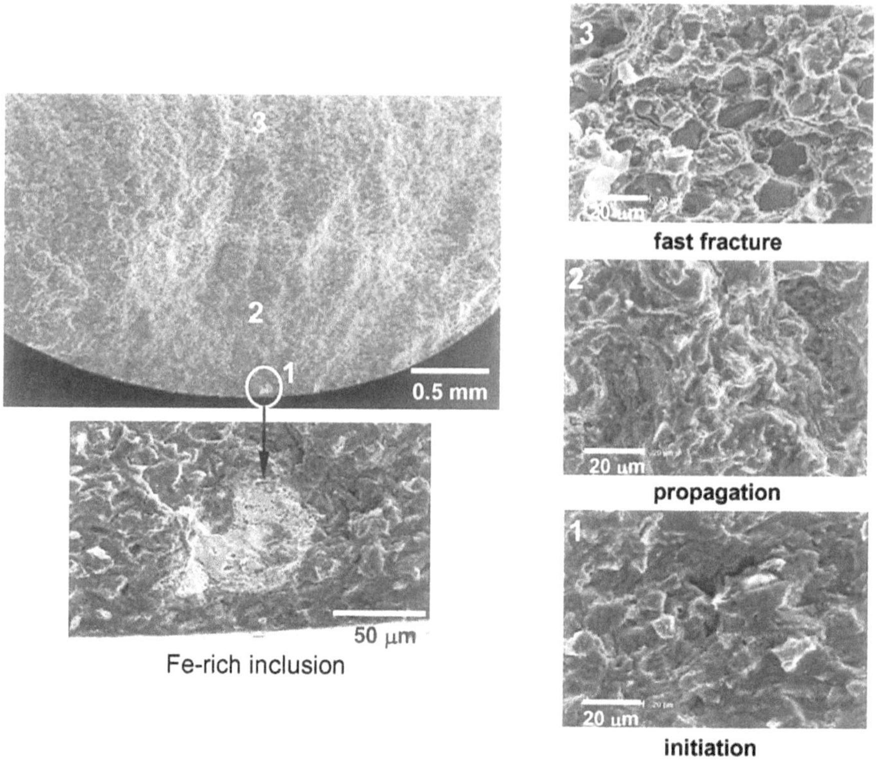

Fig. 8.15 Fatigue fracture morphology in a 2080/SiC/20_p-T6 composite consisting of initiation, propagation, and fast fracture regions (after Chawla et al., 1998a).

region of stable crack propagation, followed by fast-fracture. The last stage is characterized by a large amount of particle fracture, due to the high crack velocity. The defects (e.g., inclusions, particle clusters) act as stress concentrators and increase the local stress intensity in the material, promoting easy crack nucleation. For a given inclusion size, the stress concentration in a composite, where the inclusion is surrounded by high stiffness reinforcement particles, is lower than in the unreinforced alloy. Since more of the load is being "shared" by the high stiffness SiC particles in the composite, an inclusion in the composite will be subjected to lower stress than a similar inclusion in the unreinforced alloy.

In extruded composites, the overall size of inclusions is also lower since the ceramic reinforcement particles break the brittle inclusions into smaller sizes during extrusion, Fig. 8.16, (Chawla et al., 2000a). It is interesting to note that in the low cycle regime, cracks seem to originate relatively early in fatigue life (around 10% of total life) (Chawla et al., 1998a; Lukasak and

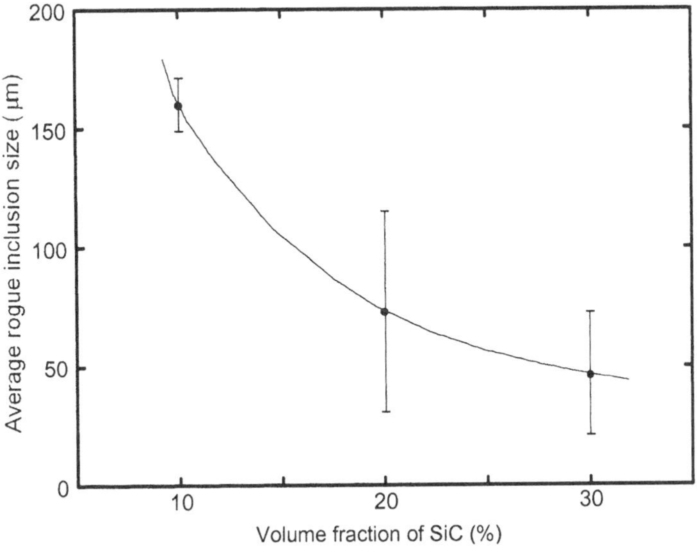

Fig. 8.16 Average "rogue inclusion" size as a function of SiC particle volume fraction. The hard particles contribute to the beneficial fracture and comminution of the Fe-rich inclusions during processing (after Chawla et al., 2000b).

Koss, 1993). In the high cycle regime, on the other hand, crack initiation can occur quite late (after about 70-90% of the life of the specimen). While crack growth is relatively unimpeded in unreinforced materials, crack growth is hindered by mechanisms such as crack deflection and crack trapping in the composite. A more detailed discussion of crack growth in these materials is given in the next section.

Several MMC applications require fatigue resistance at elevated temperatures (on the order of 150-175°C). Figure 8.17 shows the elevated temperature fatigue performance of MMCs in comparison with the room temperature data shown in Fig. 8.10 (Chawla et al., 1999). The most significant debit in fatigue strength occurred between 25 and 150°C. The incremental increase in temperature to 170°C resulted in a modest decrease in fatigue strength. In applications where elevated temperature fatigue resistance is a criterion, the temperature at which the composite is aged also becomes very important. At temperatures slightly higher than the aging temperature, a severe decrease in fatigue strength may take place because of significant matrix overaging. The fatigue strength in the composites seems to be directly proportional to the strength of the matrix, although the decrease in fatigue strength due to temperature is significantly higher than the decrease in yield strength. This may be caused by changes in matrix microstructure and decrease in matrix strength from a combination of long-term exposure and cyclic stress. The fatigue fracture morphology after

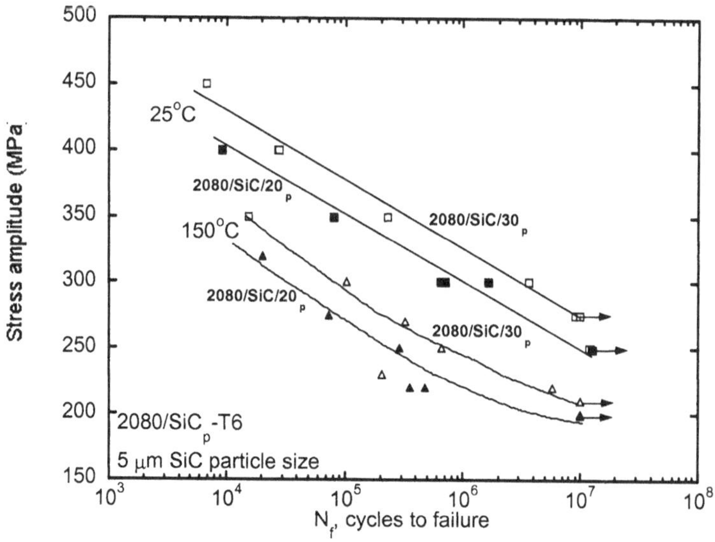

Fig. 8.17 Elevated temperature fatigue of 2080/SiC_p composites. A slight decrease in fatigue resistance is observed at elevated temperature (after Chawla et al., 1999).

elevated temperature fatigue is quite different from that at room temperature. At elevated temperature, while crack initiation takes place at the inclusions, fractography also indicates evidence of interfacial decohesion and void growth at the particle/matrix interface and particle corners, as well as in the matrix of the composite, Fig. 8.18. It appears that microvoid nucleation and coalescence in the matrix have also taken place prior at fracture.

Temperatures during metal cutting can rise to as much 1200°C (Kindermann et al., 1999). Tungsten carbide/cobalt composites, commonly referred to as cemented carbides or hardmetal, are used for cutting a range of materials under environments involving aggressive chemicals, high-temperatures, lubricants, and water-based coolants. Cyclic fatigue behavior of such materials is of importance because cyclic loading in these materials occurs during the alternating slip and stick of the metal chips during machining, vibrations of the machines, and the interrupted cutting process (Almond and Roebuck, 1980; Roebuck et al., 1984; Kindermann et al., 1999; Pugsley and Sockel, 2004). An example of an S-N curve for WC/6 wt.% Co composite in air and tannic acid (to simulate wood cutting) is shown in Fig. 8.19. The evidence of corrosion fatigue effect at low stress amplitudes is clear. Figure 8.20 shows a strong decrease in fatigue behavior at 700°C vis-à-vis room temperature. Also shown in the figure are the inert strength values of these composites, which are significantly higher than the cyclic strength values at all cycles. This indicates the loss of strength is not due to surface oxidation

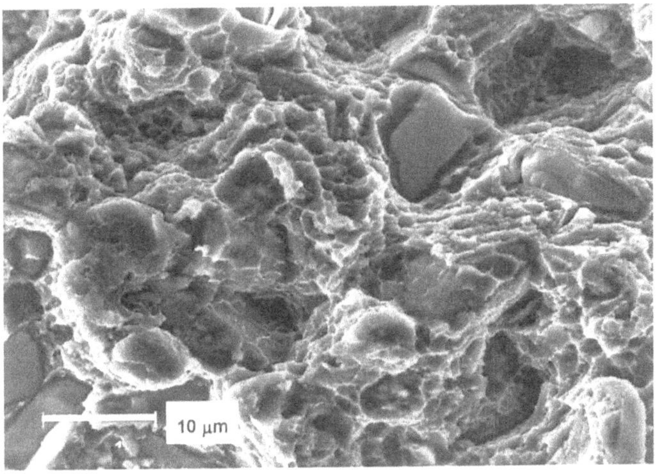

(a)

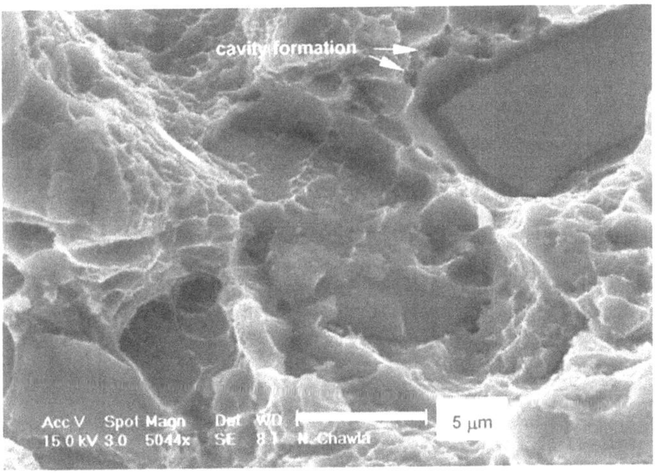

(b)

Fig. 8.18 Elevated temperature fatigue fracture morphology, showing (a) dimpled fracture and interfacial decohesion and (b) cavity formation in the matrix (after Chawla et al., 1999).

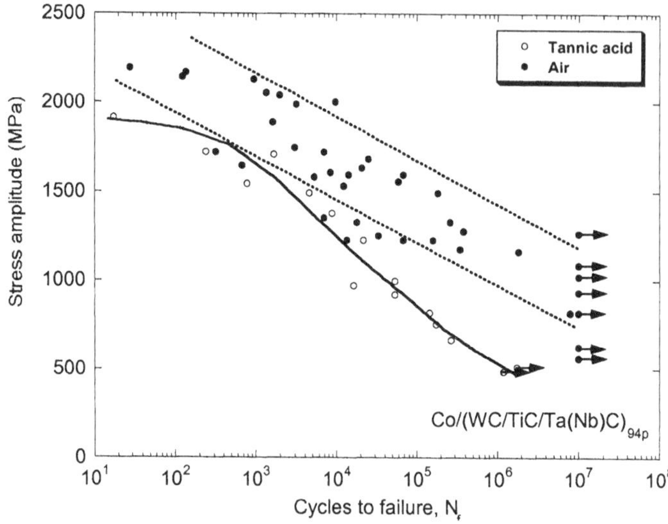

Fig. 8.19 S-N curve for WC/Co composites in air and in tannic acid at 4 Hz (after Pugley and Sockel, 2004). The effect of corrosion fatigue effect, at low stress amplitudes, is clear.

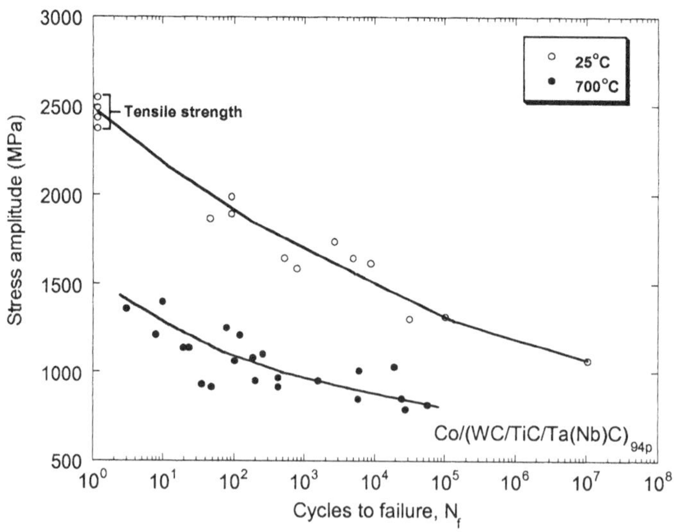

Fig. 8.20 S-N curves for WC/Co composites at room temperatures and at 700°C (after Kindermann et al., 1999).

of WC at high temperatures but a true high-temperature cyclic fatigue phenomenon.

8.2 FATIGUE CRACK GROWTH

In this section, we describe the fatigue crack growth behavior of MMCs. A brief summary of the salient features of fatigue crack growth in engineering materials, including a description of fatigue crack closure and the two parameter approach to fatigue, is given in an inset.

8.2.1 Continuous Fiber Reinforced MMCs

In general, the fibers provide a crack-impeding effect, but the nature (morphology, rigidity, and fracture strain) of the fiber, the fiber/matrix interface, and/or any reaction-zone phases at the interface can have a great influence on fatigue crack growth processes (Chawla, 1991). In composites with a relatively high interface strength, the fatigue crack propagates straight through the fibers, Fig. 8.21(a). With a weak interface, fiber/matrix debonding, crack deflection, fiber bridging, and fiber pullout take place, Fig. 8.21(b) and 8.22. This leads to a significant amount of energy-dissipation during crack growth which results in macroscopic toughening of the composite.

In general, fiber reinforced MMCs exhibit superior fatigue crack growth resistance to monolithic alloys. An example is the case of boron fiber (coated with B_4C) reinforced Ti-6Al-4V composite and the parent matrix alloy, shown in Fig. 8.23 (Harmon et al., 1987). Note that the MMC has a higher fatigue threshold and a much lower rate of crack growth through most of the fatigue life. Soumeldis et al. (1986) observed lower fatigue-crack growth

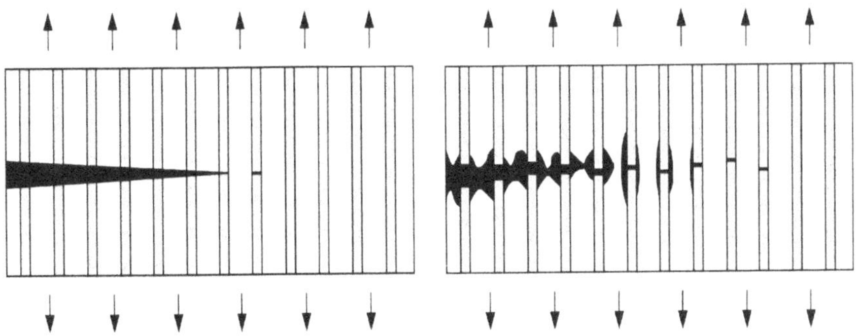

Fig. 8.21 Schematic of damage mechanisms in continuous fiber reinforced MMCs: (a) strong interface – fatigue crack propagates straight through the fibers; and (b) weak interface – fiber/matrix debonding, crack deflection, fiber bridging, and fiber pullout take place.

Fatigue Crack Growth in Engineering Materials

As described in the beginning of the chapter, the fatigue crack growth behavior of engineering materials is characterized by a plot of log da/dN versus log ΔK.. A three-region curve is obtained. Regime 1 is often denoted as the threshold regime, to indicate a threshold stress intensity factor, ΔK_{th} (typically corresponding to $\sim 10^{-10}$ m/cycle), for crack growth. With increasing ΔK, a linear portion on the log-log plot is observed, which is called the Paris regime. At even higher ΔK, the crack growth rate increases rapidly, resulting in fracture of the material.

In order to fully understand the crack growth behavior, let us define some important parameters. The nominal cyclic stress intensity factory range, ΔK, is equal to:

$$\Delta K = K_{max} - K_{min}$$

where K_{max} and K_{min} are the maximum and minimum stress intensity factors, respectively, obtained from the maximum and minimum applied stress, σ_{max} and σ_{min} and the corresponding crack lengths.

Fatigue Crack Closure

During the cyclic process, premature "closure" of the crack may take place, at a stress intensity value before reaching K_{min}. Crack closure during fatigue has been attributed to many origins, including plasticity at the crack tip, oxidation, crack surface roughness, and by wedging of asperities in the crack mouth. These are shown in the figure on the following page.

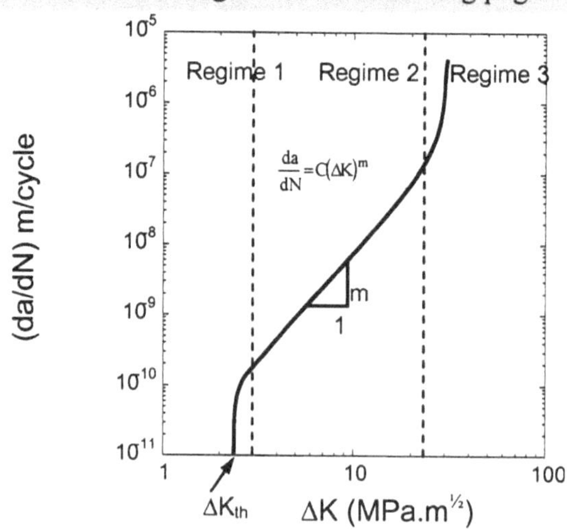

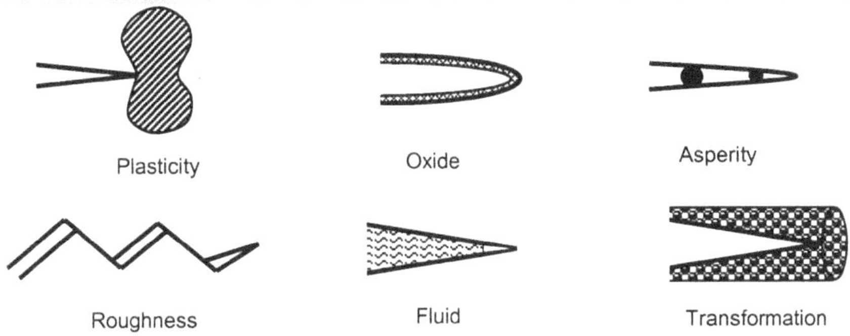

Because of premature closure of the crack faces, the driving force for crack growth is smaller than ΔK. Thus, we define an additional parameter, which corresponds to the stress intensity factor required to open the crack or K_{cl}. Now, the *effective* stress intensity factor, ΔK_{eff} (as shown schematically below), is given by:

$$\Delta K_{eff} = K_{max} - K_{cl}$$

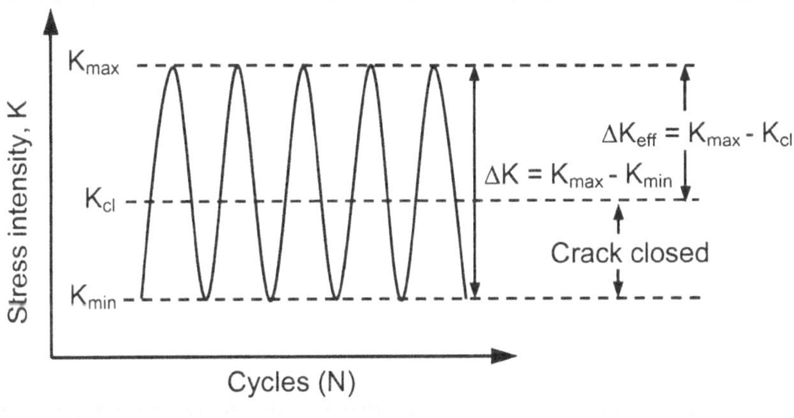

Two-Parameter Approach to Fatigue

Fatigue crack growth behavior can also be analyzed using a two-parameter approach (Doker and Marci, 1983; Vasudevan et al., 1994). The basis of this analysis is that there are two driving forces for fatigue crack growth: (a) static parameter or K_{max} and (b) cyclic parameter or ΔK. By conducting experiments at several R-ratios (K_{min}/K_{max} or $\sigma_{min}/\sigma_{max}$), the ΔK_{th} is obtained for a range of K_{max}. By plotting ΔK_{th} versus K_{max}, an 'L' shaped curve is obtained. The vertical and horizontal asymptotes of the curve correspond to ΔK^*_{th} and K^*_{max}. The physical meaning of these two parameters is that critical values of both the static and cyclic driving forces must be met *simultaneously* for fatigue crack growth to take place. This is shown by the shaded region in the curve.

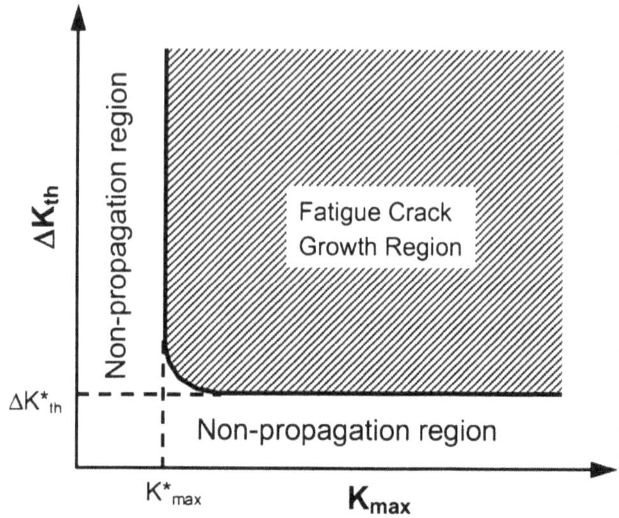

Using the two-parameter analysis, several important materials can be subdivided into different classes (Vasudevan and Sadananda, 1995). It is shown that environmental effects also play an important role on fatigue crack growth behavior.

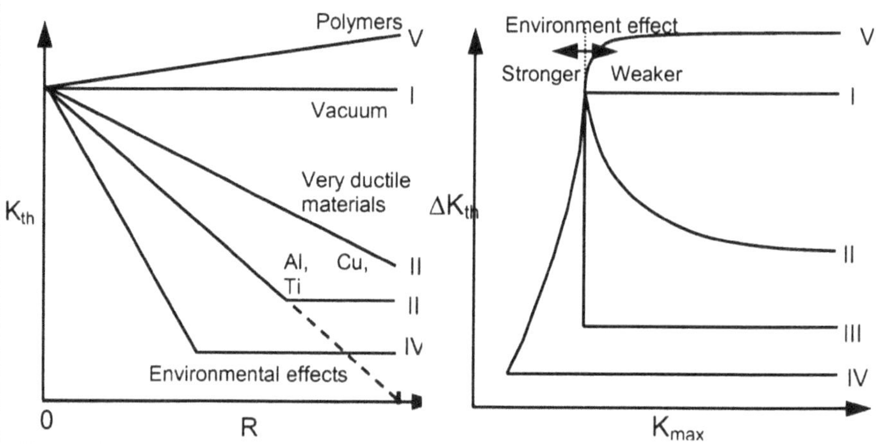

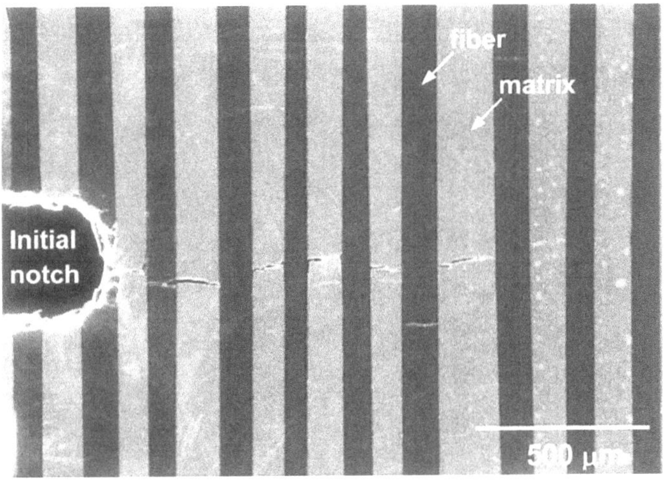

Fig. 8.22 Fatigue crack growth in a Ti-β-21s/SiC_f (SCS-6) composite exhibiting interfacial debonding, crack deflection, and fiber bridging (courtesy of J. Liu).

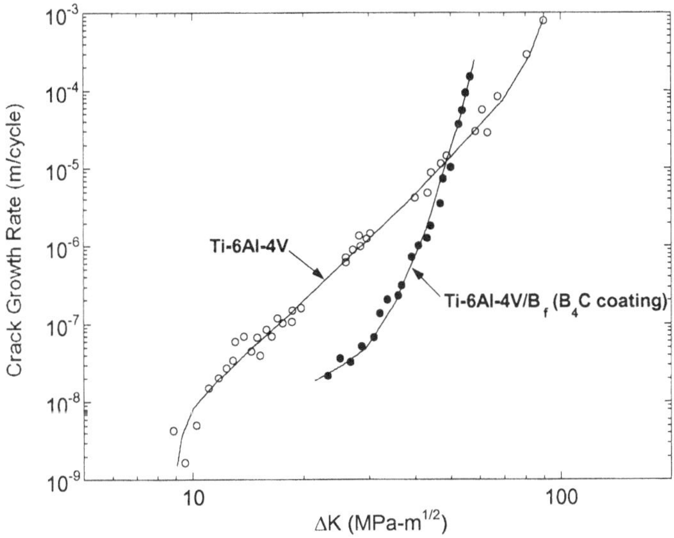

Fig. 8.23 Fatigue crack growth behavior of a boron fiber (coated with B_4C) reinforced Ti-6Al-4V composite and the parent matrix alloy (after Harmon et al., 1987). The fiber reinforced MMC has superior fatigue crack growth resistance to the monolithic alloy.

rates in fiber reinforced Ti-6Al-4V compared with that in the unreinforced alloy. Long isothermal exposures at 850°C, however, resulted in reduced crack growth resistance of the MMC. This was due to fiber degradation, fiber/matrix debonding, and an increase in matrix brittleness. Short-time isothermal exposures (up to about 10 h for Ti-6Al-4V/B_f, 30 h for Ti-6Al-4V/B_f (B_4C), and 60 h for Ti-6Al-4V/SiC_f(SCS-6)) improved the fatigue-crack resistance. This was attributed to an energy-dissipating mechanism of fiber microcracking in the vicinity of the crack tip.

Cotterill and Bowen (1993) studied the fatigue crack growth behavior of SCS-6 fiber reinforced Ti-15V-3Al-3Sn-3Cr (Ti-15-3) composites. In general, the crack growth rates decreased with increasing crack length. This was attributed to crack tip "shielding" from the bridging fibers, which are stronger and stiffer than the matrix. When a single fiber fracture took place, however, the crack growth rate increased by two orders of magnitude. This was also accomplished by a single cycle overload, or by increasing the R-ratio (e.g., an increase in K_{min}). Cotterill and Bowen (1996) examined the effects of ΔK and K_{max}, the cyclic and static stress intensity components for fatigue crack growth, respectively. It was also shown that K_{max} controlled fiber fracture, but ΔK controlled the matrix crack growth rate. Thus, if fiber fracture takes place, K_{max} is the parameter that controls the crack growth rate. The crack growth rate was also shown to be dependent on ΔK. Below a critical value of ΔK, the crack arrested, but above that value, fiber fracture took place and the crack was able to propagate, Fig. 8.24. The transition from crack arrest to catastrophic failure is also affected by fiber strength distribution and specimen geometry (Liu and Bowen, 2002).

At very high temperatures in S-N fatigue, frequency plays a role in fatigue crack growth behavior. At lower frequencies, aging of the matrix and weakening of grain boundaries contribute to fracture in the composite (Cotterill and Bowen, 1993). Stress relaxation of the matrix during creep or due to hold times (or lower cyclic frequency) may also result in crack closure and aid in crack bridging (Zhang and Ghonem, 1995).

Rao et al. (1993) examined the fatigue crack growth behavior of $6061/SiC/40_f$ (SCS-8) composites in longitudinal and transverse orientations. Their results, shown in Fig. 8.25, indicate that the crack growth resistance of the composite was higher than that of the unreinforced alloy in the longitudinal orientation, but lower in the transverse orientation. Fiber debonding and crack deflection mechanisms toughened the composites in the L-T orientation, due to the relatively weak bonding between fiber and matrix provided by the C-rich coating on the SCS-8 fibers. These authors also proposed a mechanistic model for fatigue crack growth in the transverse orientation, whereby fiber debonds grew until a critical point where individual debonds were linked, resulting in fracture, Fig. 8.26. These authors also attempted to quantify the extent of bridging by the *ductile phase*, during transverse loading. The contribution from bridging, ΔK_b, was given by:

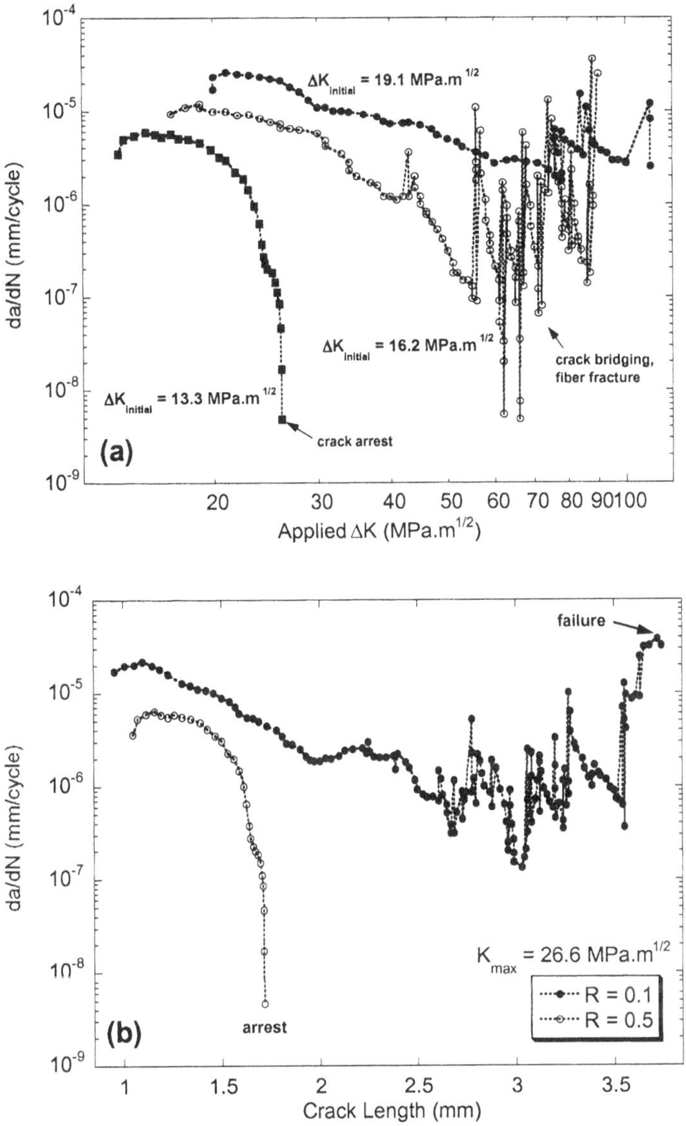

Fig. 8.24 Fatigue crack growth behavior of a SiC fiber reinforced Ti alloy composite: (a) influence of increasing ΔK and (b) influence of R-ratio. Critical values of K_{max} and ΔK are required to propagate the crack (after Cotterill and Bowen, 1996).

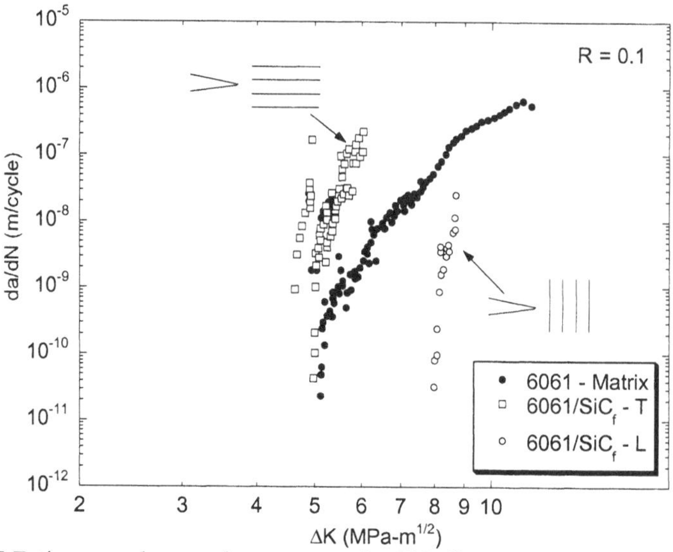

Fig. 8.25 Fatigue crack growth response of a SiC fiber reinforced Al 6061 matrix composite. The fatigue threshold in the longitudinal orientation is superior to that in the transverse orientation and that of the unreinforced alloy (after Rao et al., 1993).

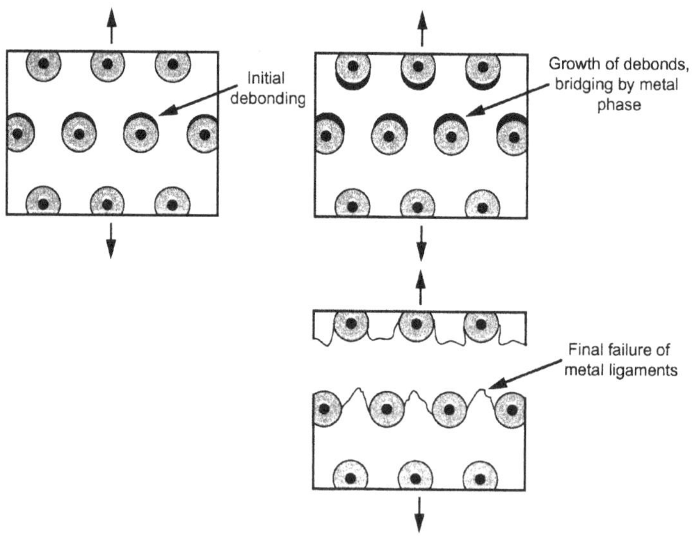

Fig. 8.26 Schematic of mechanisms for fatigue crack growth under transverse loading. Fiber debonds grow until a critical point where individual debonds are linked, resulting in fracture (after Rao et al., 1993).

$$\Delta K_b = \frac{2}{\sqrt{\pi a}} \int_0^{\ell} \sigma(x) F\left(\frac{x}{a}, \frac{a}{W}\right) dx$$

where a is the crack length, ℓ is the bridging zone length, W is the specimen width, σ(x) is the stress distribution over the bridging zone, and x is the distance behind the crack tip. The total toughness, then, is given as:

$$K = K_o + \Delta K_b$$

where K_o is the intrinsic toughness of the composite. Assuming that σ(x) = $f\sigma_o$, where f is the fraction of ductile phase, and σ_o is the flow stress of the matrix,

$$\Delta K_b = 2f\sigma_o\sqrt{\frac{2\ell}{\pi}}$$

Using this relation the steady-state toughness of the composite was predicted to be around 12.4 $\text{MPa}\sqrt{\text{m}}$, which was slightly lower than that measured experimentally. The authors attributed the difference to an additional contribution from thermally-induced dislocation punching at the interface, which strengthened the matrix, as well as residual strains from plasticity.

Davidson et al. (1989) examined the fatigue crack growth behavior of an Al_2O_3 fiber reinforced Mg alloy composite (made by liquid infiltration). A simple model was developed to predict the crack growth rate as a function of ΔK. The cyclic crack opening displacement (COD), δ_c, is given by:

$$\delta_c = \frac{c_o \Delta K^2}{E_i}$$

where E_i is the elastic modulus of the matrix (E_m) or the composite (E_c). If the crack growth rate is assumed to be proportional to COD, then the Paris law can be written as:

$$\frac{da}{dN} = C\left[\left(\frac{E_m}{E_c}\right)^{\frac{1}{2}} \Delta K\right]^n$$

By incorporating shielding from the fibers, a lower bound was obtained, while an upper bound was obtained by incorporating the interaction between the main crack and microcracks. By normalizing the crack growth data with respect to the modulus of the composite, good agreement was obtained for composites with varying fiber orientation, with respect to the loading axis, as well as with the unreinforced Mg alloy, Fig. 8.27.

Fatigue-crack propagation studies have also been conducted on aligned eutectic or *in situ* composites. Since many of these in situ composites are meant for high-temperature applications in turbines, their fatigue behavior has been studied at temperatures ranging from room temperature to 1100°C. The general consensus is that the mechanical behavior of *in situ* composites, i.e., static and cyclic strengths, is superior to that of the conventional cast superalloys (Stoloff, 1978).

8.2.2 Particle Reinforced MMCs

The crack growth behavior in particle reinforced MMCs is also very much dependent on reinforcement characteristics (Shang et al., 1988; Lukasak and Bucci, 1992; Allison and Jones, 1993; LLorca et al., 1994; Vasudevan and Sadananda, 1995), and on matrix microstructure (Bonnen et al., 1990; Sugimura and Suresh, 1992). In general, higher threshold values, ΔK_{th}, are observed for composites than for monolithic materials, Fig. 8.28. In addition,

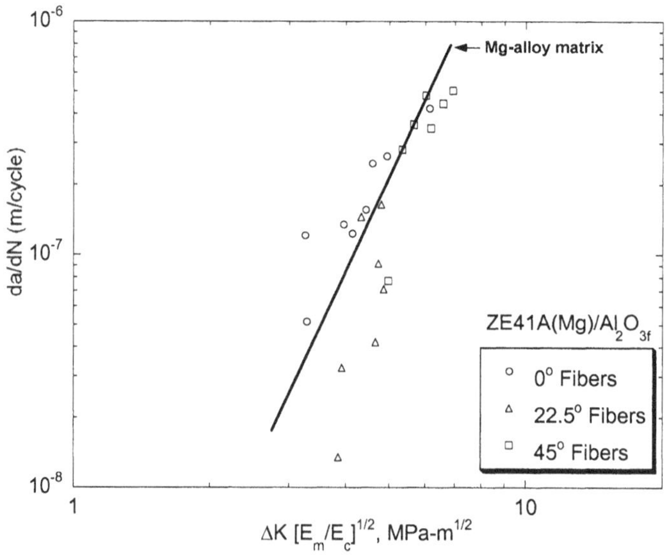

Fig. 8.27 Fatigue crack growth behavior of Mg(ZE41A)/$Al_2O_{3,f}$ composite, normalized by the Young's moduli of matrix and composite. Good agreement was obtained for composites with varying fiber orientation, as well as with the unreinforced Mg alloy (after Davidson et al., 1989).

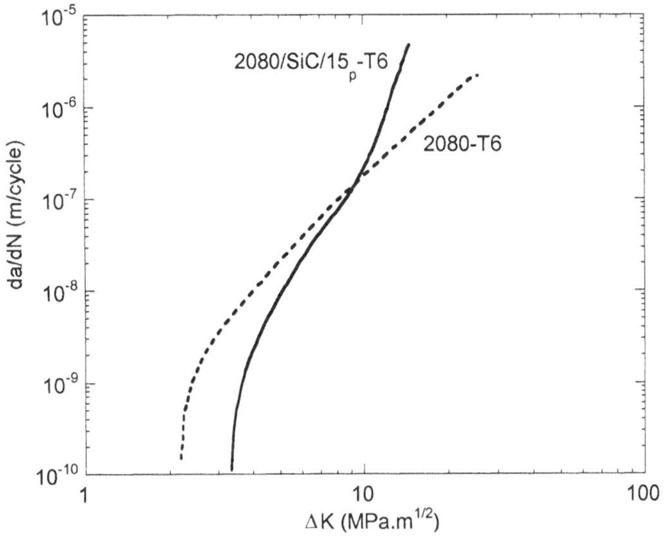

Fig. 8.28 Fatigue crack growth behavior of 2080-T6 alloy versus 2080/SiC/15$_p$-T6 composite. The composite exhibits a higher fatigue threshold and slightly higher Paris law slope (after Lukasak and Bucci, 1992).

increasing volume fraction of particles results in an increase in ΔK_{th}, Fig. 8.29. This can be attributed to the higher modulus of the composite, which results in a lower crack opening displacement for a given applied stress intensity factory. The Paris-law slope of da/dN vs ΔK curve for the composites is generally comparable to that of unreinforced alloys, although at very high ΔK, the crack propagation in the composites is much higher. This is due to lower fracture toughness of the composite, relative to the unreinforced alloy.

It is very important to note, however, that these trends are only true for composites with a constant matrix microstructure. Sugimura and Suresh, (1992), for example, examined the effect of SiC volume fraction on fatigue crack growth of cast MMCs. Because of the casting process, the grain size in the matrix was inversely dependent on the volume fraction of particles (see Chapter 4), i.e., the matrix microstructure changed with increasing volume fraction. This was because a higher volume fraction of particles resulted in a larger number of possible grain nucleation sites, and a finer overall grain size in the matrix of the composite. Thus, in their study, an increase in volume fraction of particles (i.e., a decrease in matrix grain size) contributed to a decrease in fatigue crack growth resistance, Fig. 8.30.

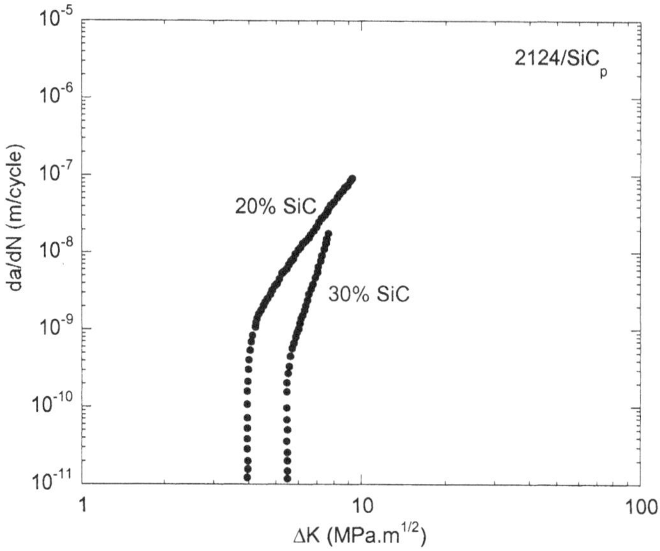

Fig. 8.29 Effect of volume fraction on fatigue crack growth of 2124/SiC_p composites (after Bonnen et al., 1990). Increasing volume fraction results in an increase in fatigue threshold (the matrix microstructure of this powder metallurgy-processed composite was relatively constant).

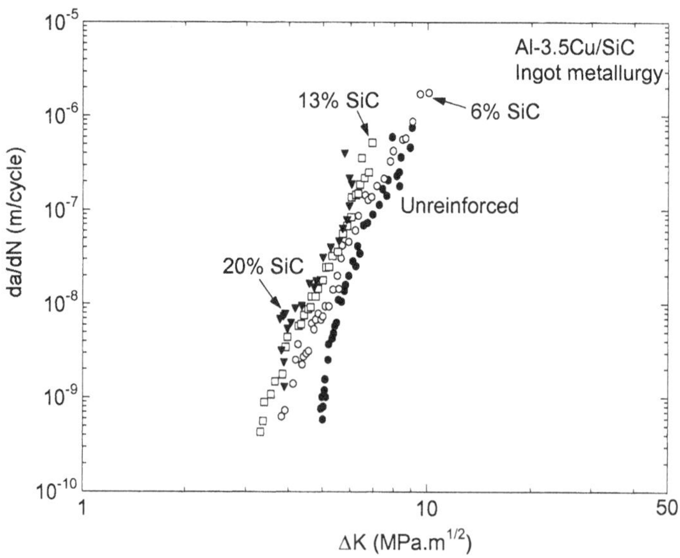

Fig. 8.30 Decrease in fatigue crack growth resistance with increasing volume fraction of SiC particles (cast composite). Because of the casting processing route, the matrix grain size at higher volume particle fractions decreased significantly, yielding lower fatigue resistance (after Sugimura and Suresh, 1992).

It is interesting to note that coarser particles provide better fatigue crack growth resistance, than finer particles, because of a large extent of roughness-induced crack closure (Shang et al., 1988), Fig. 8.31. Furthermore, experiments on overaged and underaged composites, showed little difference in fatigue crack growth resistance (Bonnen et al., 1990), indicating that the controlling deformation mechanisms are related to the particle characteristics and spatial distribution, Fig. 8.32. The fatigue crack growth behavior is also very much affected by the load ratio R ($K_{min}/K_{max} = \sigma_{min}/\sigma_{max}$). Figure 8.33(a) shows the fatigue crack growth behavior for a 2080/SiC/20$_p$ composite at various R-ratios, ranging from −2 to 0.8. With increasing R-ratio, the fatigue threshold decreases and the Paris law slope increases. These data can be further represented by using the two-parameter approach of ΔK versus K_{max}, by Vasudevan et al. (1994), shown in Fig. 8.33(b). This approach can be understood simply by the fact that critical values of both ΔK and K_{max} must be met for the crack to propagate. Thus, with increasing crack growth rate, at higher R-ratios, for a given value of ΔK a much smaller value of K_{max} needs to be applied for the crack to propagate.

The effects of ΔK and K_{max} can be explained in terms of the plastic zone size at the crack tip, and reinforcement/crack interactions (Shang et al., 1988; Chawla and Chawla, 2004). Figure 8.34 shows a schematic of the observed behavior during fatigue crack growth (Chawla and Chawla, 2004). At low K_{max} and/or ΔK the fatigue crack propagates around the particles, while at high K_{max} and/or ΔK particle fracture takes place ahead of the crack tip, resulting in relatively planar crack growth. This behavior is can be explained by the size of the plastic zone, relative to the size and fracture strength of the particles. At low K_{max} and/or ΔK, the plastic zone at the crack tip is on the order of the particle size. Thus, crack growth takes place around the SiC particles, Fig. 8.35, and crack arrest and crack deflection mechanisms contribute to enhanced resistance to crack growth. Ayyar et al. (2005) modeled the crack growth behavior in an SiC particle reinforced composite, by using the two-dimensional microstructure of the material as a basis for the finite element model. A modified crack closure criterion was used to calculate the stress intensity at the crack tip, while the maximum circumferential tensile stress criterion was used to determine the direction of crack propagation. The predicted response qualitatively shows the same behavior as that observed in the experiment, i.e., the cracks grow around the SiC particles in a somewhat tortuous fashion, Fig. 8.36. Similar models have been developed by Boselli et al. (2001), for perfectly circular particles of single size.

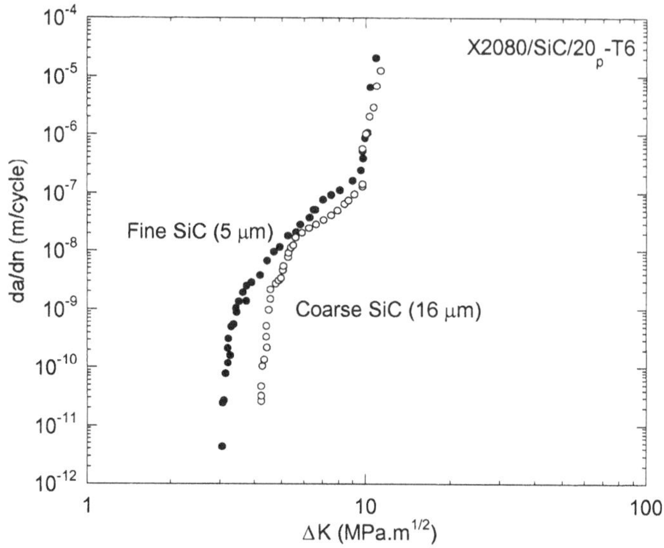

Fig. 8.31 Effect of particle size on fatigue crack growth behavior in a 2080/SiC/20p-T6 composite. The larger particles results in a larger degree of roughness-induced closure, thus increasing the fatigue threshold (after Shang et al., 1998).

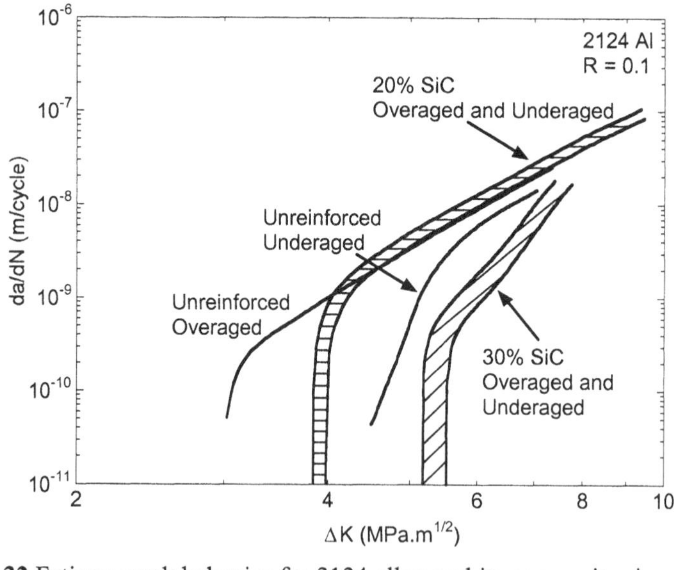

Fig. 8.32 Fatigue crack behavior for 2124 alloy and its composites in underaged and overaged conditions. While the unreinforced alloy is sensitive to aging, the composites are not, indicating that the particles control fatigue crack growth behavior (after Bonnen et al., 1990).

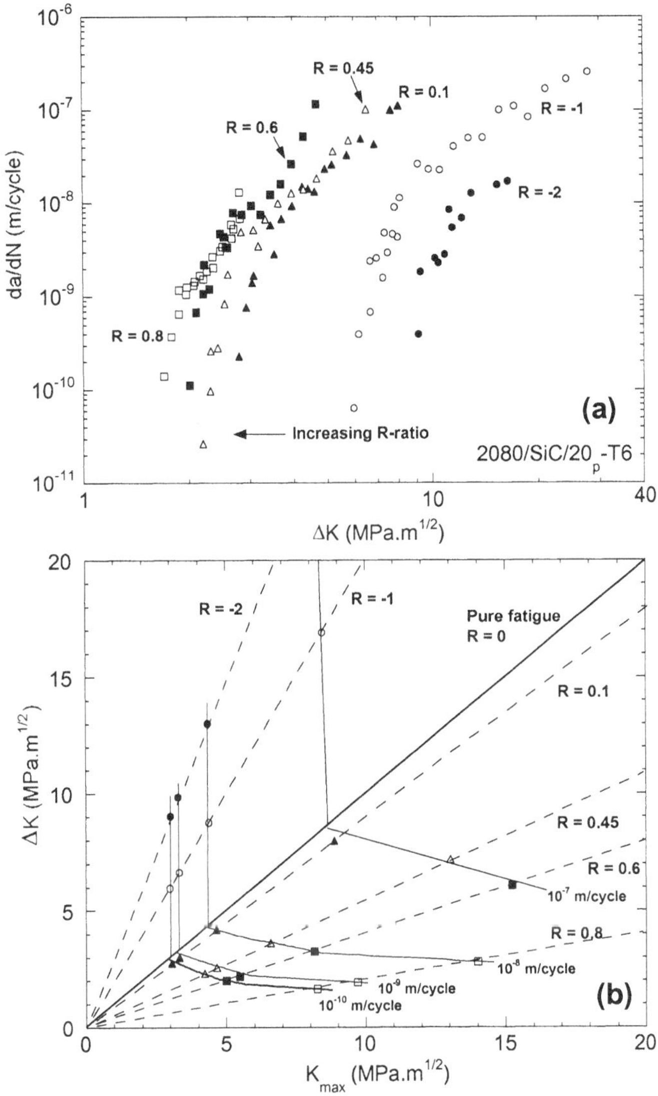

Fig. 8.33 Effect of R-ratio on fatigue crack growth of 2080/SiC/20$_p$: (a) da/dN vs. ΔK behavior, showing decreasing fatigue threshold and increasing Paris law slope with increasing R; and (b) same data plotted using ΔK vs. K_{max} approach, showing decreasing resistance to fatigue crack growth with increasing K_{max} (after Ganesh and Chawla, 2004).

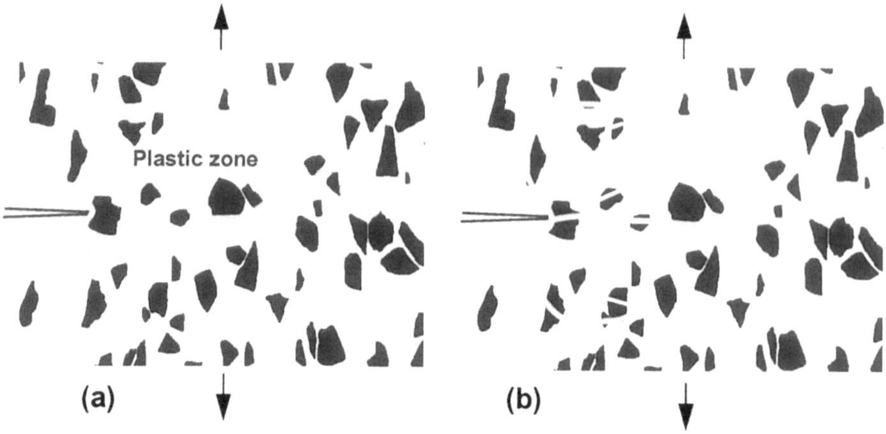

Fig. 8.34 Schematic of fatigue damage in particle reinforced MMCs: (a) low R-ratio, plastic zone is on the order of the particle size, and crack growth is tortuous, and (b) high R-ratio, plastic zone is much larger than the particle size resulting in particle fracture ahead of the crack tip, and planar crack growth (after Chawla and Chawla, 2004).

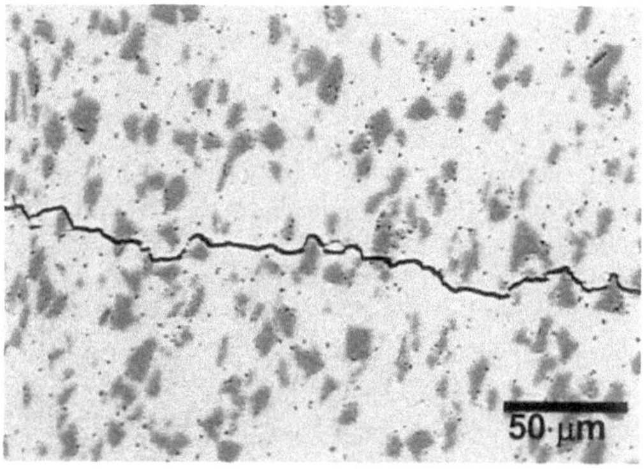

Fig. 8.35 Fatigue crack growth behavior of 2080/SiC/20_p; the crack propagates around the SiC particles.

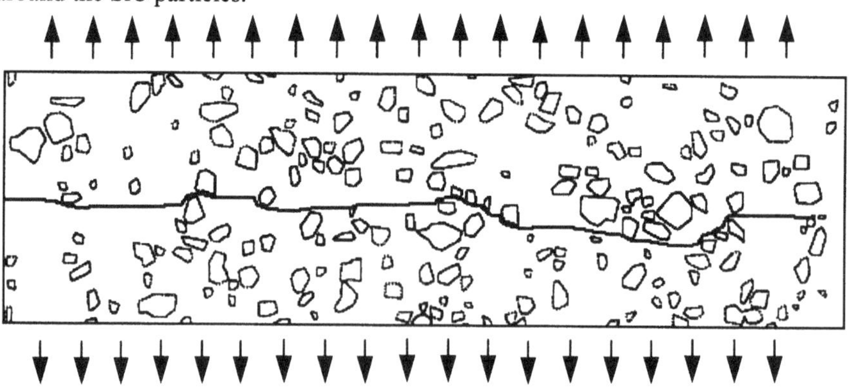

Fig. 8.36 Microstructure-based finite element model prediction of crack growth in an SiC particle reinforced Al matrix composite, showing tortuous crack growth around the reinforcement particles (after Ayyar et al., 2005).

At high K_{max} and/or ΔK, the plastic zone is large enough to engulf several particles, so particle fracture ahead of the crack tip takes place, and the crack growth is relatively linear. This behavior is confirmed by the degree of experimentally measured particle fracture at R-ratios of 0.1 and 0.8, respectively, Fig. 8.37. Note that the degree of particle fracture is much higher at the higher R-ratio, i.e., it is driven by K_{max}.

We end this section by discussing the fatigue damage mechanisms in compression-compression fatigue. Here, the nature of damage is quite different from that observed in tension-tension fatigue. Fig. 8.38 shows the fracture surface of a 2080/SiC/20$_p$ composite after fatigue growth for an R-ratio of −2. Observe the wear markings on the SiC particle surface, and that spherical particles are present on the surface. Energy-dispersive spectroscopy (EDS) analysis showed these particles to be Al_2O_3, indicating that matrix particles likely oxidized were wedged into the SiC particle. A similar behavior has been observed in steels containing carbide particles (Aswath et al., 1988).

8.2.3 Hybrid and Laminated Composites

Composites containing more than one type of reinforcement are called hybrid composites. Such composites, by using two or more type of reinforcements, extend the idea of tailor-making a composite material to

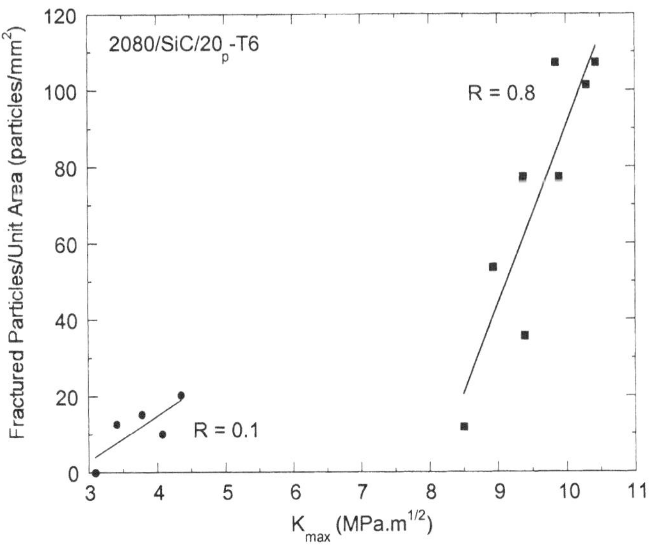

Fig. 8.37 Degree of particle fracture in 2080/SiC/20$_p$ versus applied K_{max}, at R-ratios of 0.1 and 0.8. Higher K_{max} results in a significantly larger degree of particle fracture.

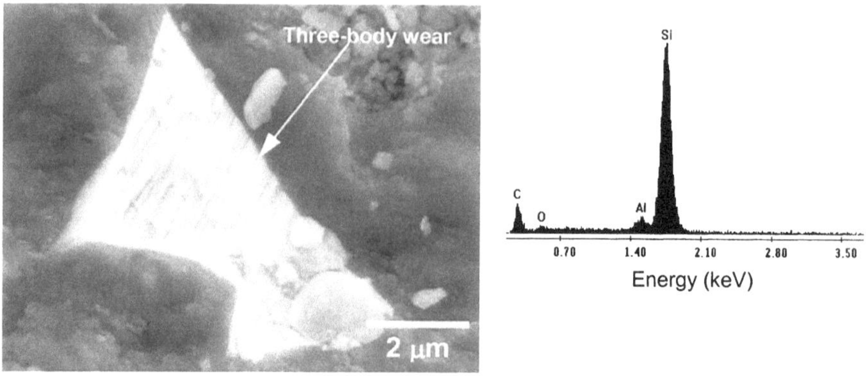

Fig. 8.38 Fatigue fracture surface after compression fatigue (R = −2). Debris from the fatigue process (Al_2O_3, from EDS spectra on the right) is embedded into the SiC particles resulting in three-body wear.

meet specific property requirements. Partial replacement of expensive fibers by cheaper but adequate fiber types is another attractive feature of hybrid composites.

Metallic sheet laminated composites have been shown to have improved fatigue resistance over their monolithic counterparts (Chawla and Liaw, 1979). There are two possible crack geometries: Crack arrester and crack divider geometries. In the crack arrest geometry, with the crack growing perpendicular to the thickness of the composite, the mechanism proposed by Cook and Gordon (1964) would appear to work (see also chapter 5). According to this model, if the interface is weak, then the crack bifurcates and changes its direction when it reaches the interface, and thus the failure of the composite is delayed. The improved fatigue crack propagation resistance in crack divider geometry has been verified by a number of researchers (McCartney et al., 1967, Taylor and Ryder, 1976, and Pfeiffer and Alic, 1978). This improved performance has been attributed either to interfacial separation, which removes the triaxial state of stress, or to a holding back of crack at the interface in the faster crack-propagating component by the slower crack-propagating component.

An interesting type of hybrid composite consists of alternating layers of high-strength aluminum alloy sheets and layers of unidirectional aramid or glass fibers in an epoxy matrix. There are two types of such laminates: ARamid ALuminum Laminate (ARALL) and GLAss REinforced (GLARE) laminate. Improved fatigue resistance of ARALL or GLARE over that of monolithic aluminum structures is the main attractive feature. Cracks in aluminum can grow only a short distance before being blocked by the aramid or glass fibers spanning the crack tip, Fig. 8.39 (Wu and Yang,

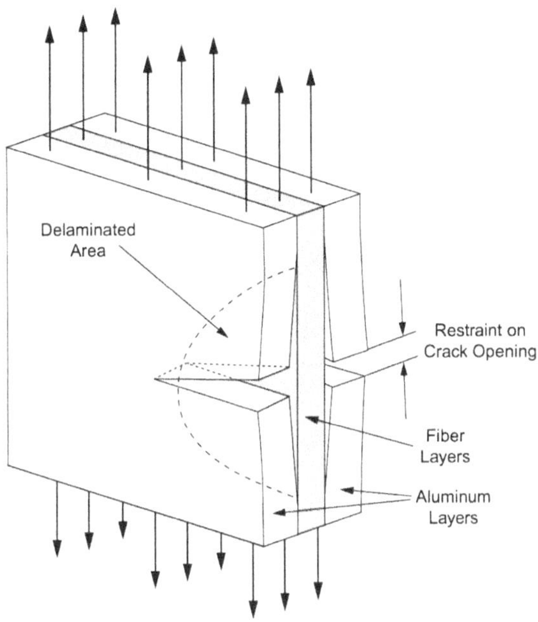

Fig. 8.39 Schematic of fatigue damage in ARALL or GLARE. Cracks can grow only a short distance before being blocked by the aramid or glass fibers spanning the crack tip (after Wu and Yang, 2005).

2005). Figure 8.40 shows the slow fatigue-crack growth characteristics of ARALLs compared with two monolithic aluminum alloys (Mueller and Gregory, 1988). ARALL has been used in tension-dominated fatigue structures such as aircraft fuselage, lower-wing, and tail skins. The use of ARALL resulted in 15-30% weight savings over conventional construction. GLARE has been chosen for use in the fuselage of the twin-deck, 550-seat, Airbus 380 aircraft. Besides having superior fatigue behavior, Glare is also lighter than aluminum, cutting the weight of the Airbus A380 by 1,000 kg. Another version of hybrid laminated MMCs is a family of layered composites consisting of metallic outer skins with a viscoelastic core material (for example, polyethylene, nylon, polypropylene, paper, or cork). Such composites are useful where sound and vibration dampening are required. The viscoelastic layer provides a high loss factor, i.e., a high capacity to convert vibrational energy to heat.

Other innovative approaches to engineering hybrid composites include laminated composites of SiC particle reinforced Al layers diffusion bonded to pure Al alloy layers, Fig. 8.41(a) (Hassan et al., 2004). It was shown that incorporation of the ductile metallic layers resulted in lower crack growth rate in the Paris law regime, higher fracture toughness, and slightly lower fatigue crack threshold, compared to conventional particle reinforced

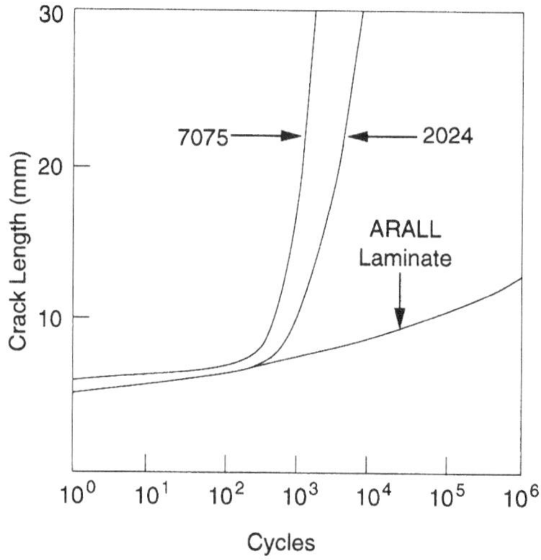

Fig. 8.40 Fatigue crack growth behavior of ARamid ALuminum Laminate (ARALL) compared to conventional 7075 and 2024 aluminum alloys (after Mueller and Gregory, 1988). The ARALL has significantly better crack growth resistance.

MMCs, Fig. 8.41(b). The authors attributed the resistance to crack growth to the higher Young's modulus of the laminate, which resulted in a lower crack opening displacement, for a given applied stress intensity factor. The elastic mismatch also resulted in a crack tip shielding effect at the interface between the layers.

8.3 THERMAL FATIGUE

The problem of thermal expansion mismatch between the components of a composite is a very general and serious one (Chawla, 1997). Thermal stresses arise in composite materials because of the large differences in the thermal expansion coefficients (α) of the fiber and the matrix. It should be emphasized that thermal stresses in composites will arise even if the temperature change is uniform throughout the volume of the composite. Such thermal stresses can be introduced in composites during cooling from high fabrication, annealing, or during any temperature excursions (inadvertent or by design) during service. Turbine blades, for example, are very much susceptible to thermal fatigue.

We described the origin and analysis of thermal stresses in chapter 6. To recapitulate, the magnitude of thermal stresses in composites is proportional to the thermal strain, $\Delta\alpha\Delta T$, where $\Delta\alpha$ is the difference in the expansion coefficients of the two components, and ΔT is the amplitude of the thermal cycle. In MMCs, the matrix generally has a much higher coefficient of

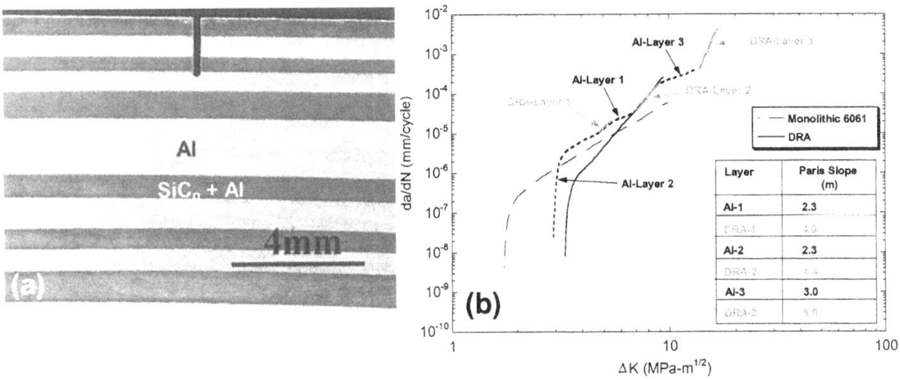

Fig. 8.41 (a) Notched hybrid composite of pure Al and SiC particle reinforced Al composite (courtesy of J. Lewandowski), and (b) fatigue crack growth behavior, showing enhanced crack growth resistance in the Paris law regime (after Hassan et al., 2004).

thermal expansion than the reinforcement. Rather large internal stresses can result when fiber reinforced composites are heated or cooled through a temperature range. When this happens in a repeated manner, we have *thermal fatigue,* because the cyclic stress is thermal in origin. Thermal fatigue can cause plastic deformation in a ductile metallic matrix (Chawla, 1973a; Chawla 1973b). Cavitation in the matrix and fiber/matrix debonding are the other forms of damage observed due to thermal fatigue in composites (Kwei and Chawla, 1992; Xu et al., 1995). Figure 8.42 shows a scanning electron micrograph of an alumina fiber reinforced magnesium alloy matrix that was subjected to different thermal cycles between room temperature and 300°C.

Xu et al. (1995) examined the thermal cycling behavior of a ZE41A Mg alloy reinforced with 35% Al_2O_3 fibers. They quantified the damage during thermal cycling by measuring the change in Young's modulus and density as a function of thermal cycles. The damage during thermal cycle may also be quantified by a change in damping behavior of the composite (Carreno-Morelli et al., 2001). A damage parameter was defined, where the damage in Young's modulus, for example, is given by:

$$D_E = 1 - \left(\frac{E_n}{E_o}\right)$$

where E_n is the Young's modulus after n cycles, and E_o is the initial Young's modulus prior to cycling. A similar approach was used to monitor damage in density, ρ. In order to correlate D_E and D_ρ, Mackenzie's equation was invoked (Mackenzie, 1950):

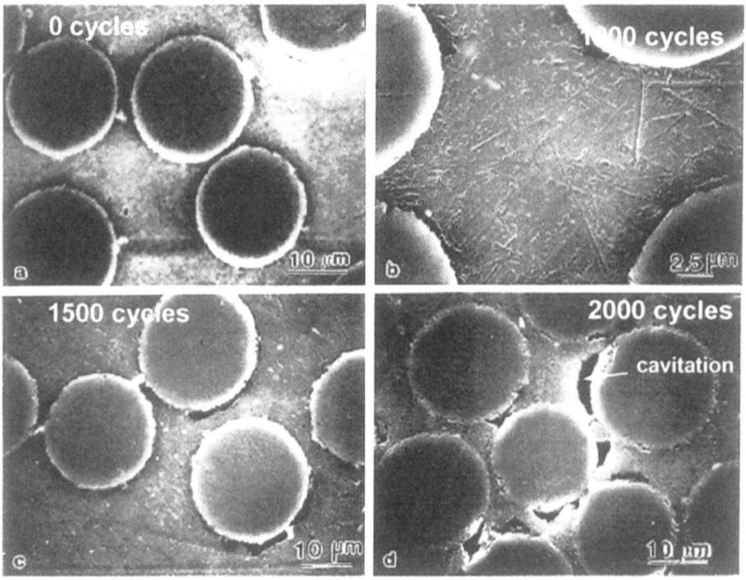

Fig. 8.42 Evolution of thermal fatigue damage in an Mg/Al_2O_{3f} composite. With increasing cycles interfacial cavitation takes place due to internal stress mismatch between fiber and matrix (after Xu et al., 1995).

$$E_n = E_o\left(1 - aV_v - bV_v^2\right)$$

where V_v is the volume fraction of voids, and *a* and *b* are constants of a fully dense material. For materials with Poisson's ratio of 0.3, a = 1.91 and b = −0.91. By incorporating these equations and using the rule of mixtures for the density of the composite, $\rho_c = \rho_f V_f + \rho_m V_m$, the following relation is obtained:

$$D_E = b\left(\frac{\rho_o}{\rho_m}\right)D_n + b_1\left[\left(\frac{\rho_o}{\rho_m}\right)D_\rho\right]^2$$

where ρ_m is the density of the matrix. Figure 8.43 shows a plot of experimental and predicted values of D_E versus D_ρ, showing very good agreement. In general, one can reduce the damage in the matrix by choosing a matrix material that has a high yield strength and a large strain to failure (i.e., ductility). The eventual fiber/matrix debonding can only be avoided by choosing the components such that the difference in the thermal expansion characteristics of the fiber and the matrix is low.

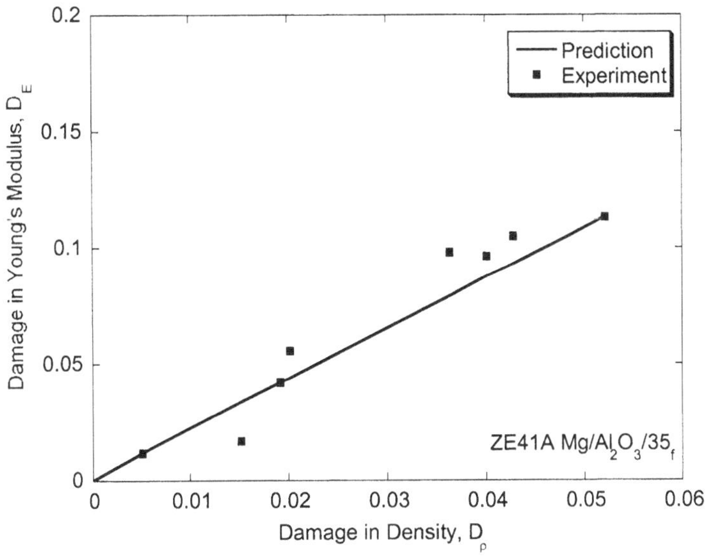

Fig. 8.43 Damage in Young's modulus versus damage in density for a $Mg/Al_2O_3/35_f$ composite showing good correlation between predicted and experimental results (Xu et al., 1995).

One type of thermal fatigue test involves cycling temperature of the sample while its gage length is kept constant. This constraint results in an internal stress on the sample. Thus, measurement of the internal stresses generated when subjecting a composite to thermal cycling can be used to evaluate study thermal fatigue. Figure 8.44 shows the results of cycling between 300°C and 500°C for alumina fiber/Al-Li composite (Kwei and Chawla, 1992). The alumina fibers were unidirectionally aligned parallel to the stress axis, and the fiber volume fraction was 35%. The plot shows the variation of the maximum stress in tension and compression as a function of number of cycles. The initial rise in the maximum tensile stress curve is due to the work hardening of the aluminum alloy matrix caused by the thermal stresses. With continued cycling, the microstructural damage sets in, and a plateau in the maximum tensile stress versus cycles curve is obtained as long as the strength increase due to work hardening of the matrix is balanced by the strength decrease due to microstructural damage, e.g., voiding at the interface. The maximum compressive stress, however, decreases with cycling to a plateau value. This may be due to the Bauschinger effect, i.e., a higher strength in tension results in a concomitant lowering of the strength in compression. Eventually, the alumina fibers suffer fracture, which causes the tensile stress curve in Fig. 8.44 to decrease.

Superimposed effects of mechanical and thermal loading are important in a variety of applications such as in an automotive engine or gas turbine. Two

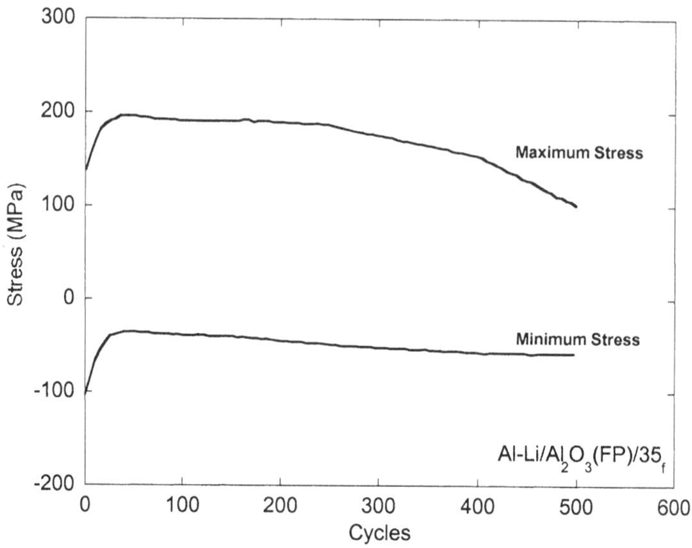

Fig. 8.44 Evolution of stess during strain-controlled thermal cycling of Al-Li/Al_2O_3/35_f composite (after Kwei and Chawla, 1992). With increasing thermal cycles, relaxation processes such as interfacial cavitation result in stress relaxation.

types of thermomechanical loading may be applied: (a) in-phase loading, where the maximum and minimum stress take place at maximum and minimum temperature, respectively; and (b) out-of-phase loading, where the maximum stress and minimum stress are opposite to the maxima and minima in temperature. The nature of thermomechanical loading has a direct influence on the internal stress in the fiber and matrix. Consider the case of out-of-phase loading. During cooling, the ceramic fiber is in a state of residual compression, and the metal matrix is in tension. Thus, at the minimum temperature, where the applied stress is a maximum, the total stress on the matrix will be enhanced, while that in the fiber will be lower. In in-phase loading, at the maximum temperature, the fiber is in tension while the matrix is in compression. Since the maximum temperature corresponds to the point of maximum stress, the stress on the fiber is maximized.

The effects of the internal stress during out-of-phase and in-phase TMF for a Ti-6242/SiC_f composite are shown in Fig. 8.45(a) (Bettge et al., 2003; Peters et al., 2004). In general, the fatigue life under TMF loading is significantly lower than in pure mechanical low cycle fatigue. In-phase loading results in lower overall fatigue resistance, while out-of-phase loading exhibits an abrupt decrease in fatigue life at about 10^3 cycles. Due to the different internal stress states, the mechanisms for fracture in TMF are also quite different. Under in-phase loading, matrix relaxation takes place which results in a gradual increase in the stress in the fiber. Thus, fatigue life is creep-controlled, Fig. 8.45. In out-of-phase loading, the strain range in the matrix is very large, so the damage is primarily matrix-controlled. With matrix crack, oxygen ingression takes place, and when the cracks reach the fiber/matrix interface, oxidation and fracture of the fiber take place, Fig. 8.45(b). Thus, at higher stress the fracture is fiber-controlled, with decreasing stress multiple matrix cracks and fiber bridging were observed. Thermal cycling damage in discontinuously reinforced MMCs is somewhat lower than that of continuous fiber reinforced MMCs, because diffusional relaxation of the matrix is easier (the constraint on the matrix is lower for

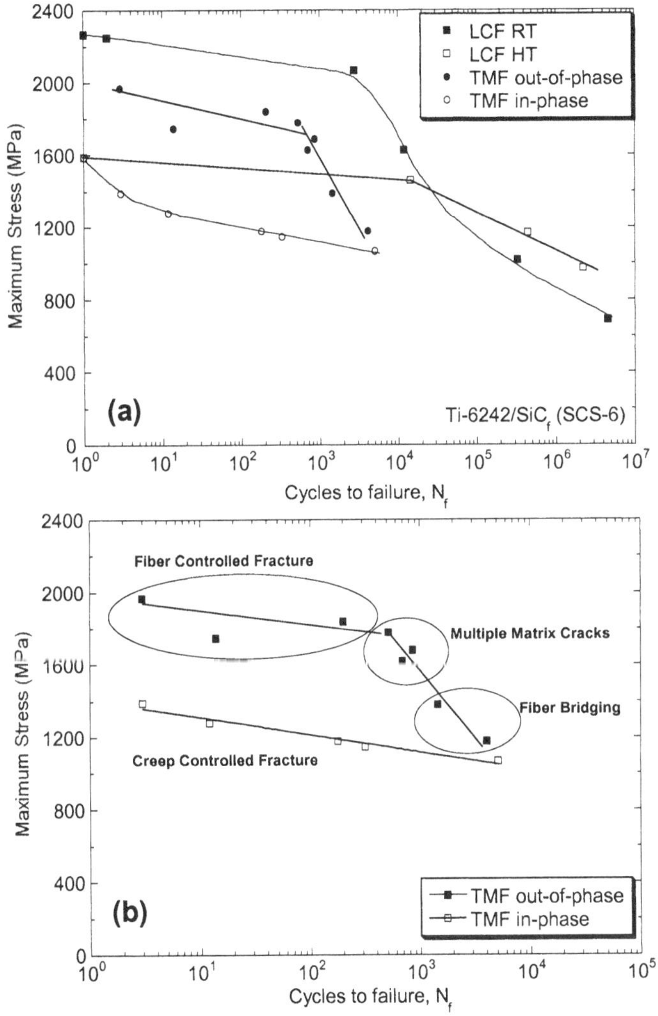

Fig. 8.45 Comparison of thermomechanical fatigue and low cycle fatigue of a Ti/SiC_f composite: (a) stress versus cycles behavior, showing the more damaging effect of thermal fatigue, and (b) damage mechanisms (after Bettge et al., 2003).

discontinuous reinforcement). Very large plastic strains (on the order of 150% for Al composites with 20 and 30% SiC) have been observed during thermal cycling (Pickard and Derby, 1990). In addition, a significant increase in surface roughness of the composite has been observed, by Rezai-Aria et al. (2003). They studied the thermal cycling behavior of an $Al/Al_2O_3/15_{sf}$ (Saffil) composite. The increase in surface roughness was attributed to the higher mobility of dislocations near the free surface. Within the bulk of the material, a cellular dislocation structure evolved during thermal cycling.

References

Allison, J.E., and J.W. Jones, (1993) in *Fundamentals of Metal Matrix Composites*, S. Suresh, A. Mortensen, and A. Needleman, eds., Butterworth-Heinemann, Stoneham, MA, p. 269.

Almond, E.A., and B. Roebuck, (1980) *Metals Technology*, **7**, 83-85.

Arsenault, R.J., and U.T.S. Pillai, (1996) *Metall. Mater. Trans.*, **27A,** 995-1001.

Arsenault, R.J., and S.B. Wu, (1987) *Mater. Sci. Eng.*, **96,** 77-88.

Aswath, P.B, S. Suresh, D.K. Holm, and A.F. Blom, (1988) *J. Eng. Mater. Tech.*, **110** 278-85.

Ayyar, A., J.J. Williams, and N. Chawla, (2005) *Comp. Sci. Tech.*, submitted.

Baker, A.A., D.M. Braddick, and P.W. Jackson, (1972) *J. Mater. Sci.*, **7**, 747-762.

Bauschinger, J., (1886) *Mitt: Mech-Tech Lab.*, XIII Munchen.

Bettge, D., B. Gunther, W. Wedell, P.D. Portella, J. Hemptenmacher, and P.W.M. Peters, (2004) in *Low Cycle Fatigue 5*, (P. D. Portella, H. Sehitoglu, K. Hatanaka, eds.), DVM, Berline, pp. 81-86.

Blatt, D., J.R. Jira, and J.M. Larsen, (1995) *Scripta Metall. Mater.*, **33**, 939-944.

Bonnen, J.J., C.P. You, J.E. Allison, and J.W. Jones, (1990) in *Proceedings of the International Conference on Fatigue,* Pergamon Press, New York, pp. 887–892.

Boselli, J., P.D. Pitcher, P.J. Gregson, and I. Sinclair, (2001) *Mater. Sci. Eng.*, **A300**, 113-124.

Calabrese, C., and C. Laird, (1974a) *Mater. Sci. Eng.,* **13**, 141 - 157.

Calabrese, C., and C. Laird, (1974b) *Mater. Sci. Eng.,* **13**, 159 - 174.

Carreno-Morelli, E., N. Chawla, and R. Schaller, (2001) *J. Mater. Sci. Lett.*, **20**, 163-165.

Champion, A.R., W.H. Krueger, H.S. Hartman, and A.K. Dhingra, (1978), in *Proc. 1978 Intl. Conf. Composite Materials* (ICCM/2), TMS-AIME, New York, p. 883.

Chawla, K.K, (1973a) *Metallography*, **6**, 155.

Chawla, K.K., (1973b) *Phil. Mag.*, **28**, 401.

Chawla, K.K., (1998) *Composite Materials: Science & Engineering*, Springer-Verlag, New York.

Chawla, K.K., (1975) *Fiber Sci. Tech.*, **8**, 49.

Chawla, K.K., (1991) in *Metal Matrix Composites: Mechanisms and Properties*, R.K. Everett and R.J. Arsenault, eds., Academic press, pp. 235-253.

Chawla, K.K., and N. Chawla, (2004) in *Kirk-Othmer Encyclopedia*, John-Wiley and Sons, New York.

Chawla, K.K., and P.K. Liaw, (1979) *J. Mater. Sci.*, **14,** 2143.

Chawla, K.K., and M. Metzger, (1972) *J. Mater. Sci.*, **7**, 34.

Chawla, N., (1997) *Metall. Mater. Trans.*, **28A**, 2423.

Chawla, N., and J.E. Allison, (2001) in *Encyclopedia of Materials: Science and Technology*, vol. 3, (B. Ilschner and P. Lukas, eds.), Elsevier Science, pp. 2969-2974.

Chawla, N., C. Andres, J.W. Jones, and J.E. Allison, (1998a) *Metall. Mater. Trans.*, **29A**, 2843.

Chawla, N., C. Andres, J.W. Jones, and J.E. Allison, (1998b) *Scripta Mater.*, **38**, 1596.

Chawla, N., L.C. Davis, C. Andres, J.E. Allison, J.W. Jones, (2000a) *Metall. Mater. Trans.*, **31A**, 951-957.

Chawla, N., U. Habel, Y.-L. Shen, C. Andres, J.W. Jones, and J.E. Allison, (2000b) *Metall. Mater. Trans.*, **31A**, 531-540.

Chawla, N., J.W. Holmes, and R.A. Lowden, (1996) *Scripta Mater.*, **35**, 1411.

Chawla, N., J.W. Jones, and J.E. Allison, (1999).in *Fatigue '99* (X.R. Wu and Z.G. Wang, eds.), EMAS/HEP.

Chawla, N., M. Kerr, and K.K. Chawla, (2005) *J. Am. Ceram. Soc.*, **88**, 101-108.

Chawla, N., and Y.-L. Shen, (2001) *Adv. Eng. Mater.*, **3**, 357-370.

Cook, J., and J.E. Gordon, (1964) *Proc. Roy. Soc. Lond.*, **A282**, 508.

Cotterill, P.J., and P. Bowen, (1993) *Composites*, **24**, 214-221.

Cotterill, P.J., and P. Bowen, (1996) *Mater. Sci. Tech.*, **12,** 523-529.

Couper, M.J., and K. Xia, (1991) in *Metal Matrix Composites - Processing, Microstructure and Properties*, (N. Hansen et al., eds.), Riso National Laboratory, Roskilde, Denmark, p. 291.

Davidson, D.L., K.S. Chan, A. McMinn, and G.R. Leverant, (1989) *Metall. Trans.*, **20A**, 2369-2378.

Doker, H., and G. Marci, (1983) *Int. J. Fatigue*, **5**, 187-191.
Foulk III, J.W., D.H. Allen, and K.L.E. Helms, (1998) *Mech. Mater.*, **29**, 53-68.
Ganesh, V.V., and N. Chawla, (2004) *Metall. Mater. Trans.*, **35A**, 53-62.
Gomez, J.P., and F.E. Wawner, (1988) personal communication.
Gouda, M., K.M. Prewo, and A.J. McEvily, (1981) in *Fatigue of Fibrous Composite Materials*, p. 101, ASTM STP, 723, Amer. Soc. Testing and Materials, Philadelphia.
Hack, J.E., R.A. Page, and G.R. Leverant, (1987) *Metall. Trans.,* **15A**, 1389.
Hall, J., J.W. Jones, and A. Sachdev, (1994) *Mater. Sci. Eng.,* **A183**, 69.
Harmon, D.M., C.R. Saff, and C.T. Sun, (1987) AFW AL- TR-87-3060. Air Force Wright Aeronautical Labs., Dayton, Ohio.
Hassan, H.A., J.J. Lewandowski, and M.H. Abd El-latif, (2004) *Metall. Mater. Trans.*, **35A**, 45-52.
Johnson, W.S., (1982) in *Damage in Composite Materials*, *ASTM STP 775*, American Society for Testing and Materials, Philadelphia, p. 83.
Johnson, W.S., (1988) in *Mechanical and Physical Behavior of Metallic and Ceramic Composites*, 9th Risø Intl. Symp. on Metallurgy and Materials Science, Rise Nat. Lab., Roskilde, Denmark.
Johnson, W.S., and R.R. Wallis, (1986) in *Composite Materials: Fatigue and Fracture, ASTM STP 907*, American Society for Testing and Materials, Philadelphia, p. 161.
Kerr, M., N. Chawla, and K.K. Chawla, (2005) *JOM*, **2**, 67-70.
Kindermann, P., P. Schlund, H.-G. Sockel, M. Herr, W. Heinrich, K. Görtring, and U. Schleinkofer, (1999) *Int. J. Refractory & Hard Materials*, **17**, 55
Kwei, L.K, and K.K. Chawla, (1992) *J. Mater. Sci.*, **27**, 1101-1106.
Lewandowski, J.J., (2000) in *Comprehensive Composite Materials*, vol. 3, (A. Kelly and C. Zweben, eds.), Elsevier Press, pp. 151-187.
Li, C., and F. Ellyin, (1996) *Mater. Sci. Eng.*, **A214**, 115.
Liu, J., and P. Bowen, (2002) *Acta Mater.*, **50**, 4205-4218.
Liu, J., and P. Bowen, (2003) *Metall. Mater. Trans.*, **34A**, 1193-1202.
LLorca, J., (1994) *Acta Metall. Mater.*, **42**, 151-162.
LLorca, J., (2002) *Prog. Mater. Sci.*, **47**, 283-353.
LLorca, J., A. Needleman, and S. Suresh, (1990) *Scripta Metall. Mater.*, **24**, 1203.
LLorca, J., J. Ruiz, J.C. Healy, M. Elices, and C.J. Beevers, (1994) *Mater. Sci. Eng.*, **A185**, 1-15.
LLorca, J., S. Suresh, and A. Needleman, (1992) *Metall. Mater. Trans.*, **23A,** 919-933.
Lukasak, D.A., and R.J. Bucci, (1992) Alloy Technology Div. Rep. No. KF-34, Alcoa Technical Center, Alcoa, PA.

Lukasak, D.A., and D.A. Koss, (1993) *Composites*, **24**, 262.
Mackenzie, J.K., (1950) *Proc. Phys. Soc.*, **B63**, 2.
Mall, S., and B. Portner, (1992) *J. Eng. Mater., Tech.*, **114**, 409-415.
McCartney, R.F., R.C. Richard, and P.S. Trozzo, (1967) *Trans. ASM*, **60**, 384.
McGuire, M.A., and B. Harris, (1974) *J. Phys., Appl. Phys.*, **7**, 1788.
Majumdar, B.S., and G.M. Newaz, (1995) *Mater. Sci. Eng.*, **A200**, 114-129.
Meyers, M.A., and K.K. Chawla, (1999) *Mechanical Behavior of Materials*, Prentice-Hall.
Mott, N.F., (1952) *Phil.Mag.*, **43**, 1151.
Mueller, L.R., and M. Gregory, (1988) paper presented at First Annual Metals and Metals Processing Conf., SAMPE, Cherry Hill, NJ.
Orowan, E., (1959) in *Internal Stresses and Fatigue in Metals*, (G.M. Rassweiler and W.L. Grube, eds.), Elsevier Press, New York.
Page, R.A., J.E. Hack, R. Sherman, and G.R. Leverant, (1987) *Metall. Trans.*, **15A**, 1397.
Paris, P.C., M.P. Gomez, and W.P. Anderson, (1961) *The Trend in Engineering*, **13**, 9.
Paris, P.C., and F. Erdogan, (1963) *J. Basic. Eng. Trans. ASME*, **85**, 528.
Pfeiffer, N.J., and J.A. Alic., (1978) *J. Eng. Mater. Tech.*, **100**, 32.
Peters, P.W.M., J. Hemptenmacher, B. Gunther, D. Bettge, and P.D. Portella, (2004) in *Proc. ECCM-11*.
Pickard, S.M., and B. Derby, (1990) *Acta Metall. Mater.* **38**, 2537-2552.
Poza, P., and J. LLorca, (1999) *Metall. Mater. Trans.*, **30A**, 857.
Pugsley, V.A., and H.-G. Sockel, (2004) *Mater. Sci. Eng.*, A**366**, 87.
Rao, K.T. Venkateshwara, S.C. Siu, and R.O. Ritchie, (1993) *Metall. Trans.*, **24A**, 721-734.
Rezai-Aria, F., T. Liechti, and G. Gagnon, (1993) *Scripta Metall. Mater.*, **28**, 587-592.
Roebuck, B., E.A. Almond, and A.M. Cottenden, (1984) *Mater. Sci. Eng*, **66**, 179.
Salazar, A., J.Y. Pastor, and J. LLorca, (2004) *IEEE Trans. Appl. Supercon.*, **14**, 1941-1947.
Sanders, B.P., and S. Mall, (1996) *J. Comp. Tech. Res.*, **18**, 15-21.
Seeger, A., J. Diehl, S. Mader, and H. Rebstock, (1958) *Phil. Mag.*, **2**, 323.
Shang, J.K., W.K. Yu, and R.O. Ritchie, (1988) *Mater. Sci. Eng.* **A102,** 181–192.
Shen, Y.-L., M. Finot, A. Needleman and S. Suresh, (1995) *Acta Metall. Mater.*, **43**, 1701.
Starke, E.A., and G. Luetjering, (1979) in *Fatigue and Microstructure,* J.T. Stayley and E.A. Starke, eds., American Society for Metals, pp 205-243.

Stoloff, N.S., (1978) in *Advances in Composite Materials*, p. 247. Applied Sci. Pub., London.

Sugimura, Y., and S. Suresh, (1992) *Metall. Trans.*, **23A**, 2231-2242.

Suresh, S., (1998) *Fatigue of Materials*, 2nd Ed., Cambridge University Press, Cambridge,UK.

Taylor, L.G., and D.A. Ryder, (1976) *Composites*, **1**, 27.

Vasudevan, A.K., K. Sadananda, and N. Louat, (1994) *Mater. Sci. Eng.*, **A188**, 1-22.

Vasudevan, A.K., and K. Sadananda, (1995) *Metall. Mater. Trans.*, **26**, 1221-1234.

Vyletel, G.M., D.C. Van Aken, and J.E. Allison, (1991) *Scripta Metall. Mater.*, **25**, 2405-2410.

Walls, D.P., G. Bao, and F.W. Zok, (1993) *Acta Metall. Mater.*, **41**, 2061-2071.

Walls, D.P., and F.W. Zok, (1994) *Acta Metall. Mater.*, **42**, 2675-2681.

Wu, G., and J.-M. Yang, (2005) *JOM*, **57**, 72-79.

Xu, Z.R., K.K. Chawla, A. Wolfenden, A. Neuman, G.M. Liggett, and N. Chawla, (1995) *Mater. Sci. Eng.*, **A203**, 75.

Zhang, T., and H. Ghonem, (1995) *Fatigue. Fract. Eng. Mater. Struc.*, **18**, 1249-1262.

Zhang, W., M. Gu, J. Chen, Z. Wu, F. Zhang, H.E. Deve′, (2003) *Mater. Sci. Eng.*, **A341**, 9-17.

Chapter 9

CREEP

The creep behavior of MMCs is of great significance, since in many structural and non-structural applications, these materials will be subjected to constant stress (or strain) for long periods of time, at temperature above half of the homologous temperature (homologous temperature is the temperature of interest divided by melting point, both in K; i.e., T/T_m). Most materials exhibit three distinct stages of creep: (i) primary creep, (ii) secondary or steady-state creep, and (iii) tertiary creep. In primary creep, the strains are relatively small. In the secondary or steady-state regime, a linear relationship exists between the strain and time (constant strain rate). This is believed to be a result of the combination of hardening and recovery mechanisms during creep. Finally, in the tertiary regime, the material undergoes cavitation and void growth, which is manifested in terms of a very rapid increase in strain with time.

In general, the steady creep rate, $\dot{\varepsilon}_s$, is described by a general expression called the Mukherjee-Bird-Dorn relation (Mukherjee et al., 1964):

$$\dot{\varepsilon}_s = \frac{AGbD}{kT}\left(\frac{b}{d}\right)^p\left(\frac{\sigma}{G}\right)^n$$

where σ is the applied stress, T is the temperature in kelvin, G is the shear modulus, b is the Burgers vector, d is the average grain size, p is the inverse grain size exponent, n is the stress exponent, D is the diffusion coefficient of the material, k is the Boltzmann's constant, and A is a dimensionless constant. The value of the stress exponent can usually be correlated with a particular creep mechanism (e.g., $n \sim 4$-5 for dislocation climb). The diffusion coefficient, D, is given by:

$$D = D_o \exp\left(\frac{-Q_D}{RT}\right)$$

where D_o is the pre-exponential constant, R is the universal gas constant, and Q_D is the activation energy for creep, which is often equal to the activation energy for diffusion. For a detailed discussion on the fundamentals of creep, the reader is referred to texts such as Evans and Wilshire (1993) and Meyers and Chawla (1999).

In general, the addition of high stiffness reinforcement greatly increases the creep resistance over that of unreinforced alloys. The addition of the reinforcement also changes the creep deformation mechanisms, relative to the pure matrix. A schematic of typical creep curves (strain versus time) for MMCs with continuous and discontinuous reinforcement, is shown in Fig. 9.1 (Lilholt, 1991). Composites with continuous fibers exhibit a short primary creep regime, followed by a long steady-state regime, Fig. 9.1. This behavior can be predicted by simple viscoelastic models, such as an isostrain model, where the matrix is modeled as the viscous component and the fiber is elastic. In the case of discontinuous reinforcement (short fibers or particles), a more typical creep curve is observed, with three distinct creep regimes, Fig. 9.1, since the degree of load transfer is not as high as that for continuous fibers.

9.1 CONTINUOUS FIBER REINFORCED MMCs

The creep strength of continuous fiber MMCs is usually significantly higher than that of the unreinforced alloy. Figure 9.2(a) shows creep strain versus time for Ti-6Al-4V/SiC$_f$ composite at temperatures ranging between 430 and 650°C (Leyens et al., 2003). As mentioned above, the creep curves exhibit

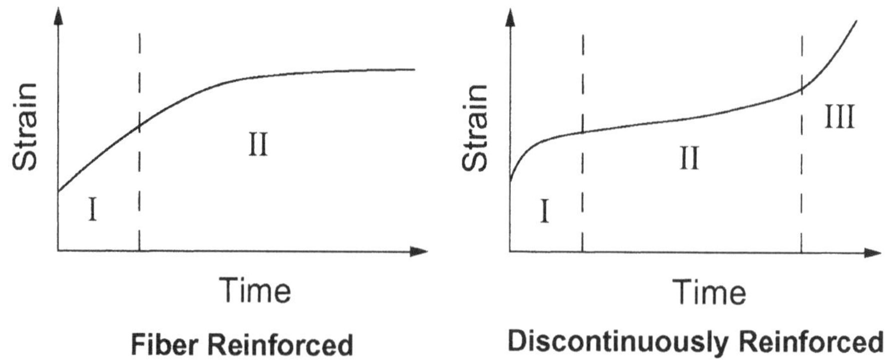

Fig. 9.1 Schematic of creep strain versus time for fiber reinforced and discontinuously reinforced MMCs (after Lilholt, 1991). The three stages of creep are (I) primary creep, (II) steady-state creep, and (III) fast fracture.

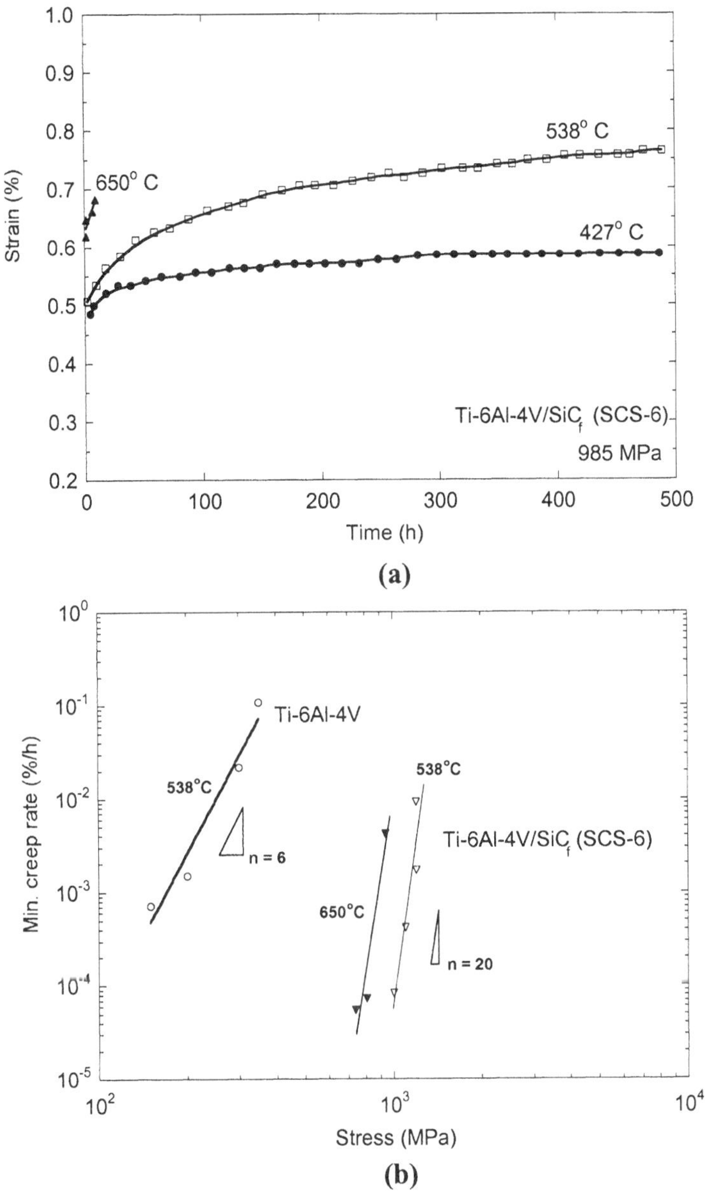

Fig. 9.2 Creep behavior of Ti-6Al-4V/SiC$_f$ composite versus the unreinforced alloy: (a) creep strain versus time plots for temperatures ranging between 430 and 650°C and (b) steady-state creep rate versus stress. The MMC exhibits significantly higher creep resistance, but much higher creep stress exponent, n, (after Leyens et al., 2003).

two stages during creep. At the highest temperature, 650°C, the creep strain increases exponentially with time. A plot of creep strain rate, $\dot{\varepsilon}$, versus stress, σ, Fig. 9.2(b), shows that a much higher stress is required to induce a given creep strain in the composite, relative to the unreinforced alloy. The stress exponent, for the composite, however, is much larger (n ~ 20 for the composite, vis-à-vis n ~ 6 for the unreinforced alloy). The reasons for the anomalously high values of the stress exponent in the composite are explained later in this chapter. In general, off-axis creep results in poorer creep resistance (Ohno et al., 1994). At 45° large-scale shear deformation takes place in the matrix, while at 90° (transverse loading) interfacial debonding results in creep fracture.

Bullock et al. (1977) studied the behavior of directionally solidified *in situ* composites of $Ni\text{-}Ni_3Al\text{-}Cr_3C_2$. The creep rate was found to be inversely proportional to the scale of the eutectic microstructure, i.e., the mean fiber radius, λ. The Hall-Petch relation was modified for creep rate and written as:

$$\log\dot{\varepsilon} = \log\dot{\varepsilon}_{\infty} + K\lambda^{-\frac{1}{2}}$$

where $\dot{\varepsilon}$ is the creep rate, $\dot{\varepsilon}_{\infty}$ is the creep rate for $\lambda = \infty$, and K is a constant. A comparison of the experimentally-determined creep rates and the model prediction is presented in Fig. 9.3, showing reasonable agreement.

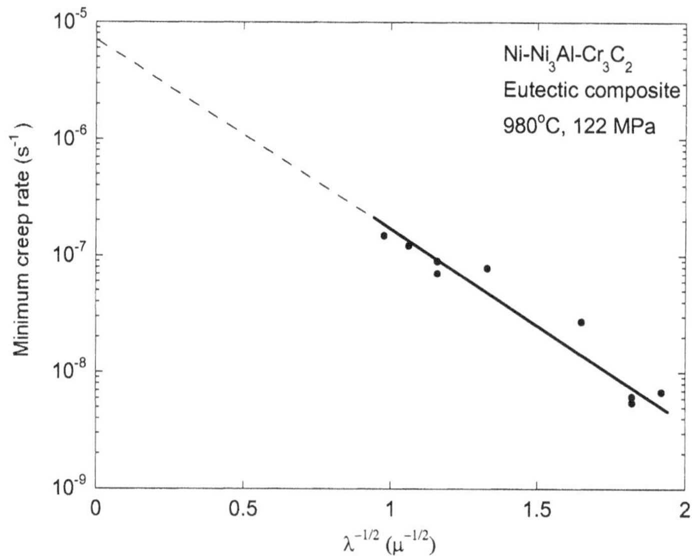

Fig. 9.3 Minimum creep rate versus inverse square root of fiber radius of directionally solidified *in situ* composites of $Ni\text{-}Ni_3Al\text{-}Cr_3C_2$ (after Bullock et al., 1977). The creep rate follows a modified Hall-Petch relationship.

The evolution of load transfer during creep, from the creeping matrix to the rigid reinforcement (most fibers do not creep at temperatures at which the metal matrix creeps), has an important influence on the creep of the composite. As the matrix creeps, an increasing fraction of the load is transferred to the fibers, Fig. 9.4(a). Upon unloading, the load in the fiber and matrix decrease. The stress in the matrix, however, dips below zero and then increases slightly. This phenomenon is termed creep recovery. A three dimensional finite element model of a fiber reinforced MMC (E_m = 60 GPa, E_f = 470 GPa, 15 vol.% of fibers, with a matrix deforming by power-law creep) illustrates this behavior very nicely, Fig. 9.4(b) (Sørensen et al., 1992).

The gradual transfer of load from the matrix to the fibers can be modeled by assuming that the fiber behaves elastically and that the matrix follows power law creep ($\dot{\varepsilon} = A\sigma^n$). Then the creep rate of the composite is given by:

$$\dot{\varepsilon}_c = \frac{A\sigma^n\left[1 - \dfrac{\dot{\varepsilon}}{\dot{\varepsilon}_\infty}\right]^n}{\left[1 + \dfrac{V_f E_f}{V_m E_m}\right] V_m^n}$$

where $\dot{\varepsilon}_\infty$ is the asymptotic creep strain, which is achieved when all of the load has been transferred to the fibers. It is given by:

$$\dot{\varepsilon}_\infty = \frac{\sigma_c}{V_f E_f}$$

The creep behavior of the composite is also dependent on whether the matrix creeps faster than the matrix, $\dot{\varepsilon}_m > \dot{\varepsilon}_f$, or vice versa, Fig. 9.5. In the case of matrix creeping faster, one may expect the fibers to fracture first. In the reverse scenario, $\dot{\varepsilon}_f > \dot{\varepsilon}_m$, the matrix should crack first, allowing the fibers to bridge the crack.

Lilholt (1985) proposed analytical models for predicting the creep behavior of fiber reinforced MMCs. He considered two cases: (a) fiber elastic and matrix creeping, and (b) both and fiber and matrix creeping. Load transfer to the fibers was modeled using a modified shear-lag theory, with the matrix between the fibers deforming by shear, and matrix at the fiber ends deforming in tension. While pure metals undergo power-law creep, the local stress in the matrix of the composite is significantly higher, so an

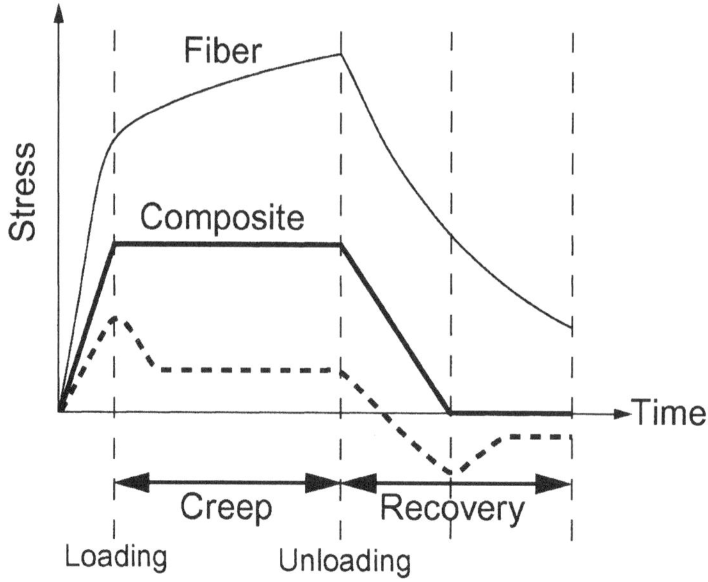

(a)

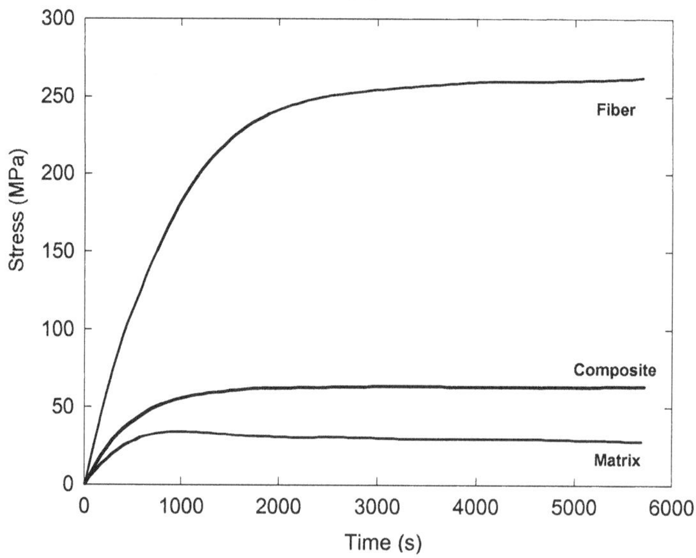

(b)

Fig. 9.4 Evolution of load transfer during creep, from the creeping matrix to the rigid reinforcement: (a) schematic of experimental behavior and (b) three-dimensional finite element model (after Sørensen et al., 1992). As the matrix creeps, an increased fraction of the load is transferred to the fibers. Upon unloading, slight creep recovery in the matrix takes place.

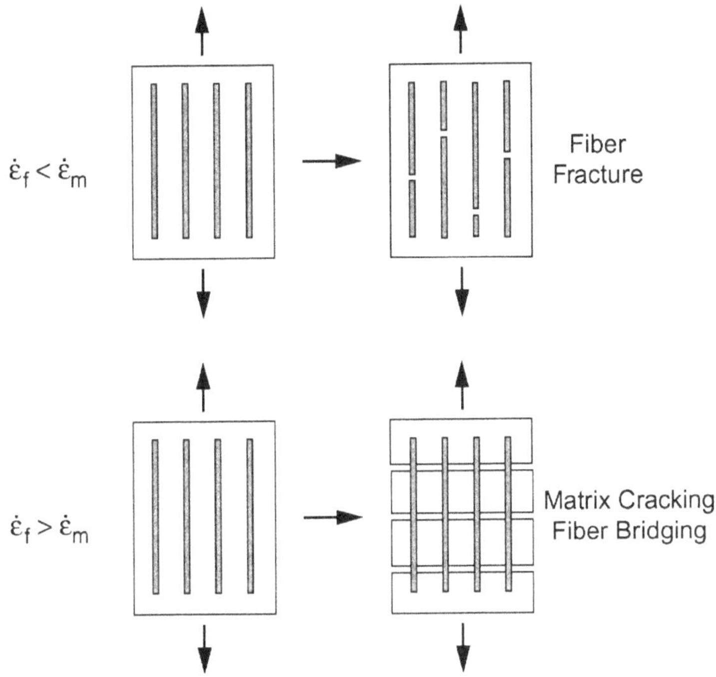

Fig. 9.5 Creep behavior of fiber reinforced MMC. When the matrix creeps faster than the matrix, $\dot{\varepsilon}_m > \dot{\varepsilon}_f$, the fibers fracture first. For $\dot{\varepsilon}_f > \dot{\varepsilon}_m$, the matrix cracks first, allowing the fibers to bridge the crack.

exponential law describing matrix creep was deemed to be more appropriate (at these stresses the power law breaks down). The exponential law was given by:

$$\dot{\varepsilon} = \dot{\varepsilon}_o \exp\left[-\frac{Q}{RT}\left(1-\frac{\sigma}{\sigma_o}\right)\right]$$

where σ is the applied stress, $\dot{\varepsilon}$ is the creep rate, $\dot{\varepsilon}_o$ is a constant, Q is the activation energy for creep (taken here as the activation energy for dislocation glide), and σ_o is the strength of the glide obstacles at 0 K. The total composite strength was modeled as a sum of the following components:

$$\sigma_c = \sigma_m + \sigma_{th} + \langle\sigma\rangle$$

where σ_m is the creep strength of the matrix, σ_{th} is a threshold stress for Orowan bowing of dislocations, and $\langle\sigma\rangle$ is a mean stress that is directly proportional to the applied strain. Good agreement was obtained with experimental results on model composites, such as Ni/W_f and $Ni/NiAl_f,Cr_3C_{2f}$.

Goto and McLean (1989, 1991) modeled the creep behavior of continuous and short fiber reinforced MMCs, based on the nature of the fiber/matrix interface, Fig. 9.6. If the fiber/matrix interface is completely incoherent, Orowan loops will stop at the interface and climb parallel to the fiber length. This results in extensive recovery and very little hardening at the interface. It also results in extensive strain relaxation at the interface and contributes to slipping of the boundary. For the case of the completely coherent boundary, two cases exist. If the modulus of the matrix is greater than that of the fiber,

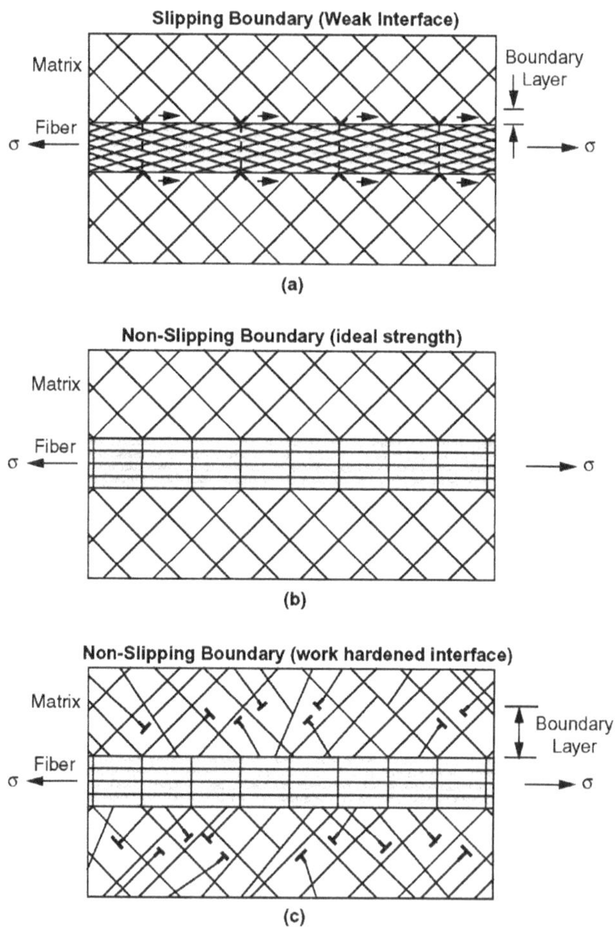

Fig. 9.6 Types of fiber/matrix interface used in modeling creep behavior (after Goto and McLean, 1989, 1991): (a) weak interface, (b) ideal strength interface, and (c) work hardened interface.

i.e., $E_m > E_f$, the dislocations are attracted to the boundary, the coherency is retained, although some slippage may take place. For the case of $E_m < E_f$, the dislocation loops are repelled from the fiber and a work hardened zone is created that is not conducive to interface slipping. The results of their model showed that the work hardened boundary temporarily carries a large fraction of the load, prior to transferring it to the fibers. This results in an enhancement in creep life. The case for a weak interface did not appear to have a significant effect on creep behavior in continuous fibers, although it did affect the short fiber behavior.

Aspects of fiber fracture and interfacial debonding were also analyzed by Lee et al. (1995). These authors conducted a parametric analysis of the creep behavior of SCS-6 fiber reinforced Ti matrix composites. They used an iterative computer simulation to determine the stress in the fiber and matrix at a given time. Models with single and multiple fibers were considered. Figure 9.7 shows two examples of numerical model predictions. Figure 9.7(a) shows the case of composites with (i) no fiber fracture, (ii) fiber fracture but no interfacial debonding, and (iii) fiber fracture with interfacial debonding. As expected, the material with fiber fracture and interfacial debonding exhibits the highest creep strain, for a given time, while the composite with no fiber fracture is the most creep resistant. The effect of interfacial strength is illustrated in Fig. 9.7(b). Note that with increasing interfacial strength, creep of the matrix is more constrained, so the overall composite creep rate is lower.

9.2 DISCONTINUOUSLY REINFORCED MMCs

Dlouhy et al. (1993, 1995) examined the creep behavior of Al_2O_3 short fiber reinforced Al7Si3Cu alloy matrix composites, processed by squeeze casting Three major mechanisms for creep damage were proposed: (i) load transfer to the fibers through a work hardened zone (WHZ) some distance from the fiber/matrix interface; (ii) diffusional/recovery mechanisms that result in a decrease in dislocation density in the WHZ; and (iii) multiple fiber fracture. The WHZ develops in the primary creep regime, which contributes to significant load transfer to the fibers. The recovery process results from dislocations moving to the fiber ends by combined climb and glide processes, Fig. 9.8. Figure 9.9 shows a comparison of the dislocation structure, in the matrix of the composite, in the as-processed condition and after creep deformation at 623 K, stress of 40 MPa, and rupture at 1.7% total strain. Note the much higher dislocation density at the fiber/matrix interface after creep. The addition of Mg to the alloy resulted in the formation of

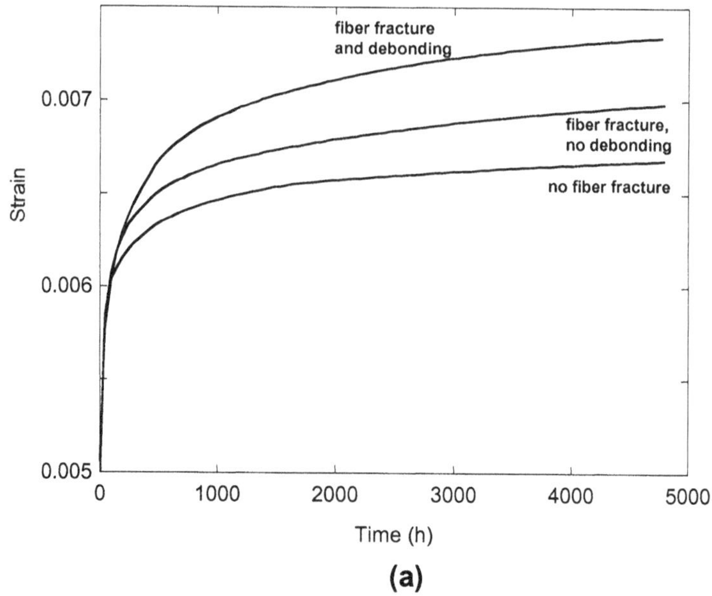

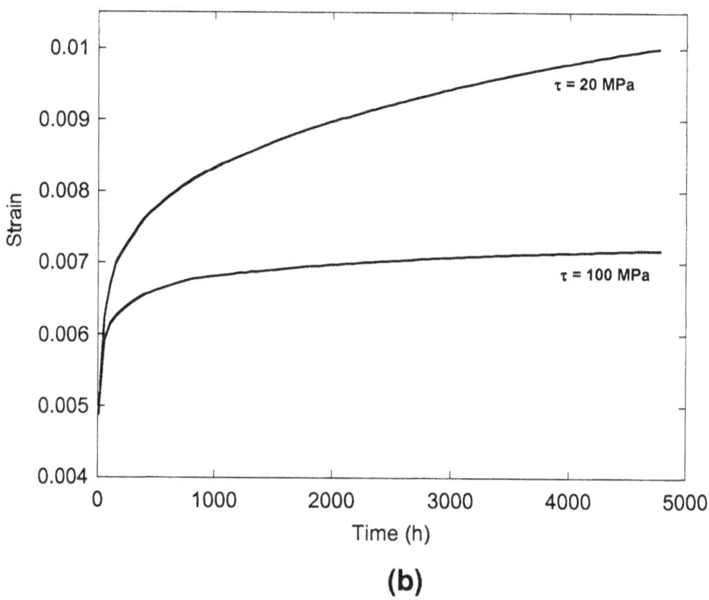

Fig. 9.7 Numerical analysis of creep behavior of SCS-6 fiber reinforced Ti matrix composite (after Lee et al., 1995): (a) effect of fiber fracture and debonding – fracture and debonding increase the creep rate of the composite; and (b) effect of interfacial shear strength, τ – increasing interface strength increases the constraint on the matrix, which lowers the creep rate of the composite.

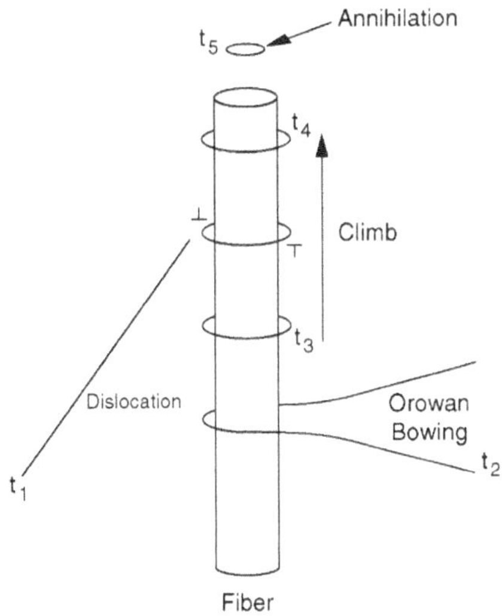

Fig. 9.8 Proposed mechanism for creep deformation in a fiber reinforced MMC (after Dlouhy et al., 1993, 1995). A dislocation bows around the fiber. The dislocation segment at the interface climbs parallel to the fiber axis and is annihilated at the fiber ends.

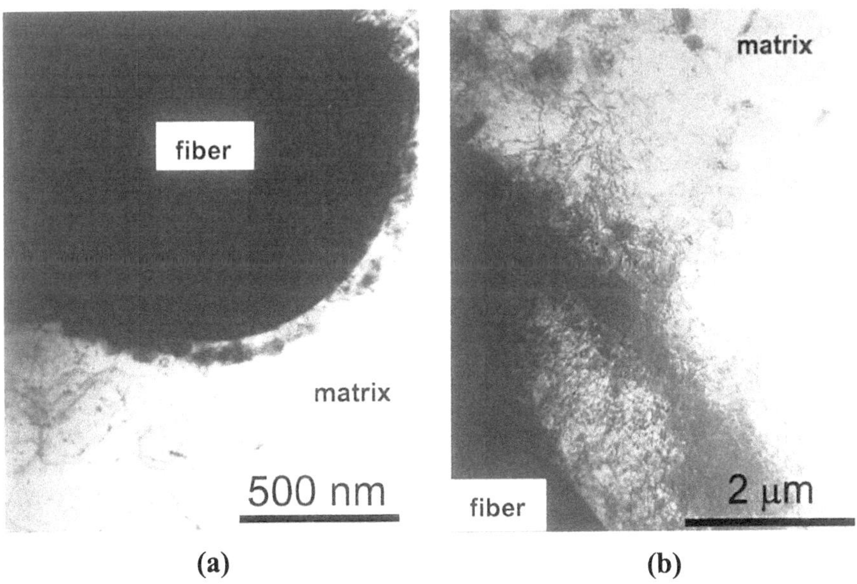

Fig. 9.9 Comparison of dislocation in the matrix of the composite in (a) the as-processed condition and (b) after creep deformation at 623 K, 40 MPa, and rupture at 1.7% total strain (courtesy of A. Dlouhy and G. Eggeler). Note the enhanced dislocation density at the fiber/matrix interface after creep.

intermetallic particles at the fiber/matrix interface (Dlouhy et al., 1993). The proposed effect of the particles at the interface was a longer effective path for recovery, and thus, an increase in creep strength. Eventually, the stress concentration caused by dislocation pile-up at the fiber/matrix interface results in fiber fracture, see Figs. 9.10 and 9.11. This also aids in recovery because of matrix diffusion at the fiber cracks. Fiber fracture and the increase in recovery contribute to the onset of tertiary creep. By using springs (elastic behavior) and dashpots (viscoelastic behavior) to represent individual components of the material behavior, excellent correlation with the experiment was obtained.

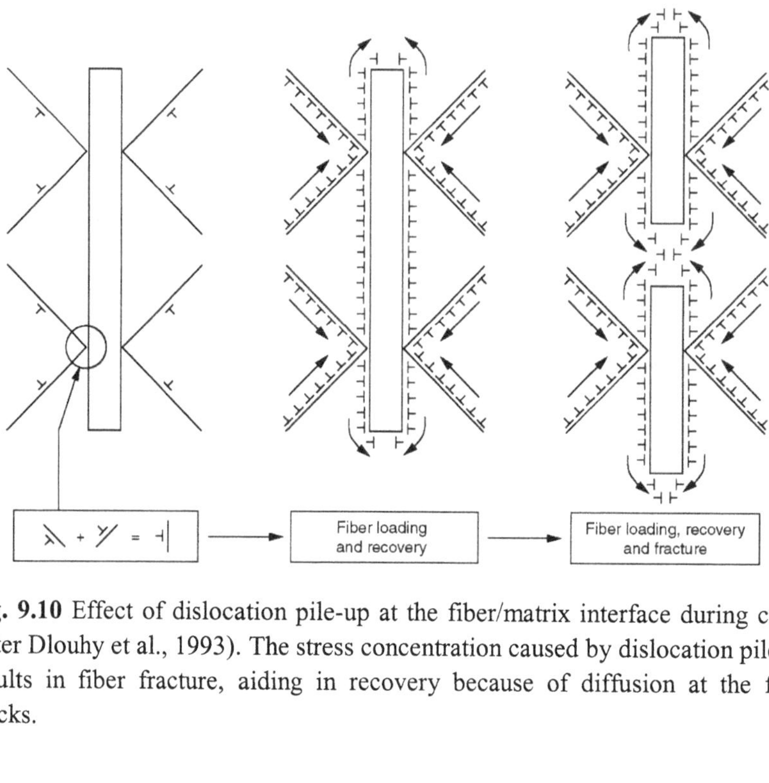

Fig. 9.10 Effect of dislocation pile-up at the fiber/matrix interface during creep (after Dlouhy et al., 1993). The stress concentration caused by dislocation pile-up results in fiber fracture, aiding in recovery because of diffusion at the fiber cracks.

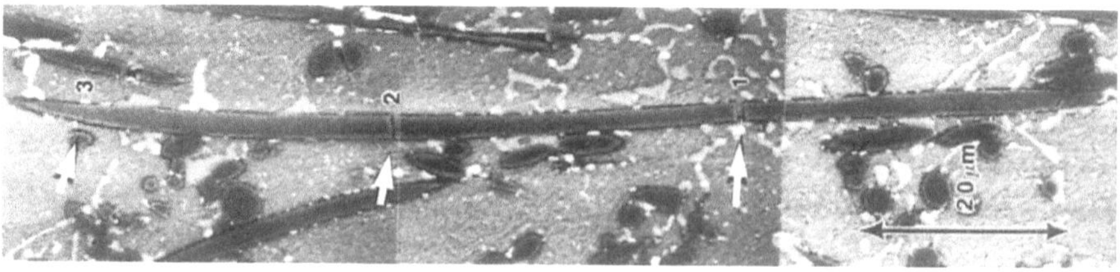

Fig. 9.11 Fiber fracture (indicated by arrows) in Al_2O_3 short fiber reinforced Al7Si3Cu alloy matrix composite during creep (courtesy of A. Dlouhy and G. Eggeler). The loading axis is horizontal.

While short fibers or whiskers provide significant creep strengthening, particulate reinforcement has also been used to enhance creep resistance. Nieh (1984) studied the creep behavior of 6061/SiC/20_p and 6061/SiC/20_w, and compared it to that of the unreinforced 6061 alloy, Fig. 9.12. He noted a much higher creep resistance in the composite with an accompanying higher sensitivity to applied stress (much higher stress exponent n). Enhanced creep resistance was also obtained with the higher aspect ratio whiskers than with particles, presumably due to more effective load transfer from the matrix to the whisker of high stiffness and large aspect ratio. Webster (1982) also characterized the creep behavior of the matrix alloy and its whisker reinforced composite with increasing temperature. At intermediate temperatures (500-720K) the strength was controlled by the whiskers, as the load was transferred to the high modulus and high aspect ratio reinforcement. The strength becomes matrix-controlled at very high temperatures (720-900 K), perhaps due to increasingly lower interfacial shear strength and lower efficiency in load transfer to the whisker reinforcement. Evidence of creep cavitation, predominantly at reinforcement particle clusters, has also been observed (Whitehouse et al., 1998).

Krajewski et al. (1993, 1995) studied the creep behavior of 2219/TiC/15_p-T6 composites, and compared it to that of the unreinforced alloy. They found that the precipitate structure in the matrix of the composite had a dominant effect on controlling creep rate. The composite had a finer interprecipitate

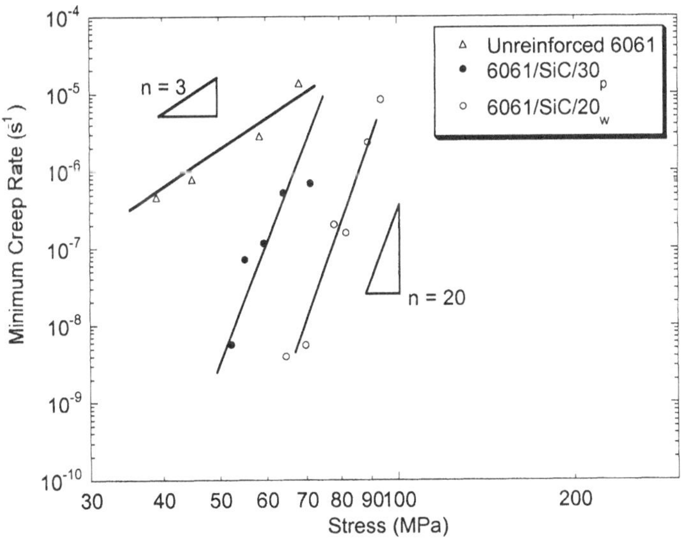

Fig. 9.12 Creep behavior of 6061/SiC particle and whisker reinforced composites (after Webster, 1982). Notice the higher creep resistance of the whisker reinforced material, due to more effective load transfer to the reinforcement.

spacing than that of the unreinforced alloy, due to indirect strengthening (see Chapter 7). Figure 9.13(a) shows the unreinforced precipitate structure in the matrix of the composite. The precipitates exhibit characteristic needle-like morphology observed in Al-Cu and Al-Cu-Mg alloys. After creep deformation at 250°C and 75 MPa, the precipitates serve as barriers for dislocation motion, Fig. 9.13(b).

In unreinforced aluminum alloys, Sherby et al. (1977) showed that the creep behavior was proportional to the substructure grain size, λ, to the third power, i.e., λ^3. They also concluded that in materials where a subgrain size during creep was relatively constant, the steady-state creep rate was better described by a stress exponent of 8, rather than the conventional value of 5 for dislocation creep. Thus, the following equation was used to describe the steady state creep rate in pure aluminum:

$$\dot{\varepsilon} = S\left(\frac{D_{eff}}{b^2}\right)\left(\frac{\lambda}{b}\right)^3\left(\frac{\sigma}{E}\right)^8$$

where S is a constant, E is the Young's modulus of the materials, and D_{eff} is the effective diffusivity for creep. Krajewski et al. (1993, 1995) also found that the creep rate in 2219/TiC/15_p composites was proportional to the interprecipitate spacing, in the matrix, to the third power. They rationalized this behavior by postulating that a substructure is formed due to the presence of the reinforcement, and the size of the substructure may be controlled by the interprecipitate spacing.

The anomalously high values of the stress exponent, n, and activation energy, Q, can be rationalized by using the concept of a threshold stress (Webster, 1982; Nieh, 1984, Nardone and Strife, 1987). Nardone and Strife (1987) used the concept of a threshold stress, σ_R, for creep deformation in composites. This theory was originally used to explain the high values for Q and n in dispersion-strengthened alloys (Davies et al., 1973; Parker and Wilshire, 1975; Nardone and Tien, 1986, Kerr and Chawla, 2004). By introducing the threshold stress, the general steady-state creep rate is modified to:

$$\dot{\varepsilon}_{ss} = A\left(\frac{\sigma - \sigma_R}{E}\right)^n \exp\left(\frac{-Q}{RT}\right)$$

where A is a constant, E is the elastic modulus of the composite, and Q is the activation energy. A methodology for determining the threshold stress from

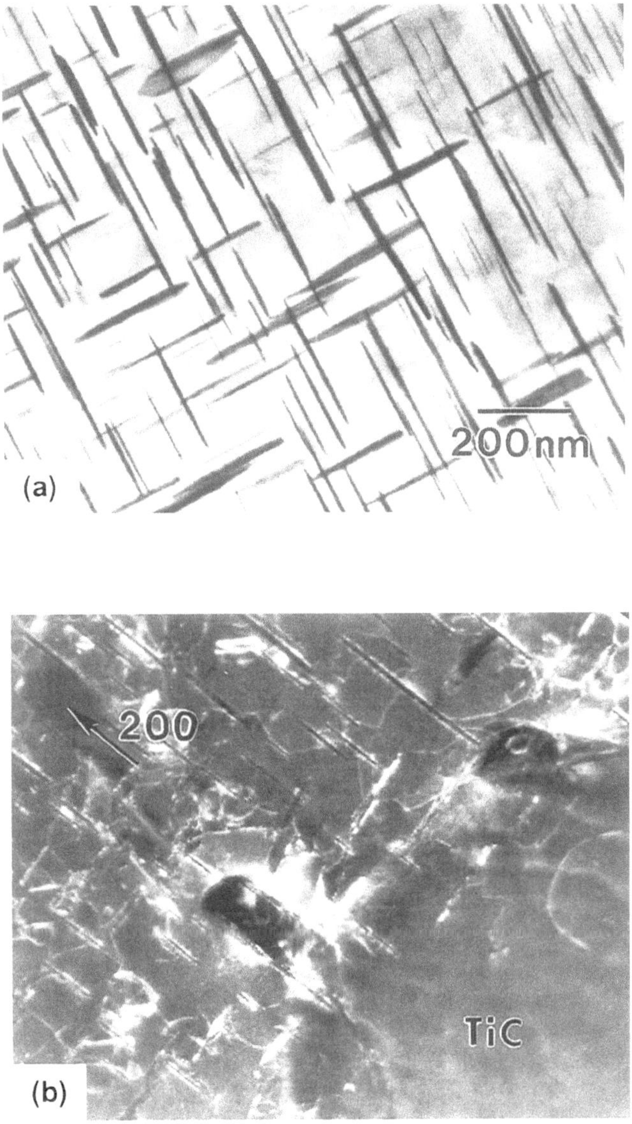

Fig. 9.13 Precipitate structure in the matrix of a 2219/TiC/15_p-T6 composite (after Krajewski et al., 1993): (a) as-processed and (b) after creep deformation at 250°C and 75 MPa. Note the interactions between dislocations and precipitates (courtesy of P. Krajewski).

experiments is shown in an inset.

The physical explanation for the threshold stress in discontinuously reinforced composites can be attributed to a variety of reasons (Dunand and Derby, 1993; Pandey et al., 1992): (1) Orowan bowing between particles, (2) back-stress associated with dislocation climb, and (3) attractive force between dislocations and particles, resulting from relaxation of the strain field of dislocations at the particle/matrix interface (Arzt and Wilkinson, 1986). Due to the higher work hardening rate of the matrix, i.e., the addition of particle decreases the volume of matrix material, increasing the word hardening rate relatively to the unreinforced alloy, the enhancement of dislocation/dislocation interactions can contribute to σ_R, although this mechanism is more plausible at ambient temperature.

A threshold stress approach cannot always be used to explain the high stress exponents observed in MMCs. Load transfer to the reinforcement, despite the lower aspect ratios of particles and whiskers, is significant. With increasing load transfer to the reinforcement, the resolved shear stress on dislocations in the matrix may be lowered significantly below that required for Orowan bowing. Dragone and Nix (1992) studied the creep behavior of Al_2O_3 short fiber reinforced Al-5% Mg alloy between 200-400°C. They also observed anomalously high values of stress exponent (n ~ 12-15) in the composites, while the unreinforced alloy exhibited much lower, typical values (n ~ 3), Fig. 9.14. The measured activation energy (225 kJ/mol) for the composites was also anomalously high. A threshold stress analysis showed that the contribution from Orowan bowing was very small. Using a model consisting of randomly oriented short fibers in the Al alloy matrix, and considering the progressive damage to the fibers during creep, they were able to predict the experimentally-observed high values of stress exponent and activation energy. Dragone and Nix (1990) also noted that the arrangement of fibers had a significant effect on the degree of matrix constraint. A decrease in effective stress (increase in matrix constraint) was observed with increasing volume fraction, fiber aspect ratio, and degree of overlap between fibers. Figure 9.15 shows a decrease in the von Mises stress in the matrix with increasing volume fraction and aspect ratio of the short fibers. The stress and strain distributions in fiber and matrix with increasing time are shown in Fig. 9.16. The stress in the fiber increases steadily with creep, indicating load transfer from the matrix to the fiber. Localization of plastic stain begins at the sharp corner of the fiber and progresses at the fiber/matrix interface normal to the loading axis. The normal and shear stresses at the fiber /matrix interface are also quite large, indicating that void growth or debonding may take place during the creep process.

Threshold Stress Analysis in Creep

The threshold stress analysis can be illustrated as follows. The figure below in part (a) (after Li and Langdon, 1998a), below, shows the experimental creep data for an Al 7005/Al_2O_3/20_p composite, in terms of shear strain rate, $\dot{\gamma}$, versus shear stress, τ, between 573-773 K. As described above, the composite exhibits an increasing stress exponent with decreasing applied stress.

The next step is to find the "true" stress exponent. This is obtained by plotting the data in terms of $\dot{\gamma}^{1/n}$ for several values of n. The plot with the best linear fit is usually a good indication of the true stress exponent. In the current case, a stress exponent of ~ 4 was obtained, part (b) (after Li and Langdon, 1998a). The exponent calculated from linear regression should be similar to that of the unreinforced alloy, where pure metallic behavior applies (unless the matrix itself has some oxide dispersions, and behaves as a dispersion strengthened material). Linear extrapolation of the linear fits to the x-axis yields the threshold stress of the material at a given temperature, τ_o.

In order to complete the analysis, the normalized creep rate is plotted versus the effective stress, part (c) (after Li and Langdon, 1998a). The normalized creep rate, from the Mukherjee-Bird-Dorn equation, is given by:

$$\frac{\dot{\gamma}kT}{D_cGb}$$

The effective stress is the applied stress minus the threshold stress, normalized by the temperature-dependence of the shear modulus:

$$\frac{(\tau - \tau_o)}{G}$$

If the threshold stress values are correct, the above plot should "collapse" the data for all temperatures onto a single line with a slope equal to the true stress exponent.

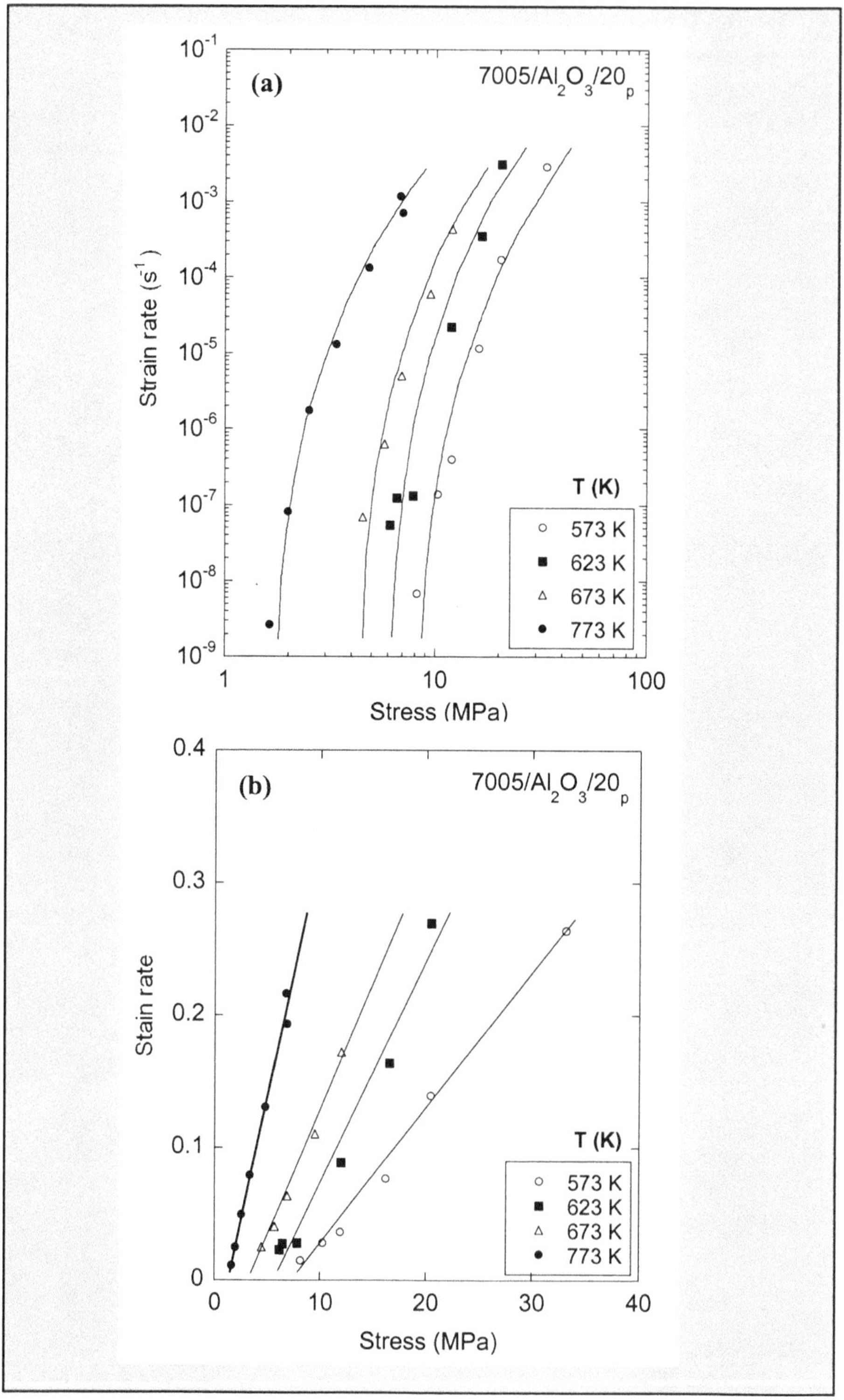
(a)
7005/Al2O3/20p
Strain rate (s-1)
Stress (MPa)
T (K)
573 K
623 K
673 K
773 K
(b)
7005/Al2O3/20p
Stain rate
Stress (MPa)
T (K)
573 K
623 K
673 K
773 K

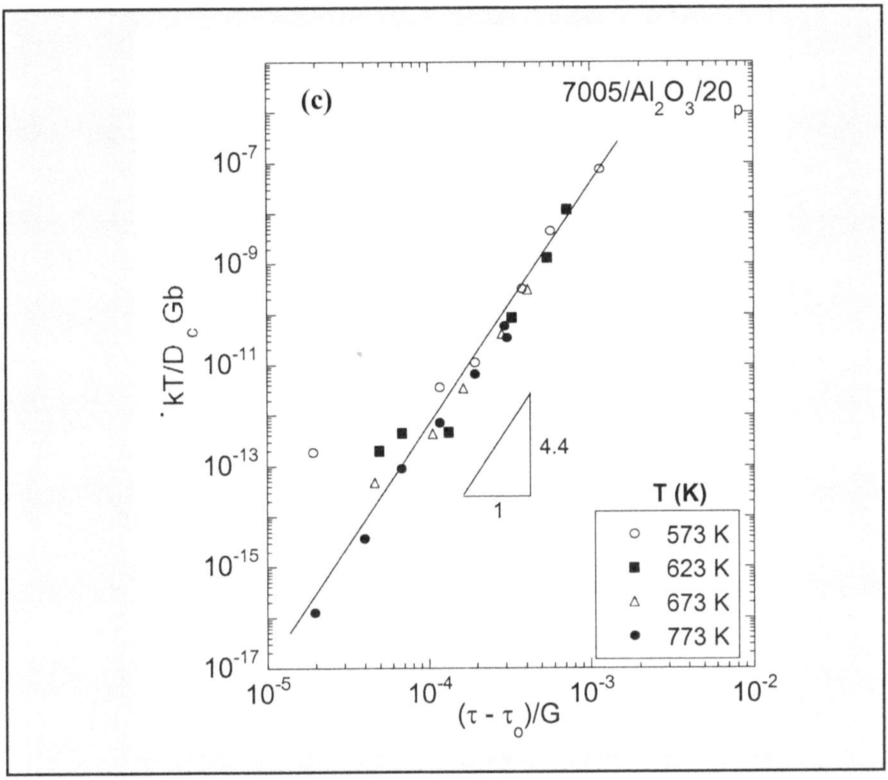

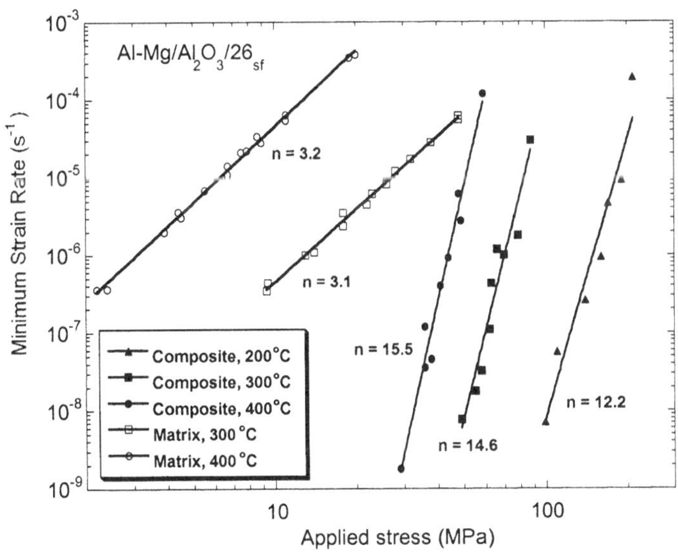

Fig. 9.14 Comparison of creep behavior of Al-5Mg/Al_2O_3/26_{sf} versus the unreinforced alloy (after Dragone and Nix, 1992). Anomalously high values of the stress exponent, n, were observed in the composite.

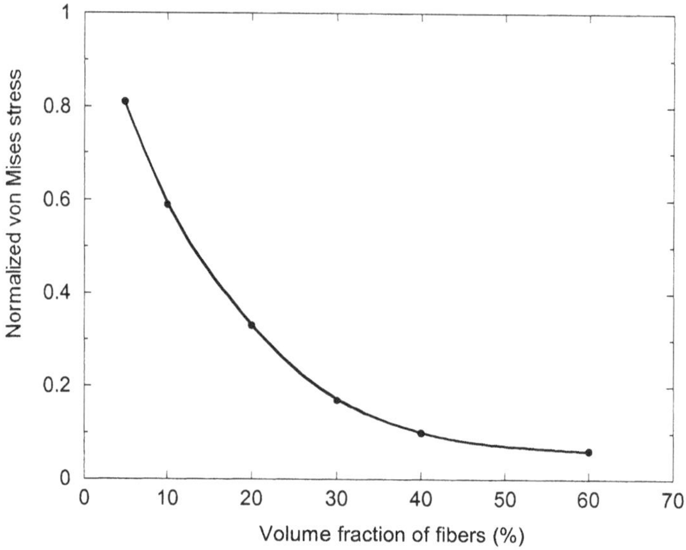

(a)

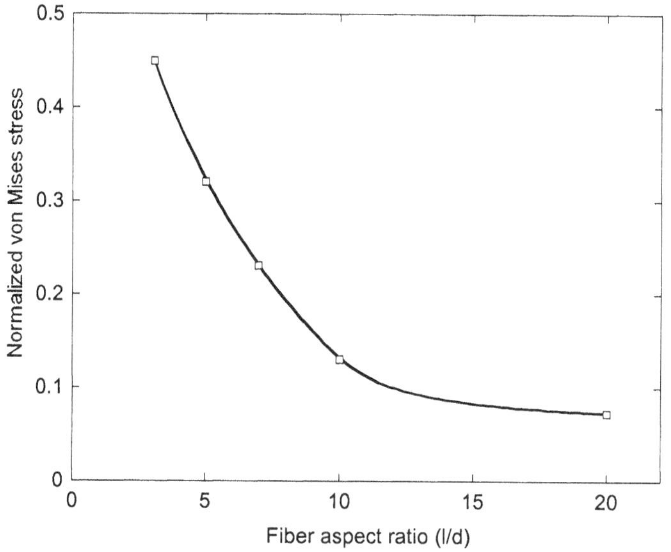

(b)

Fig. 9.15 Finite element model of creep in an Al_2O_3 short fiber reinforced Al alloy matrix composite (after Dragone and Nix, 1990). The von Mises stress in the matrix decreases with: (a) increasing volume fraction and (b) aspect ratio of the short fibers.

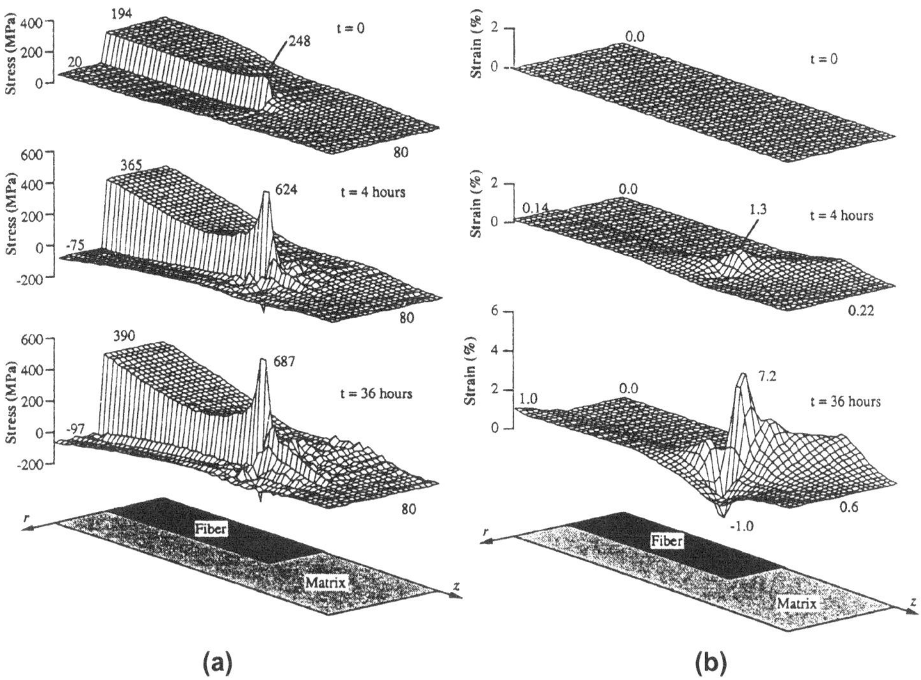

Fig. 9.16 Finite element model of creep in an Al_2O_3 short fiber reinforced Al alloy matrix composite: (a) stress distribution and (b) strain distribution during, with increasing time (after Dragone and Nix, 1990; courtesy of W.D. Nix). The stress in the fiber increases steadily with creep, indicating load transfer from the matrix to the fiber. Localization of plastic stain begins at the sharp corner of the fiber and progresses at the fiber/matrix interface normal to the loading axis.

Other continuum-based unit-cell approaches have also been used to model creep of discontinuously reinforced metal matrix composites (Bao et al., 1991; Davis and Allison, 1995; Atkins and Gibeling, 1995; Biner, 1996). Assuming a spherical particle and the power-law creep formulation for the metal matrix, Davis and Allison (1995) showed that the ratio of composite to matrix steady-state creep rates depends primarily on the volume fraction and geometry of the reinforcing phase, with the stress exponent of the composite and that of the matrix remaining relatively constant. The higher resistance to creep in the composite is largely attributed to the constrained matrix flow, leading to a reduced creep rate in the composite. Changes in modulus ratio of reinforcement to matrix affected the initial stress distribution and creep rates, but did not really affect the final creep rates. Higher modulus mismatch between reinforcement and matrix resulted in higher initial creep rates. Residual stresses due to thermal expansion mismatch also resulted in higher initial creep rates in the composite.

Other aspects of creep deformation, such as grain boundary sliding have also been included (Biner, 1996). It was shown that when the grains in the matrix were allowed to slide, that incorporation of the reinforcement resulted in an enhancement in stress in the matrix (described above) which can result in inhomogeneous grain boundary sliding. Figure 9.17(a) shows the evolution of grain boundary cavitation and sliding in a short fiber reinforced metal matrix composite. As described above, the strain concentration takes place at the pole of the reinforcement, which coincides with the onset of matrix grain cavitation. With increasing time, grain boundary sliding takes place at the cavities or facet cracks. This effect is accentuated with an increase in reinforcement aspect ratio, Fig. 9.17(b), as the local strain concentration in the matrix, immediately above the reinforcement, increases. Figure 9.18 shows a comparison of predicted creep rates for (i) matrix exhibiting cavitation and sliding, (ii) composite exhibiting matrix cavitation and sliding, and (iii) composite exhibiting matrix cavitation without sliding. The unreinforced alloy has a higher creep rate than the composite. The composite with matrix sliding but no cavitation exhibits slower creep rates at lower stress, but the creep rates approach that of the matrix at relatively large stress.

It has also been observed that below a critical strain rate diffusional relaxation around the SiC particles is the rate-controlling mechanism, while above this point, a greater degree of load is carried by the high stiffness particles (Zong and Derby, 1997). In addition to the threshold stress, additional proposed mechanisms for the anomalously high values of Q and n include power law break-down of the matrix (Zong and Derby, 1997; Lilholt, 1985) and interfacial decohesion at the particle/matrix interface (Taya and Lilholt, 1986). The reinforcement may contribute to changes in the matrix during creep by localized recrystallization at corners or interfaces and precipitate coarsening at the particle/matrix interface, where the density of precipitates is the highest (because of the greatest thermal mismatch stress upon cooling being at the interface).

It is interesting to note that in powder processed composite materials, oxide dispersions (not present in the unreinforced alloy) may also contribute to extremely high "anomalous" values of n and Q (Park et al., 1990; Li and Langdon, 1998b). Park et al. (1990) suggested that the presence of fine oxide particles, incoherent with the matrix, arising from the powder metallurgy process used to fabricate the composite, served as effective barriers for dislocation motion and gave rise to a threshold creep stress. The high creep stress exponent and increase in exponent with decreasing applied stress were attributed to the oxide particles in the matrix. The work of Li and Langdon

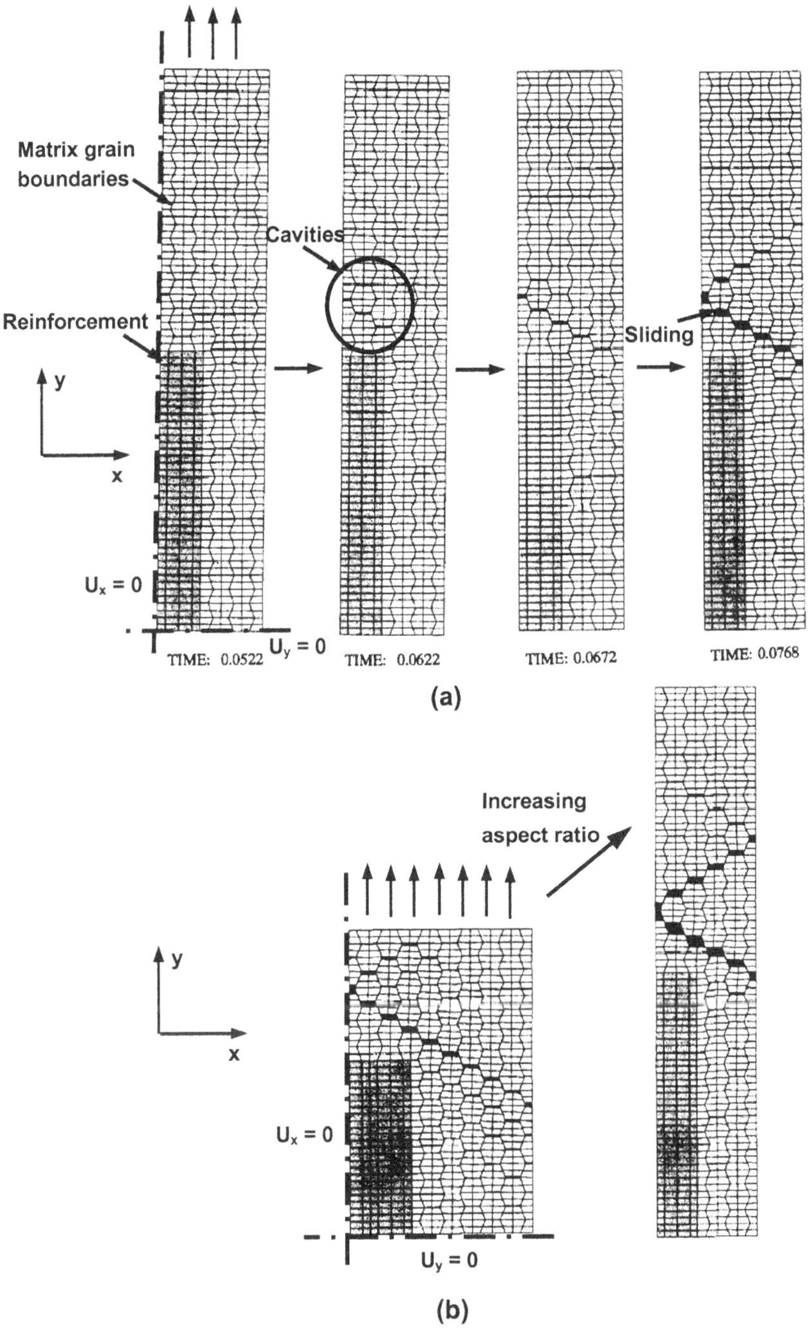

Fig. 9.17 (a) Finite element modeling of grain boundary cavitation and sliding in a short fiber reinforced metal matrix composite (after Biner, 1996; courtesy of S. Biner). Strain concentration takes place at the pole of the reinforcement, which coincides with the onset of matrix grain cavitation. With increasing time, sliding takes place at the cavities or facet cracks. (b) Increase in reinforcement aspect ratio accentuates local strain concentration in the matrix.

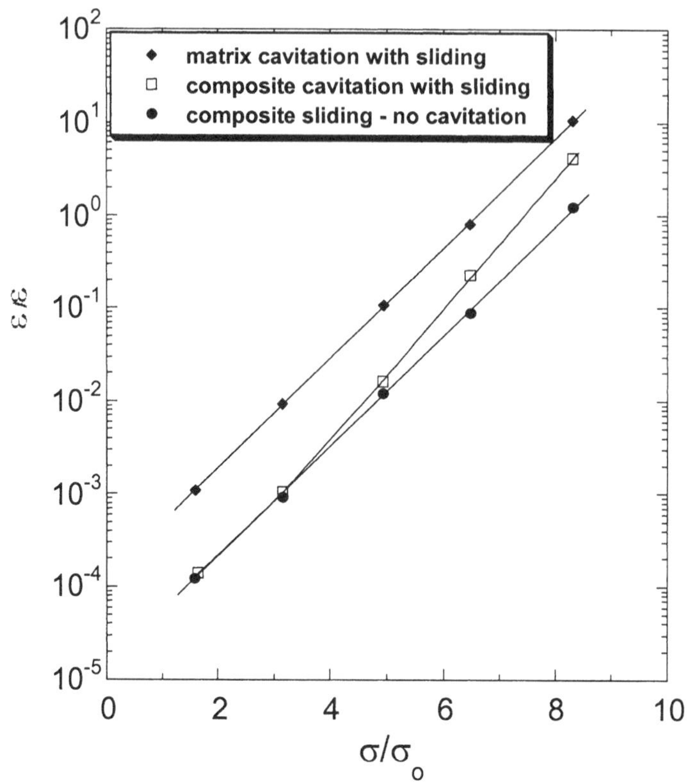

Fig. 9.18 Finite element model comparison of predicted creep rates for (i) matrix exhibiting cavitation and sliding, (ii) composite exhibiting matrix cavitation and sliding, and (iii) composite exhibiting matrix cavitation without sliding (after Biner, 1996). The unreinforced alloy has a higher creep rate than the composite. The composite with matrix sliding but no cavitation exhibits slower creep rates at lower stress, but the creep rates approach that of the matrix at relatively large stress.

(1998a) supports this conclusion. They add that in composites processed via ingot metallurgy, compared to powder metallurgy-processed composites of the same composition, viscous glide is the rate-controlling mechanism because of the absence of oxide particles.

Li and Langdon (1998a) also proposed two separate classes of creep behavior in metal matrix composites. In class M (pure metal type) materials dislocation climb is the rate-controlling mechanism, with a stress exponent of around 5 and activation energy similar to the value for self-diffusion in the matrix. In class A (alloy type) metals, viscous dislocation glide is the rate-controlling mechanism, with a stress exponent of around 3 and an activation energy associated with the viscous drag of the solute atmospheres.

Exceptionally high creep rates were observed at the highest stress levels, perhaps due to dislocations breaking away from solute atom atmospheres (Li and Langdon, 1998b). It should be noted that in unreinforced solid solution alloys a transition occurs between class M behavior at low stresses to class A behavior at higher stresses (Yavari et al., 1981).

Li and Langdon (1998a) derived an expression which can be used to determine the transition in material behavior from class M to class A:

$$\alpha\left(\frac{kT}{ec^{1/2}Gb^3}\right)^2 = B\left(\frac{\Gamma}{Gb}\right)^3\left(\frac{D_c}{D_g}\right)\left(\frac{\sigma_e}{G}\right)^2$$

where σ_e is the effective stress, β is a constant associated with the relative contributions from various viscous glide processes, e is the solute-solvent size mismatch, Γ, is the stacking fault energy of the matrix, B is a constant, and D_c and D_g, are the diffusion coefficients for dislocation climb and glide, respectively. A graphical representation of this equation, for several material systems, is shown in Fig. 9.19.

9.3 SUPERPLASTICITY

Superplasticity can be defined as the ability of a material to undergo uniform very large plastic strains (> 100% strain) (Meyers and Chawla, 1999). The

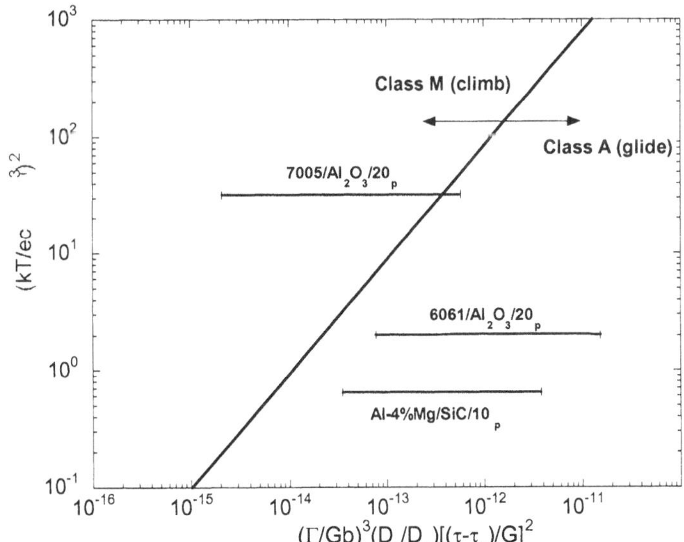

Fig. 9.19 Graphical representation of the trasition from class M (pure metal type) behavior class A (alloy type) behavior (after Li and Langdon, 1998a).

stress (σ)-strain rate $(\dot{\varepsilon})$ behavior of most materials can be described by the following equation:

$$\sigma = K\dot{\varepsilon}^m$$

where K and m are constants, and m is termed the strain-rate sensitivity. For a newtonian viscous solid, m = 1. Thus, any enhancement in m will contribute to an enhancement in superplasticity. In most alloys, the microstructural requirement for a high value of m is a very fine grain size. This is because the large plastic strains attained in superplasticity are accommodated by grain boundary sliding (Ahmed and Langdon, 1977; Mohamed et al., 1977).

Superplasticity has been demonstrated in particle reinforced metal matrix composites, such as SiC particle or whisker reinforced Al. Because of the lower ductility and higher strain hardening rates in these materials, conventional superplasticity by reduction in grain size is not feasible. Wu and Sherby (1984) used thermal cycling to generate internal stresses in the composite, due to thermal expansion mismatch between reinforcement and matrix. The internal stresses assisted plastic flow and increased the m exponent. Nieh et al. (1984) conducted isothermal forming in the solid-liquid region of the material, and were able to obtain 300% strain at relatively high strain rates (~ 3×10^{-1}/s). Mahoney and Ghosh (1987) studied the superplastic behavior of an Al-Zn-Mg-Cu matrix composite with SiC particles (about 5 μm in diameter). They were able to achieve superplastic strains of 500% in the composite, compared to 800 % in the unreinforced alloy. Fig. 9.20 shows a plot of flow stress versus strain rate during superplasticity at 516°C. An increase in SiC volume fraction resulted in an increase in the flow stress necessary for superplasticity. The flow stress versus strain rate behavior was divided into three regions, Fig. 9.21. The expected behavior of the composite is given by the dashed line. The measured flow stress, however, particularly in Region I, is much higher than the expected behavior. This was attributed to a threshold stress due to pinning of grain boundaries by the SiC particles. In Region II, the strain rate sensitivity is the highest (highest value of m), while in Region III, superplasticity is controlled by dislocation creep in the matrix.

The measured activation energies during superplasticity are often higher than for lattice diffusion or grain boundary diffusion. Li and Langdon (1998c) showed that by incorporating the contribution of load transfer to the particles (through a threshold stress approach), the true activation energy was similar to that for grain boundary diffusion. Mishra et al. (1997) studied the

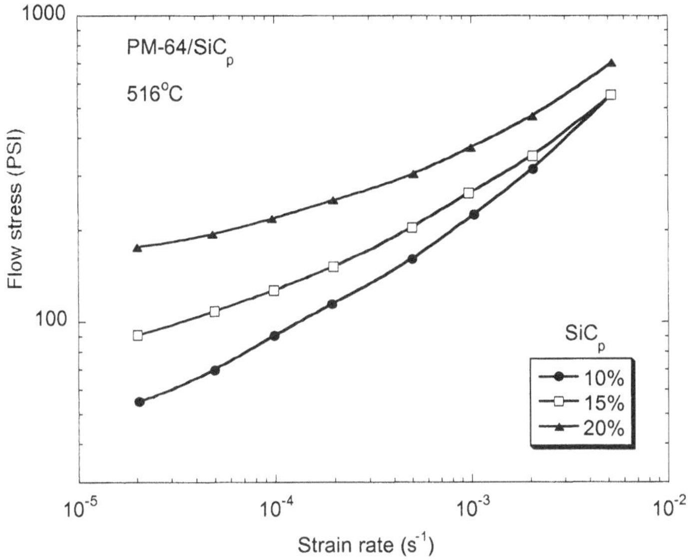

Fig. 9.20 Slow stress versus strain rate during superplasticity at 516°C of an Al alloy matrix composites reinforced with SiC particles (after Mahoney and Ghosh, 1987). An increase in SiC volume fraction resulted in an increase in the flow stress necessary for superplasticity.

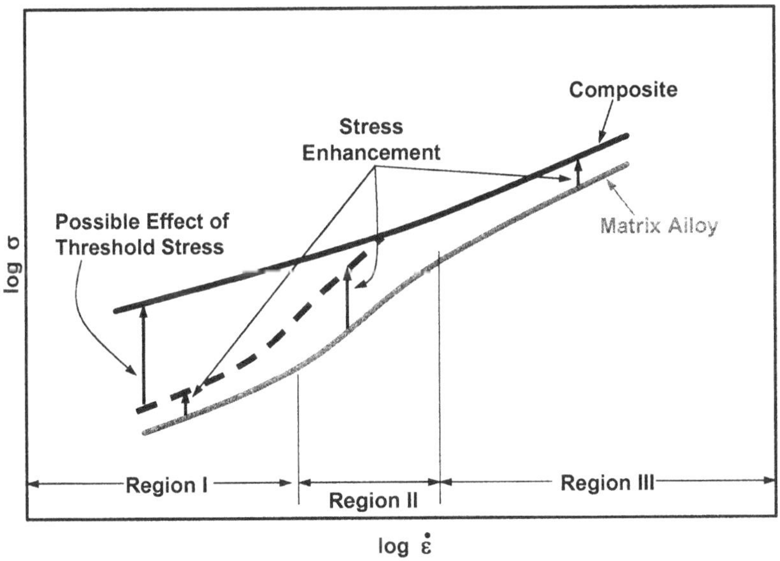

Fig. 9.21 Schematic of flow stress versus strain rate behavior for MMCs and unreinforced alloy (after Mahoney and Ghosh, 1987). The expected behavior of the composite is given by the dashed line. The measured flow stress, particularly in Region I, is much higher than the expected behavior. In Region II, the strain rate sensitivity is the highest (highest value of m), while in Region III, superplasticity is controlled by dislocation creep in the matrix.

mechanisms for superplasticity in Si_3N_4 particle reinforced 2124 Al matrix composites. In single phase materials, slip accommodation is provided by grin boundary sliding. In dispersion strengthened systems, where fine second phase particles are introduced, they pin the grain boundaries, so local diffusional relaxation around the particles must take place. In particle reinforced MMCs, however, the particles are larger than that in dispersion strengthened systems, so sliding is controlled by diffusional accommodation at the particle/matrix interface. If strain accommodation does not take place, then cavitation will occur. Mahoney and Ghosh (1987) showed that for an increase in reinforcement fraction, the fraction of voids increases, for a given applied strain, Fig. 9.22(a). This can be explained by the increase in triaxiality of stress in the matrix due to the presence of rigid, non-deforming particles (see chapter 7).

The onset of cavitation can be delayed by the superposition of a compressive hydrostatic pressure, which counteracts the triaxial tensile stress (Vasudevan et al., 1989; Lewandowski and Lowhaphandu, 1998). Figure 9.22(b) shows the delay in the onset of void growth, during superplasticity, by the superposition of a hydrostatic pressure. Mabuchi and Higashi (1999) suggested that a liquid phase during superplasticity my also decrease local stress concentrations and delay the onset of superplasticity. A large fraction of liquid phase is not desirable, however, since it will form a brittle intermetallic between the particle and matrix (see chapter 5), and cavitation will take place (Mishra et al., 1997).

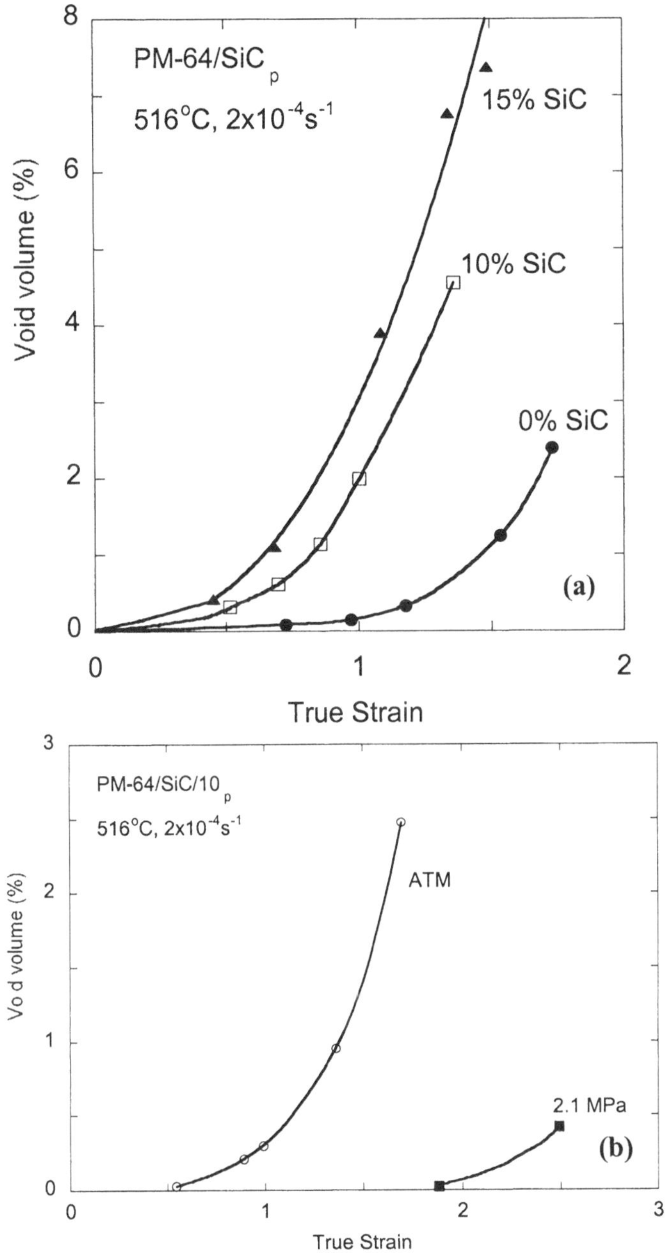

Fig. 9.22 (a) Increase in degree of cavitaiton (void volume) with increase in SiC particles in an Al alloy matrix composite (after Mahoney and Ghosh, 1987). If strain accommodation does not take place, then cavitation will occur. The increase in traixiality of stress in the matrix due to the presence of rigid, non-deforming particles results in cavitation. (b) Delay in the onset of cavitation results when a compressive hydrostatic pressure is superimposed, which counteracts the triaxial tensile stress.

References

Ahmed, M.M.I., and T.G. Langdon, (1977) *Metall. Trans.*, **8**, 1832.

Arzt, E., and D.S. Wilkinson, (1986) *Acta Metall Mater..*, **34**, 1893-1898.

Atkins, S.L., and J.C. Gibeling, (1995) *Metall. Mater. Trans.*, **26A**, 3067-3079.

Bao, G., J.W. Hutchinson, and R.M. McMeeking, (1991) *Acta Metall. Mater.*, **39**, 1871-1882.

Biner, S.B., (1996) *Acta Mater.*, **44**, 1813-1829.

Bullock, E., M. McLean, and D.E. Miles, (1977) *Acta Metall.*, **25**, 333-344.

Davies, P.W., G. Nelmes, K.R. Williams, and B. Wilshire, (1973) *Metal Sci. J.*, **7**, 87-92.

Davis, L.C., and J.E. Allison, (1995) *Metall. Mater. Trans.*, **26A**, 3081-3089.

Dlouhy, A., N. Merk, and G. Eggeler, (1993) *Acta Metall. Mater.*, **41**, 3245-3256.

Dlouhy, A., G. Eggeler, and N. Merk, (1995) *Acta Metall. Mater.*, **43**, 535-550.

Dragone, T.L., and W. D. Nix, (1990) *Acta Metall. Mater.*, **38**, 1941.

Dragone, T.L., and W. D. Nix, (1992) *Acta Metall. Mater.*, **40**, 2781.

Dunand, D.C., and B. Derby, (1993) in *Fundamentals of Metal Matrix Composites*, (S. Suresh, A. Mortensen, and A. Needleman, eds.), Butterworth-Heinneman, Boston, pp. 191-214.

Evans, R.W., and B. Wilshire, (1993) *Introduction to Creep*, The Institute of Materials, London.

Goto, S., and M. McLean, (1989) *Scripta Mater.*, **23**, 2073-2078.

Goto, S., and M. McLean, (1991) *Acta Metall. Mater.*, **39**, 153-164.

Kerr, M., and N. Chawla, (2004) *Acta Mater.*, **52** 4527-4535.

Krajewski, P.E., J.E. Allison, and J.W. Jones, (1993) *Metall. Trans.*, **24A**, 2731-2741.

Krajewski, P.E., J.E. Allison, and J.W. Jones, (1995) *Metall. Mater. Trans.*, **26A**, 3107-3118.

Lee, S., S.M. Jeng, and J.-M. Yang, (1995) *Mech. Mater.*, **21**, 303-312.

Lewandowski, J.J, and P. Lowhaphandu, (1998) *Int. Mater. Rev.*, **43**, 145-187.

Leyens, C., J. Hausmann, and J. Kumpfert, (2003) *Adv. Eng. Mater.*, **5**, 399-410.

Li, Y., and T.G. Langdon, (1998a) *Acta Mater.*, **46**, 1143-1155.

Li, Y., and T.G. Langdon, (1998b) *Mater. Sci. Eng.*, **A245**, 1-9.

Li, Y., and T.G. Langdon, (1998c) *Acta Mater.*, **46**, 3937-3948.

Lilholt, H., (1985) *Comp. Sci. Tech.*, **22**, 277-294.

Lilholt, H., (1991) *Mater. Sci. Eng.*, **A135**, 161-171.

Mabuchi, M., and K. Higashi, (1999) *Acta Mater.*, **47**, 1915-1922.

Mahoney, M., and A.K. Ghosh, (1987) *Metall. Trans.*, **18A**, 653-661.
Meyers, M.A., and K.K. Chawla, (1999) *Mechanical Behavior of Materials*, Prentice Hall, Upper Saddle River, NJ, pp. 540-591.
Mishra, R.S., T.R. Bieler, and A.K. Mukherjee, (1997) *Acta Mater.*, **45**, 561-568.
Mohamed, F.A., M.M.I. Ahmed, and T.G. Langdon, (1977) *Metall. Trans.*, **8**, 933.
Mukherjee, A.K., J.E. Bird, and J.E. Dorn, (1964) *Trans. ASM*, **62**, 155.
Nardone, V.C., and J.R. Strife, (1987) *Metall. Trans.*, **18A**, 109-114.
Nardone, V.C., and J.K. Tien, (1986) *Scripta Mater.*, **20**, 797-802.
Nieh, T.G., (1984) *Metall. Trans.*, **15A**, 139-146.
Nieh, T.G., C.A. Henshall, and J. Wadsworth, (1984) *Scripta Metall.*, **18**, 1405-1408.
Ohno, N., K. Toyoda, N. Okamoto, T. Miyake, and S. Nishide, (1994) *Trans. ASME*, **116**, 208-214.
Pandey, A.B., R.S. Mishra, and Y.R. Mahajan, (1992) *Acta Metall. Mater.*, **40**, 2045-2052.
Park, K.-T., E.J. Lavernia, and F.A. Mohamed, (1990) *Acta Metall. Mater.*, **38**, 2149-2159.
Parker, J.D., and B. Wilshire, (1975) *Metal Sci. J.*, **9**, 248-252.
Sherby, O.D., R.H. Klundt, and A.K. Miller, (1977) *Metall. Trans.*, **8A**, 843-850.
Sørensen, N, A. Needleman, and V. Tvergaard, (1992) *Mater. Sci. Eng.*, **A158**, 129-137.
Taya, M., and H. Lilholt, (1986) in *Advances in Composite Materials and Structures* (S.S. Wang and Y.D.S. Rajapakse, eds.), ASME, New York, 21-27.
Vasudevan, A.K., O Richmond, F. Zok, and J.D. Embury, (1989) *Mater. Sci. Eng.*, **A107**, 63 69.
Webster, D., (1982) *Metall. Mater. Trans.*, **13A**, 1511-1519.
Whitehouse, A.F., H.M.A. Winand, and T.W. Clyne, (1998) *Mater. Sci. Eng.*, **A242**, 57-69.
Wu, M.Y., and O.D. Sherby, (1984) *Scripta Metall.*, **18**, 773-776.
Yavari, P., F.A. Mohamed, and T.G. Langdon, (1981) *Acta Metall.*, **29**, 1495.
Zong, B.Y., and B. Derby, (1997) *Acta Mater.*, **45**, 41-49.

Chapter 10

WEAR

Wear can be defined as the loss of material that occurs when two surfaces rub against each other. Two common forms of wear are:

- Adhesive Wear
- Abrasive Wear

Adhesive wear is also known as scouring, galling, or seizing. Deformation in shear is the main mechanism in adhesive wear. Generally, high contact stresses are involved which lead to local plastic deformation, resulting in the formation of adhesive bonds between the two surfaces, Fig. 10.1. Continuing the sliding motion increases the shear stress in the bonded region until the yield strength of the softer material is significantly exceeded.

Abrasive wear occurs when hard ceramic particles such as WC or SiC get trapped between the sliding surfaces, Fig. 10.2. The degree of material loss (volume or mass) is a function of the relative hardness of the ceramic particles and the sliding surface. Environment can be an additional factor in wear. It can lead to fretting, cavitation, erosion, etc. During abrasive wear, the surface asperities are worn down and the contact surfaces become mated. This reduces the local contact stress because of the increase in contact area. After this initial abrasive wear, the removal of oxidized particles occurs along the surface. This is generally a steady state process inasmuch as it requires reoxidation of the denuded surface in order to continue removing oxidized particles. The final stage of wear occurs in adhesive mode and occurs if the contact pressure increases to the point of shearing the particles. The result of shear is the formation of thin plate-like wear debris sheets. Depending upon the applied load, the sheets can result in significant material loss.

It is important to remember that, more often than not, wear is a system

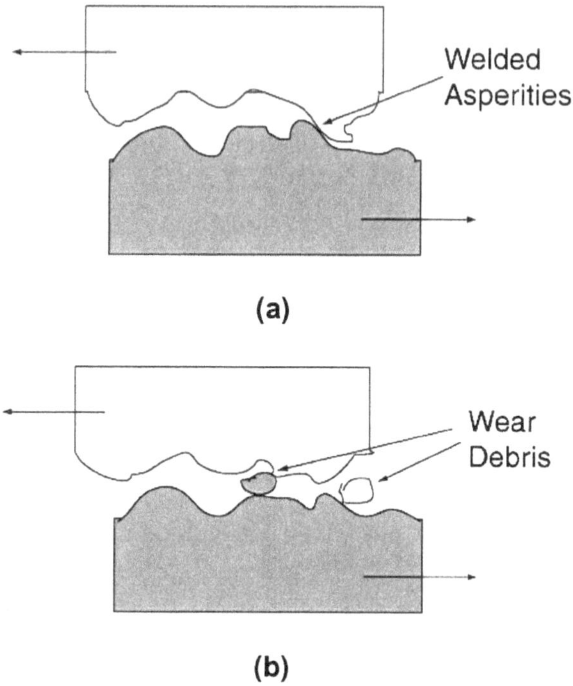

Fig. 10.1 Adhesive wear due to high contact stresses which lead to local plastic deformation, and (a) welded asperities and (b) wear debris between the two surfaces.

property. Consider the different types of variables involved: the matrix, reinforcement fibers or particles, orientation of fibers with respect to the wear direction, reinforcement/matrix interface, morphology, and size, and volume fraction of the phases in the composite. These are *intrinsic* microstructural variables. Testing or operational conditions can also affect the wear behavior. These include the type of counter surface or abrasive, force, speed, contact area, geometry, and environment (lubrication or dry). These are *extrinsic* variables. Two important wear testing techniques are described in the inset.

Under appropriate conditions, the addition of ceramic particles to a metal matrix can lead to improved wear resistance. It should be recognized, however, that if the conditions are not suitable, the addition of ceramic particles may decrease wear resistance. For example, the ceramic particles may be removed from the matrix and lead to a wear process by three body abrasive wear, Fig. 10.2(b). Another point to keep in mind is that as the volume fraction or size of the reinforcement becomes large, the fracture

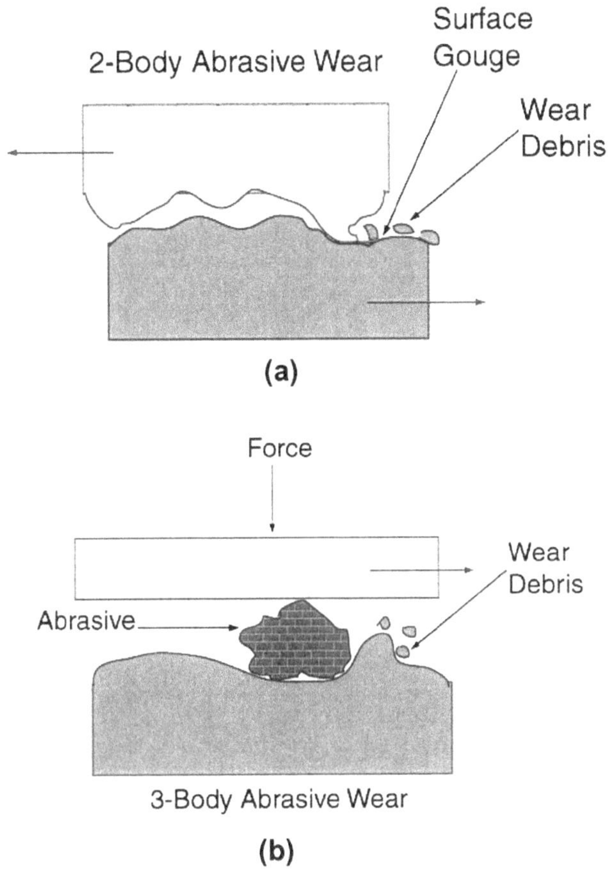

Fig. 10.2 Abrasive wear involving ceramic particles such as WC or SiC leading to debris trapped between the sliding surfaces: (a) two body wear and (c) three body wear.

toughness of the composite will be reduced significantly. If the fracture toughness is inadequate, the particles will fracture and contribute to the wear process. Increasing wear resistance without a penalty in toughness is a challenge.

Most of the work in the area of wear of MMCs has been on particle reinforced composites. Thus, in this chapter we focus mainly on the wear behavior of two important systems: silicon carbide/aluminum composites and tungsten carbide/cobalt. In the end we devote a short section on wear aspects in braking materials. A more comprehensive treatment on the applications of braking materials is given in chapter 11.

Wear Testing

Pin-on-disk wear testing (ASTM G99-95a)

In this test, a normal load is applied via pin onto a rotating disk. The sliding motion may be in one direction or reciprocating. The tests may be conducted in dry or in lubricated conditions. The coefficient of friction, μ, can be obtained from this test at any point in the test. Typically one measures the mass before and after the wear test by means of a sensitive microbalance. The wear rate is given by:

$$\text{wear rate} = \frac{\text{mass loss}}{\text{sliding distance}}$$

The wear rate may also be reported as:

$$\text{wear rate} = \frac{\text{volume loss}}{\text{sliding distance}}$$

High stress and low stress wear testing

In the high stress wear test (ASTM B611), a coarse alumina particle slurry (~590 μm) is used as an abrasive. A steel wheel is rotated at 100 rpm through the slurry bath. The test sample is pressed against the steel wheel with a known force. One measures the mass loss, after 1000 revolutions of the steel wheel, and obtains the wear number as:

$$\text{Wear number} = \frac{\text{Density}}{\text{Mass loss}}$$

Physically, the wear number gives the number of revolutions to remove a unit volume of the sample. The higher the wear number, the better the wear resistance.

In the low stress wear test (ASTM G65), dry quartz sand (200-300 μm) is used as the abrasive. It is softer than the alumina slurry used in the high stress wear test. Instead of a steel wheel, a rubber wheel is used. After 0.5 h at 200 rpm, we obtain the volume loss as:

$$\text{Volume loss} = \frac{\text{Mass loss}}{\text{Density}}$$

The lower the volume loss, the higher the wear resistance.

10.1 PARTICLE REINFORCED ALUMINUM MATRIX COMPOSITES

In the abrasive stage of wear of MMCs, the degree of abrasion is a function of the applied load, speed, size and volume fraction of reinforcement, and interfacial bond strength between the reinforcement and matrix. Abrasive wear resistance of several Al alloys containing a range of ceramic particles and volume fractions have been examined by a number of researchers (Hosking et al., 1982; Wang and Rack, 1991; Chung and Hiang, 1994; Zhang et al., 1995; Garcia-Cordovilla et al., 1996). The addition of ceramic particles, such as alumina, to aluminum alloys result in substantial reduction of the wear rate (Hosking et al., 1982). The reduction in wear rate is influenced by the size of the reinforcement particle and the volume fraction of reinforcement. The greatest wear resistance is typically achieved with the alloy containing the largest volume fraction and the largest particle size, as shown in Fig. 10.3. Figure 10.4 shows an example of the effect of SiC particles on wear resistance (Venkataraman and Sundarajan, 1996a, 1996b). The improvement of the composite over the unreinforced alloy is greatest at higher loads.

After wear, the worn surface of the unreinforced aluminum alloys typically consists of long, continuous grooves, which are the result of abrasive particles plowing the surface. The surface of composites, on the other hand, exhibit substantially reduced surface plowing (Garcia-Cordovilla et al., 1996). The wear rate of the composites decreases as the volume fraction of reinforcement increases. The wear rate usually decreases as the particle size increases. This may be attributed to the particle strength being higher than the interface strength, for smaller particles. Thus, the smaller particles are pulled out before fracturing. The larger particles remain in the matrix longer and are able to carry the surface load until they fracture into smaller particles.

One should remember that, besides hardness, the fracture toughness of the reinforcement also has an important role to play. As the stress is increased, the fracture toughness of the reinforcement becomes an important parameter. The fracture toughness determines the extent of fracture of the reinforcement. Thus, the governing factor for wear with increased stress is the fracture toughness of the reinforcement particles.

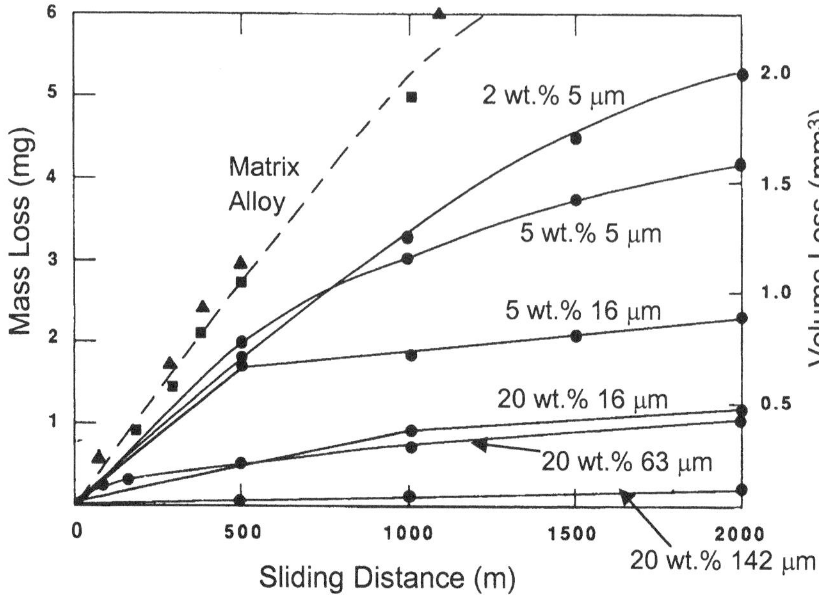

Fig. 10.3 Improved wear resistance of aluminum 2024 alloy matrix composites with increasing volume fraction and size of the alumina particles (after Hosking et al., 1982).

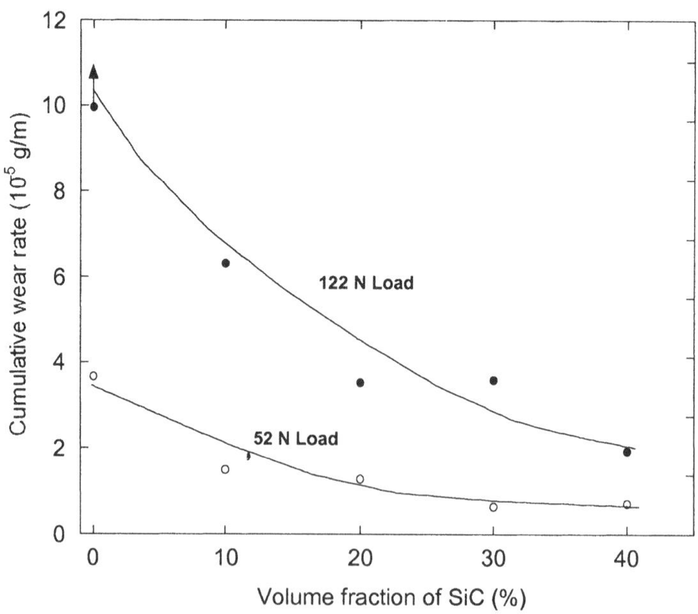

Fig. 10.4 Increase in wear resistance with increased volume fraction of SiC (after Venkataraman and Sundarajan, 1996a).

Oxidation of surface debris during wear process is another phenomenon during wear of MMCs. Abrasion transforms to oxidative wear after a certain sliding distance. With increasing load, the wear behavior changes from mild, oxidative wear to severe wear by adhesion. Examination of the worn surface and debris indicates that debris formation is more influential on wear rate, within the abrasive regime, than deformation of the material (Ludema, 1996). After a certain sliding distance, the surface asperities wear down, which reduces the contact stress because of the increase in contact area for the same load. As a result, the wear rate and coefficient of friction also decrease during this period.

Suh (1973) proposed a delamination wear mechanism wherein voids and cracks nucleate at second phase particles in the subsurface region below the contact surface. This was attributed to the large hydrostatic pressure acting near the asperity contact. Cracks during wear typically propagate parallel to the contact surface (Suh, 1973, Jahanmir and Suh, 1977). During sliding wear the surface is plastically deformed, which introduces dislocations in the subsurface of the contacting materials. The dislocations lead to eventual formation of cracks and/or voids in the subsurface region. Fragmentation of the particles can also lead to decohesion of the particle/matrix interface, resulting in more void formation. With continued sliding, the cracks and void eventually coalesce and result in larger cracks. When the crack reaches a critical length, the material between the crack and the surface is removed from the surface in the form of wear debris. The severe wear by adhesion can occur by such a delamination process initiated by subsurface cracking. Three factors influence this phenomenon: Particle size, volume fraction, and type of reinforcement. Larger particles require larger loads to initiate subsurface cracking and eventual delamination. The work of Zhang and Alpas (1993) and Chung and Hiang (1994) supports Suh's delamination theory. They examined cross-sections of worn surfaces of aluminum alloys reinforced with SiC particles, which showed evidence of wear debris in the form of flakes. These flakes of debris were produced by subsurface delamination. Microscopic analysis showed cracks, which formed at the particle/matrix interface and propagated parallel to the surface. The cracks and resultant voids coalesced with eventual delamination of material from the surface. Depending upon the rate of wear debris formation, delamination may result in a reduction of wear by the formation of a debris layer between the contacting surfaces. This layer, also termed a mechanically mixed layer (MML) is comprised of the reinforcement, matrix material, and material from the counter surface (Venkataraman and Sundarajan, 1996a, 1996b).

Venkataraman and Sundarajan (1996b) found the hardness of the MML to be approximately six times that of the bulk composite. They hypothesized that as the deformation increased, the reinforcement particles at the wearing surface fragmented and the number of voids began to increase. Once a critical void density was reached, shear instability was initiated at selected locations within the subsurface. The shear instability promoted turbulent plastic flow, which resulted in a mixture of material from both surfaces. The mixture of the material led to the formation of the MML, which appeared to have a direct influence on the wear rate. The rate of formation of the MML controlled the wear rate and was directly related to delamination wear.

Frequently, improved wear performance can also be obtained by adding a lubricant or by adding a reinforcement that provides lubricating characteristics. Rohatgi et al. (1992) observed that carbon, in the form of graphite particles, in aluminum can decrease the coefficient of friction against dry steel. Carbon in microcrystalline form, on the other hand, had practically no effect. This is understandable inasmuch as graphite flakes cleave easily and serve as lubricants. One should be cautioned that in addition to the parameters described above, the environment can also play a decisive role in the ultimate wear performance.

10.2 TUNGSTEN CARBIDE/COBALT COMPOSITES

Tungsten carbide/cobalt (WC/Co) composites, commonly known as cemented carbides, are one of the most important metal matrix composites used for their excellent wear resistance (Upadhyaya, 1998). They were originally developed for drawing tungsten incandescent filament, where wear resistance is clearly important. Now they are also widely used for cutting, machining, and drilling purposes. Cemented carbides consist mainly of fine tungsten carbide particles and the binder metal cobalt.

The wear behavior of conventional WC/Co composites is characterized by plasticity of cobalt matrix and the high strength and hardness of the WC particles. In high speed machining, the temperature at the cutting edge can exceed 1000°C leading to phenomena such as diffusion and oxidation of the cutting tool, abrasion, and adhesion, e.g., the chip can adhere to the flank face. In general, the wear behavior is directly related to hardness, which in turn is a function of volume fraction and size of tungsten carbide particles.

10.2.1 Hardness and Strength of Cemented Carbides

Hardness is a critical property in determining the wear resistance of a material. WC/Co composites have been studied extensively and some useful correlations have been obtained. Lee and Gurland (1978) proposed the following expression for the hardness of WC/Co composites:

$$H = H_{WC} C V_{WC} + H_{Co}(1 - C V_{WC})$$

where H is the hardness of the composite; H_{WC} and H_{Co} are the hardness of WC and Co, respectively; C is the contiguity of the WC particles, and V_{WC} is the volume fraction of WC in the composite. The parameter contiguity is defined as:

$$C = \frac{2S_{cc}}{2S_{cc} + S_{cm}}$$

where S_{cc} is the area between the carbide particles and S_{cm} is the area between the carbide particles and the matrix in a unit volume. The hardness of cobalt is related to the mean free path (λ) of the cobalt matrix in the composite by the following empirical relation:

$$H_{Co} = 304 + 12.7\lambda^{-0.5}$$

The important result in this case is that the hardness of cobalt matrix goes as the inverse square of the mean free path.

The homogeneous distribution of a high volume fraction of fine WC particles in a cobalt matrix, in conventional cemented carbides, enhances strength, wear resistance, thermal conductivity etc. of the composite, but at the expense of toughness. This has led to the development of a new type of composite which we describe in the next section.

10.2.2 Double Cemented Carbide (DCC)

A novel particulate composite metal matrix composite called Double Cemented Carbide has been developed (Fang and Albert, 1999; Fang et al., 1999; Deng at al., 2001). This material consists of granules of WC/Co composite embedded within Co or another metal matrix as shown in Fig. 10.5. Thus, it is a kind of "dual" composite with a "composite-within-a-composite" structure. This concept provides microstructural design variables not available in conventional particulate metal matrix composites. We describe some interesting relationships among the processing, microstructure and properties of this novel composite material. These additional degrees of freedom enable remarkably enhanced *combinations* of properties, such as

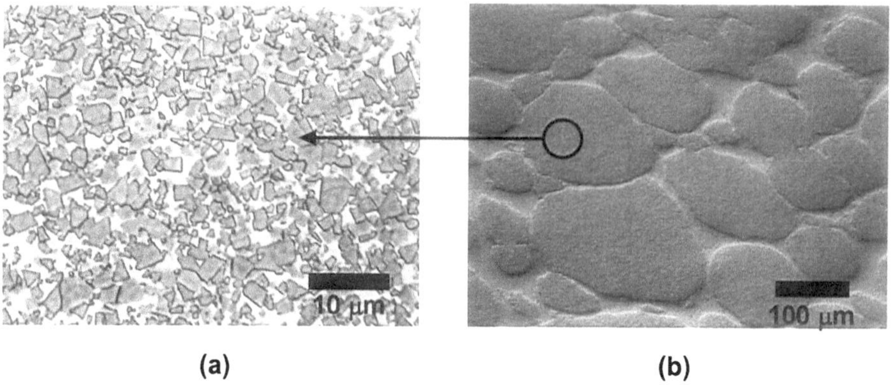

Fig. 10.5 Microstructures of cemented carbide (WC/Co) with (a) conventional and (b) DCC structure, containing the same *total* cobalt contents, 25 vol%. The coarse granules in (b) have a fine dispersed microstructure similar to that of (a).

high toughness and wear resistance.

Figures 10.5(a) and (b) show composites containing 75 vol% WC particles in a cobalt matrix but with the two different types of structure: (a) conventional particulate composite structure with homogeneously distributed particles and (b) DCC structure. The DCC material has twice the toughness and six times the high-stress abrasive wear resistance than the conventional material, at the same total particle volume fraction. These improvements in properties result from the partitioning of the matrix to produce particle-free matrix regions separating the granular regions, which have higher particle volume fractions than the overall average. These ductile intergranular regions enhance fracture resistance with little detriment to abrasive wear resistance. Wide ranges of combinations are achievable by fine tuning of granule composition and properties, granule size, matrix composition and heat treatment.

Figure 10.6 shows a plot of fracture toughness vs. wear number in terms of kilo revolutions per unit cm^3. It illustrates the difference in toughness/wear resistance behavior between conventional and DCC composites. Recall that the higher the number of revolutions, the better the wear resistance. The thicker line represents the relationship for essentially all grades of conventional cemented carbide, i.e., with wide variations in cobalt matrix fraction and WC particle size. As with most conventional particulate composites, improvement in either property comes at the expense of the other. Thus, enhanced toughness means poor wear resistance and vice versa. This restricted relationship results from the limited degrees of freedom in

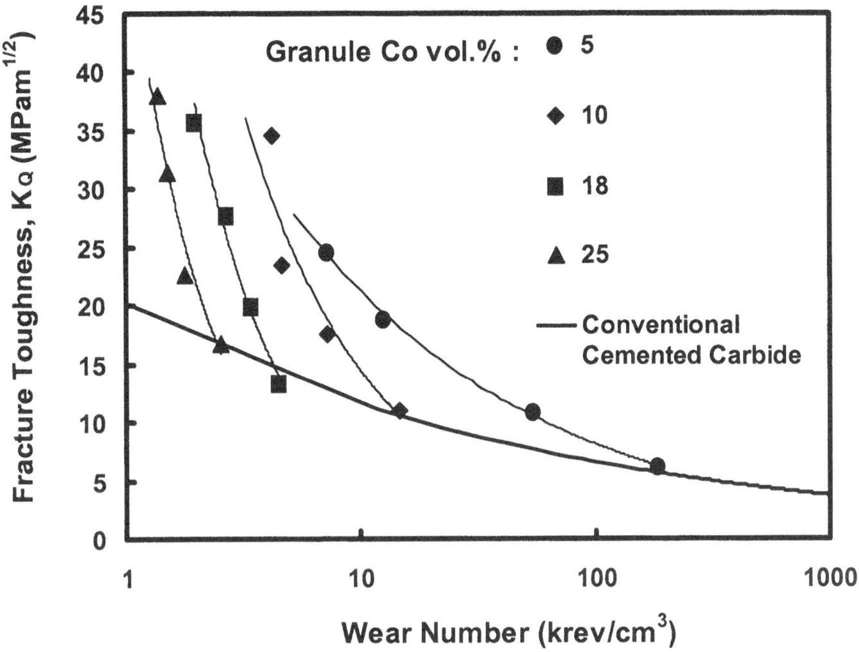

Fig. 10.6 Superior combination of toughness and wear resistance of DC Carbide vs. conventional cemented carbide (after Deng et al., 2001).

microstructural design and restricts the balance of properties in essentially all particulate MMCs. The curves for different DCCs with different granule properties (decreased cobalt yields harder granules with increased wear resistance and decreased toughness) and different volume fractions of intergranular matrix, show that at comparable wear resistance the DC structure has higher toughness than the conventional structure and at comparable toughness DC has higher wear resistance (Deng et al. 2001).

The improved toughness of the DCC composite is a function of the increased mean free path through the intergranular matrix (~20 μm vs. < 1 μm in conventional cemented carbide). The improved abrasive wear resistance of the double composite is due to a different wear resistance mechanism than with conventional particulate MMCs. Protrusion of the large, hard composite granules from the wear surface can be seen after wear, Fig. 10.7. The best wear resistance/toughness combination results from a high volume fraction of relatively large, hard granules, since increased granule size increases the intergranular mean free path, which enhances toughness. Larger granules exhibit the greatest protrusion for wear resistance and are the most difficult to remove via matrix erosion.

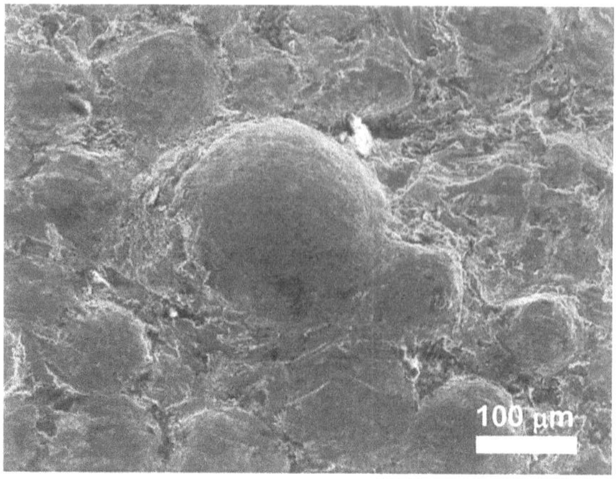

Fig. 10.7 Wear surface of DC carbide after high stress abrasive wear testing, showing protrusions (after Deng at al., 2001).

10.3 Wear in Braking Materials

Wear and frictional characteristics are very important in materials used for braking applications. In addition, one must consider physical properties such as thermal conductivity. Every time we apply a brake to stop a moving vehicle, the kinetic energy of the rotating elements of the vehicle must be absorbed by the brake. This energy is dissipated in the form of heat within the brake. The rise in temperature because of energy dissipated as heat can be simply written as:

$$E = mC\Delta T$$

where m is the mass of the device, C is the specific heat, and ΔT is the temperature rise. Conventionally, we use steel, cast iron (trains), and asbestos. The latter has been mostly replaced because of its carcinogenic nature. A material used in a braking system must be able to generate the torque sufficient for stopping the vehicle, must absorb and dissipate the heat produced during the slippage process, and withstand the high temperatures reached during stopping. It should also have superior structural properties for transferring the torque to the tires, without adding extra weight to the vehicle to make it inefficient (Chawla, 1998). MMCs with appropriate frictional characteristics such as wear, durability and ability to withstand higher temperatures are excellent candidates for braking materials. Heat resistance capability of aluminum alloys can be increased substantially by adding 30% by volume of SiC particles. Such a composite was used to make a brake rotor (Nakanishi et al., 2002).The composite brake rotor had a higher heat resistance than those than a conventional cast iron rotor, especially a more stable coefficient of friction.

Zeuner et al. (1997) developed a MMC disc brake for high-speed trains that resulted to be not only mechanically superior to conventional cast irons, but also cost effective in terms of manufacturing braking system of a train is composed by four brake discs and the necessary calipers, hand brake, and electromagnetic track brakes. The higher thermal conductivity and lower modulus of elasticity of aluminum result relatively low thermal stresses during the braking operation, which is a great advantage for Al-MMC discs over cast iron. For the same braking load, the cast iron discs showed higher thermal stresses and non uniform temperature distribution than the Al-MMC disc, resulting in cracks due to thermal stresses in cast iron brakes but not in composite, see chapter 11.

In summary, metal matrix composites, especially the particulate variety, can provide significant improvements in applications involving friction and wear. The ability to tailor their properties such as stiffness, hardness, and toughness are the main keys to such improvements. The material properties, reinforcement size and volume fraction have a profound influence on wear resistance. Increased volume fraction increases the stiffness but generally lowers the fracture toughness. Thus, increased volume fraction of ceramic particles will result in greater wear resistance at lower loads, but the fracture toughness becomes an important issue with increased load. For a given volume fraction, increased particle size provides the best wear resistance, but larger particles are also more susceptible to fracture. Fine particles are less likely to experience fracture, but are more likely to be buried in matrix and removed by delamination. The loading conditions, matrix and reinforcement properties, and reinforcement size and volume fraction influence the creation of a mechanically mixed layer between the contacting surfaces, which will help to protect the surfaces and slow the wear rate. Thus, in designing for wear resistance, it is important to select the proper combinations of matrix, reinforcement, and reinforcement size and volume fraction.

References

Chawla, K.K., (1998) *Composite Materials: Science and Engineering*, Springer-Verlag, p. 276.

Chung, S., and B. Hiang, (1994) *Tribol. Inter.*, **27**, 307-314.

Deng, X., B.R. Patterson, K.K. Chawla, M.C. Koopman, Z. Fang, G.

Lockwood and A. Griffo, (2001) *J. Refractory Metals and Hard Materials*, **19**, 547.
Fang, Z. and S.J. Albert, (1999) U.S. Patent 5,880,382.
Fang, Z., G. Lockwood, and A. Griffo, (1999) *Metall. Mater. Trans.*, **30A**, 3231
Garcia-Cordovilla, C., J. Narciso, and E. Louis, (1996) *Wear*, **192**, 170.
Hosking, F. M., F. Folgar-Portillo, R. Wundnerlin, and R. Mehrabian, (1982) *J. Mater. Sci.*, **17**, 477.
Jahanmir, S., and N.P. Suh, (1977) *Wear*, **44**, 17-38.
Lee H.C. and J. Gurland, (1978) *Mater. Sci. Eng.*, **33**, 125.
Ludema, K.C., (1996) *Friction, Wear, Lubrication: A Textbook in Tribology*, CRC Press, Boca Raton, FL., p. 139.
Nakanishi, H., K. Kakihara, A. Nakayama, and T. Murayama, (2002) *Japan Soc. Auto. Eng. Rev.*, **23**, 365-370.
Rohatgi, P.K., S. Ray, Y. Liu, (1992) *Inter. Mater. Rev.*, **37**, 192.
Suh, N.P., (1973) *Wear*, **25**, 111-124.
Upadhyaya, G.S., (1998) *Cemented Tungsten Carbides*, Noyes Pub., Westwood, NJ.
Venkataraman, B., and G. Sundarajan, (1996a) *Acta Mater.*, **44** 451-460.
Venkataraman, B., and G. Sundarajan, (1996b) *Acta Mater.*, **44** 461-473.
Wang A., and H.J. Rack, (1991) *Wear*, **146**, 337.
Zeuner, T., P. Stojanov, P.R. Sahm, H. Ruppert, and A. Engels, (1998) *Mater. Sci. Tech.*, **14**, 857-863.
Zhang, J., and A.T. Alpas, (1993) *Mater. Sci. Eng.*, **A160**, 25-35.
Zhang, Z.F., L.C. Zhang, and Y.W. Mai, (1995) *J. Mater. Sci.*, **30**, 1961-1966.

Chapter 11

APPLICATIONS

Metal matrix composites are used in a myriad of applications. The high strength-to-weight ratio, enhanced mechanical and thermal properties over conventional materials, and tailorability of properties make them very attractive in a variety of applications. Increasingly MMCs have been used in several areas including (Evans et al., 2003):

- Aerospace
- Transportation (automotive and railway)
- Electronics and thermal management
- Filamentary superconducting magnets
- Power conduction
- Recretational Products and Sporting Goods
- Wear-resistant materials

In this chapter we review some important applications of MMCs, and point out the advantages of using MMCs.

11.1 AEROSPACE

In aerospace applications, low density, tailored thermal expansion and conductivity, high stiffness and strength, are the primary drivers. In this industry, performance often outweights cost considerations in materials development.

11.1.1 Aircraft Structures

MMCs have been used in several applications in aerospace components. This stems in large part to the fact that materials with increased specific stiffness and strength can significantly enhance the performance of the aircraft. A simple example is shown in Fig. 11.1. Stringers are often used to stabilize the skin in the fuselage. Increased skin stiffness means that there is less chance of the skin buckling. Furthermore, a larger stringer can be used to stabilize the skin. Using a SiC particle reinforced Al matrix composite skin, the effective allowable stringer size increased resulting in enhanced stability of the skin.

In military aircraft, MMCs have found a large number of applications (Miracle, 2001). In the F-16 aircraft, aluminum access doors were increasingly susceptible to fatigue cracking. The aluminum doors were replaced with particle reinforced MMC (Al/SiC_p) as a weight-neutral solution. MMC was also used as a replacement for unreinforced aluminum in the ventral fins of the F-16, Figures 11.2(a) and 11.2(b). Figure 11.2 (c) shows the deflection of the ventral fin as a function of number of missions. The MMC material exhibited much lower deflection because of the higher stiffness. The incorporation of the MMC resulted in four-fold increase in life of the component and $26 million life cycle cost savings. Similar improvements were obtained by incorporation of the MMC in the F-16 fuel access door covers. Continuous fiber reinforced MMCs have also been used in military aircraft applications, due to their high specific strength, stiffness, and fatigue resistance. SiC monofilament reinforced Ti matrix composites have been used as nozzle actuator controls for the F119 engine in F-16. The MMC replaced a heavier Inconel 718 in the actuator links and stainless steel in the piston rods.

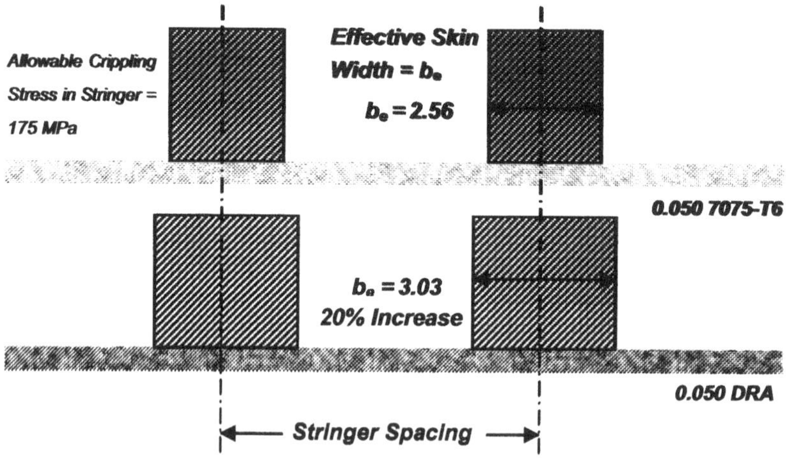

Fig. 11.1 Schematic of stringers on aircraft skin. By using a particle reinforced MMC skin, the effective allowable stringer size increases, resulting in enhanced stiffness of the skin (courtesy of D. Bowden).

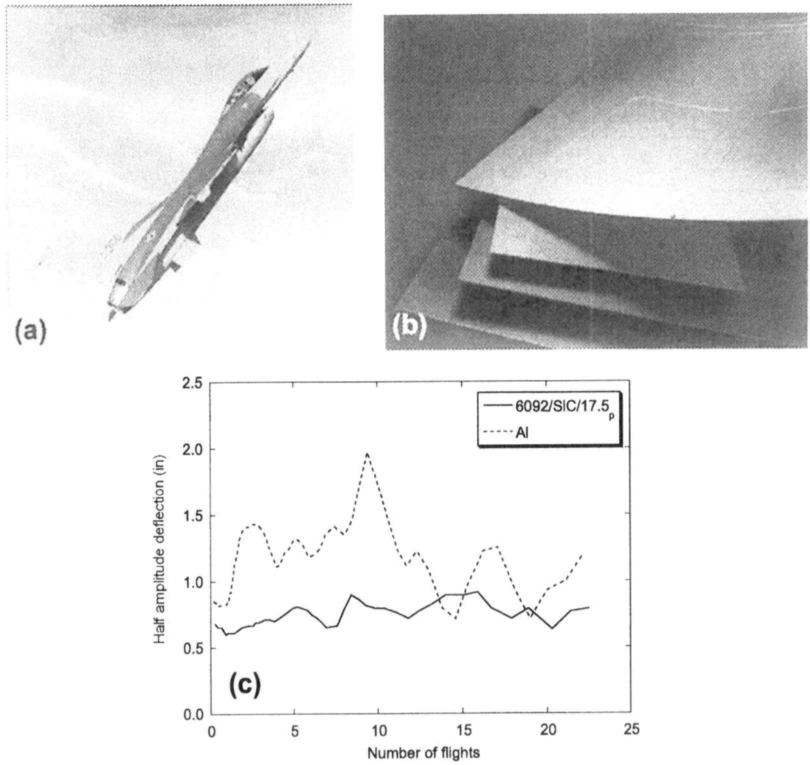

Fig. 11.2 SiC particle reinforced Al alloy MMC used as a replacement for unreinforced aluminum in the ventral fins of the F16 (courtesy of D. Miracle): (a) F16 aircraft, (b) composite ventral fin (courtesy of DWA Aluminum Composites), and (c) deflection of the ventral fin as a function of number of missions. The MMC material exhibited much lower deflection because of the higher stiffness.

MMCs have also been used in commercial aircraft. Figure 11.3 shows an application of MMC in the fan-exit guide vane of a Pratt & Whitney engine on a Boeing 777. The MMC replaced a carbon/epoxy composite that had problems with foreign object damage (FOD). The increased specific stiffness characteristics of MMCs were also exploited in a fuselage strut application considered by Airbus. The MMC replaced a carbon fiber reinforced polymer for the strut, in order to reduce cost and increase damage tolerance.

In general, the cost of particle reinforced MMCs is significantly lower than that of carbon/epoxy counterparts. The distribution of costs for each type of

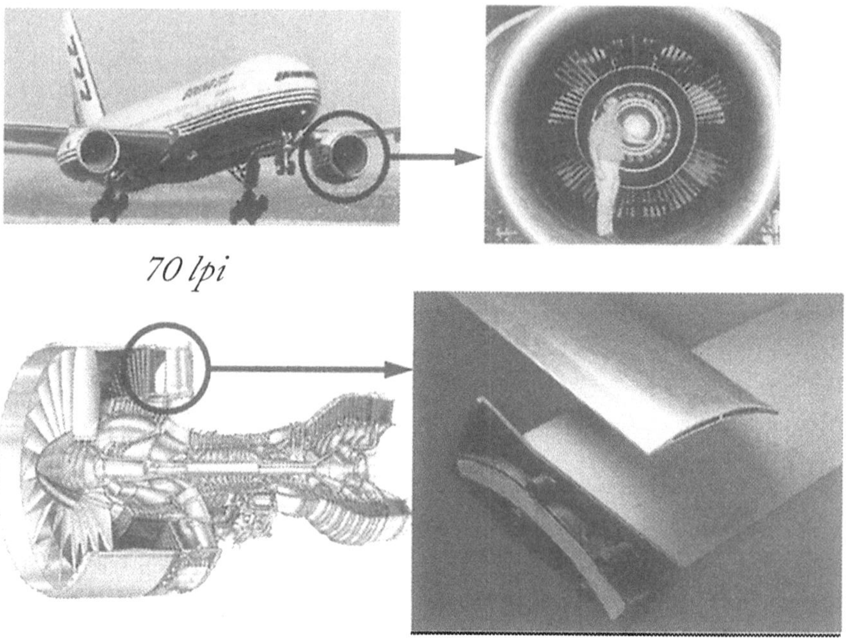

Fig. 11.3 Application of a SiC particle reinforced Al MMC in the fan-exit guide vane of a Pratt & Whitney engine on a Boeing 777 (courtesy of D. Miracle). The MMC replaced a carbon/epoxy composite that had problems with foreign object damage (FOD) and at a lower cost.

material is also quite different. A comparison of the cost breakdown for polymer matrix composites and the MMC is shown in Fig. 11.4. In polymer matrix composites, a large fraction of the cost goes into primary and secondary processing. In the MMC, the processing cost is a very small fraction of the total cost. Rather, the raw material cost of the matrix and reinforcement makes up most of the cost.

Another important application in aerospace is blade sleeves in helicopters, Fig. 11.5. The blade sleeve must be able to tolerate centrifugal loads from the blades to the rotor. Thus, fatigue life, fretting resistance, toughness, and high specific strength are required. The composite consisted of SiC particles in an Al alloy matrix. Powder metallurgy-processed billets were extruded, cut, and forged. The final component yielded only 1% in waste. The fatigue strength of the composite was around 270 MPa (assuming a fatigue runout at 10^7 cycles), which was a 50-70% increase over unreinforced Al. The fracture toughness was reasonable, at 25 MPa$\sqrt{m}$, and the specific strength and cost were lower than thos of the initial Ti alloy used in this application.

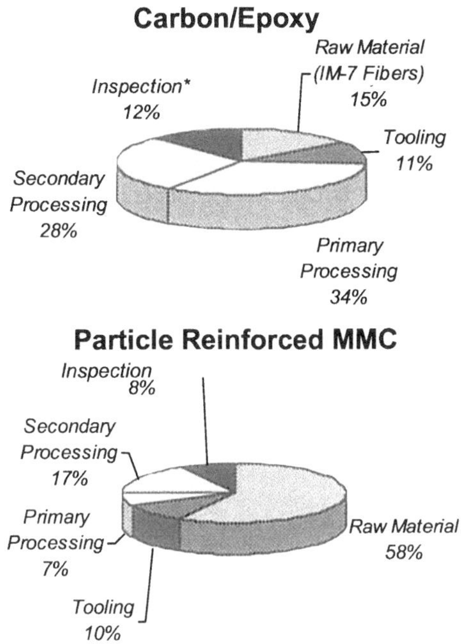

Fig. 11.4 Cost comparison between the polymer matrix composite and the particle reinforced MMC for the fan-exit guide vane (courtesy of D. Bowden). In the polymer matrix composite, a large fraction of the cost goes into primary and secondary processing. In the MMC, the processing cost is a very small fraction of the total cost. Raw material cost (matrix and reinforcement)makes up most of the cost.

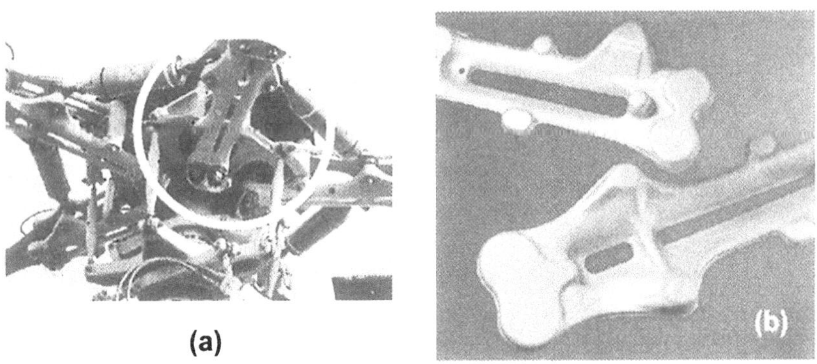

Fig. 11.5 (a) MMC blade sleeves in helicopters and (b) forged components (courtesy of D. Miracle and DWA Aluminum Composites). The blade sleeve must be able to tolerate centrifugal loads from the blades to the rotor. Thus, fatigue life, fretting resistance, toughness, and high specific strength are required. The specific strength equivalent and cost were lower than those of the monolithic Ti alloy used in this application.

11.1.2 Fiber Metal Laminates

Fiber Metal Laminates (FMLs) consist of alternately stacked thin (about a millimeter or less in thickness) sheets of metal (commonly aluminum) and fiber reinforced polymer (commonly epoxy) (Vlot and Gunnick, 2001; Vogelesang et al, 1995). The first commercial FML was *AR*amid *AL*uminum *L*aminates, ARALL. Arall was used in a few select aircraft components, but it had structural limitations that prevented wider use. GLARE or *GLA*ss *RE*inforced laminates, a glass-aluminum FML, was developed in part to overcome these limitations. Glare consists of alternating layers of aluminum and glass fiber/epoxy layers. Such a laminate is produced in an autoclave for the curing of the polymer matrix. The different layers of the laminate are stacked before curing, manually or by automated machines. There is a high degree of tailorability in these composites because both the number of layers as well as the direction of the fibers in the PMC layers can be varied depending on the application of the structural part. The main characteristics of Glare are extremely good fatigue properties, high damage tolerance capabilities, and optimal impact properties. Many structural parts (primary and secondary) can be identified where these properties play an important role. In particular, Glare has been chosen for use in the fuselage of the twin-deck, 550-seat, Airbus 380 aircraft.

Figure 11.6 shows schematically the configuration of metallic sheets and sheets of polymer matrix composite. Besides having superior fatigue behavior, Glare is also lighter than aluminum, cutting the weight of the Airbus A380 by 1,000 kg. It was discovered that bonded and laminated aluminum had favorable resistance to crack growth because cracks would grow in a single layer at a time, and the remaining layers would effectively

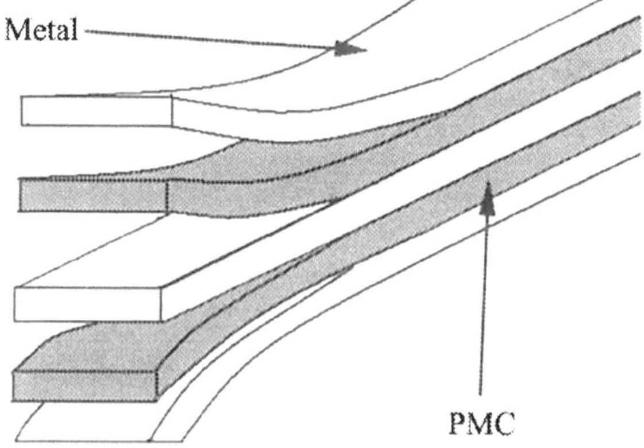

Fig. 11.6 Schematic of fiber metal laminate consisting of alternating sheets of metal and fiber reinforced polymer matrix composite.

bridge the crack. This discovery was made use of in the development of the F-27 aircraft wings, which is a very highly fatigue sensitive structure.

Fuselage construction with ARALL revealed that under loading conditions encountered in an aircraft fuselage, the aramid fibers around a fatigue crack would fracture. Once the fibers fractured, the crack was no longer inhibited and would grow. As the crack grows, delamination occurs and a compressive stress is placed on the aramid fibers, causing buckling. Low compressive stress of aramid fibers is well known (see, for example, Chawla, 1998). It was observed that aramid fibers would fracture under rather low compressive loads and it also became evident that ARALL was also sensitive to strength reductions caused by any holes that might be drilled in it. Such holes are unavoidable in a large structure such as a fuselage; they would lead to premature fatigue cracks to develop. Although use of ARALL was not suitable for fuselage construction, the weight savings by using this material were very attractive. The first commercial application of ARALL was on the C-17 military transport plane. Because of heavy weight problems toward the rear of the plane, the large aluminum cargo door was replaced with ARALL.

Glass fibers were used to replace aramid fibers in the new Glare FML, which has superior blunt notch strength and the ability to handle greater compressive stresses than Arall (Vlot and Gunnick, 2001). Glare is also more resistant to impact than Arall and for this reason the first commercial application of Glare was in the cargo floor of Boeing's 777 where the excellent impact properties of Glare were put to use. The next area of use of Glare was in the form of panels for the fuselage of Airbus's A380 aircraft.

The main consideration that underlies the design of fiber metal laminates is their superior fatigue strength. This is shown schematically in Fig. 11.7 where a plot of crack length as function of number of cycles shows Glare to be superior to aluminum under cyclic loading. Flammability is another important consideration for any aircraft material. Damage from sources such as fire has troubled early aircraft design because of the low melting point of aluminum. For example, an accident at Manchester airport resulted in loss of life from an external fire that entered the passenger cabin by burning through and melting the fuselage (Vlot and Gunnick, 2001). The introduction of GLARE to fuselage design improved the fire resistance of aircraft because of the nature of the composite. After burn through of the outer aluminum layer, Figure 11.8, the glass fiber/epoxy layer is exposed to the flames. Since the melting temperature of the glass fibers is high and the epoxy layer becomes carbonized, the prepreg effectively creates an insulation barrier,

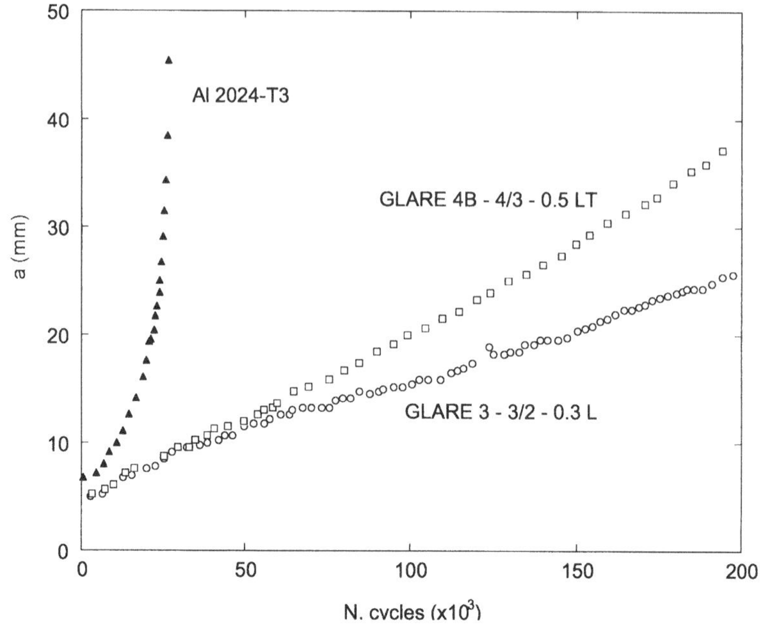

Fig. 11.7 Comparison of fatigue crack growth resistance of GLARE composites and 2024 Al. GLARE has significantly better fatigue resistance (after Vlot and Gunnick, 2001).

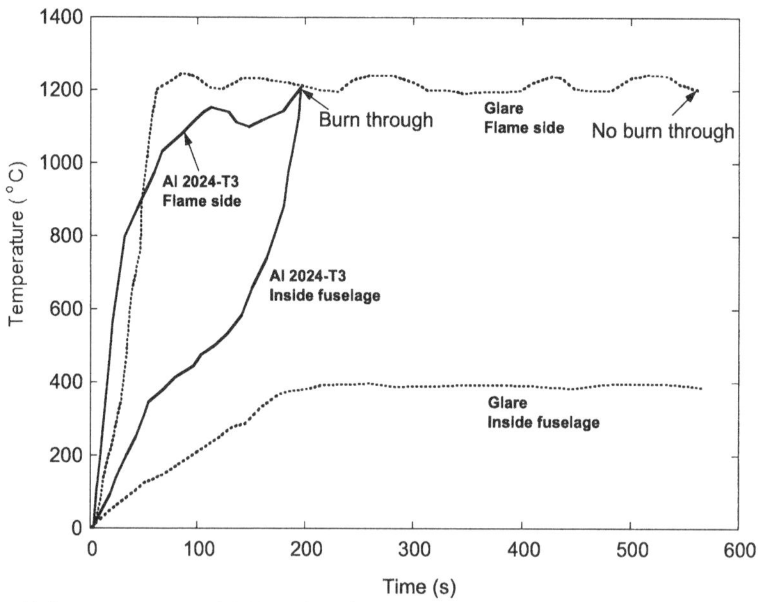

Fig. 11.8 Temperature of 2024 Al and GLARE inside of fuselage and on flame side (after Vlot and Gunnick, 2001). The introduction of GLARE to fuselage design improved the fire resistance of aircraft because of the nature of the composite.

which prevents damage to the remaining aluminum layers. No flame penetration was observed for Glare after 10 minutes, whereas the 2024-T3 aluminum samples showed burn-through within 100 s at approximately 1150 °C. The fire retardant character of GLARE allows much of its strength and fatigue properties, after burn through of the outer aluminum layer, to remain although repair may be necessary (Vogelesang et al, 1995).

Fiber metal laminates have a large degree of tailorability built into them. It is easy to see that, depending on the final application, one can choose the fiber/polymer composite, the metallic alloy type, the layer thickness and stacking sequence, fiber orientation in the PMC layer, etc. The density of the GLARE laminates depends on the relative thickness of aluminum sheet and glass fiber/epoxy layers, the number of layers in the laminate, and the fiber volume fraction. In general, the density of GLARE laminates is at least 8% lower than that of the monolithic aluminum alloy.

11.1.3 Missiles

An important application for MMCs is in missiles (Shakesheff and Purdue, 1998). With increasing performance demands for missiles, conventional aluminum alloys do not have the required strength and temperature resistance. Steel and titanium are not acceptable from a weight point of view. MMCs offer enhanced strength and stiffness with no penalty in weight. In addition the elevated temperature exposure of the missile is for a very short duration (from the time it is launched to where it meets its target). Thus, MMCs are being considered in missile wings and fins, Fig. 11.9.

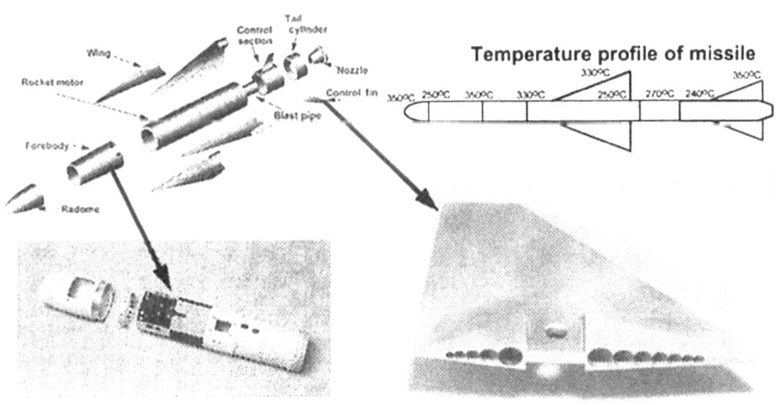

Fig. 11.9 MMC applications in missiles (after Shakesheff and Purdue, 1998; reprinted with permission from Maney Publishing). The elevated temperature exposure of the missile is for a very short duration.

11.1.4 Space Structures

Reduction in the weight of a component is a significant driving force for any application in aerospace. The first successful application of continuous fiber reinforced MMCs was boron fiber reinforced tubular struts in the frame and rib truss members of the mid-fuselage section of the space shuttle, Fig. 11.10 (Rawal, 2001). The struts provided a weight savings of 45% over unreinforced aluminum. In the Hubble telescope, a pitch-based continuous carbon fiber reinforced aluminum was used for waveguide booms because of its light weight, high elastic modulus, and low coefficient of thermal expansion, Fig. 11.11. The material is made of diffusion bonded sheet material and is 3.6 meter in length. The boom requires good stiffness and low coefficient of expansion to maintain the position of the antenna during space maneuvers. The boom also functions as a wave-guide due to its excellent electrical conductivity. Table 11.1 shows selected properties of unidirectional fiber reinforced MMCs used in space applications.

Table 11.1. Properties of unidirectional fiber reinforced metal matrix composites used in space applications (after Rawal, 2001). P100 is a pitch based carbon fiber.

	P100/6061 Al	**P100/AZ91C Mg**	**Boron/Al**
Reinforcement volume fraction (%)	42.2	43	50
Density (g/cm^3)	2.5	2.0	2.7
Poisson ratio	0.3	0.3	0.2
Specific heat (J/kg-K)	812	795	801
Longitudinal			
Young's Modulus (GPa)	343	324	235
Ultimate tensile strength (MPa)	905	710	1100
Thermal conductivity (W/m-K)	320	189	–
Coefficient of thermal expansion (10^{-6}/K)	-0.49	0.54	5.8
Transverse			
Young's Modulus (GPa)	35	21	138
Ultimate tensile strength (MPa)	25	22	110
Thermal conductivity (W/m-K)	72	32	–

Fig. 11.10 Al/B_f tubular struts in the frame and rib truss members of the mid fuselage section of the space shuttle (courtesy of S. Rawal; reprinted with permission from The Minerals, Metals, and Materials Society). The struts provide a weight savings of 45% over unreinforced aluminum.

Fig. 11.11 Carbon fiber reinforced 6061 Al matrix composite used as antenna waveguide/boom on the Hubble Space Telescope (courtesy of S. Rawal; reprinted with permission from The Minerals, Metals, and Materials Society): (a) before integration on the telescope and (b) deployed in space.

In addition to Al-based composites, copper-based composites containing Nb, Ta, or Cr as the second phase in a discontinuous form are also of interest for certain applications requiring high thermal conductivity and high strength, e.g., in high heat flux applications in rocket engine thrust chambers.

11.2 TRANSPORTATION (AUTOMOTIVE AND RAILWAY)

MMCs have been used in a variety of automotive applications. An early and successful engine application was selectively reinforced aluminum pistons in the Toyota diesel engine (Donomoto et al., 1983). In this application, an alumina-silica chopped fiber preform was incorporated into the ring groove area of the piston during pressure casting of the aluminum. The conventional diesel engine piston has an Al–Si casting alloy with a crown made of a nickel cast iron. The main property requirement was increased wear resistance in this area. The previous approach used a Ni-resist ring that increased weight and differed in coefficient of thermal expansion from the aluminum alloy piston material. Aluminum matrix materials reinforced with SiC particulate have also been used in piston applications, primarily in drag racing cars. In this case, the lower coefficient of thermal expansion of the MMC compared to conventional aluminum allowed reduced clearances between the piston and cylinder wall with improved performance.

Another early MMC application in an automotive engine was a hybrid particulate reinforced Al matrix composite used as a cylinder liner in the Honda Prelude, Fig. 11.12. The composite consisted of an Al-Si matrix with 12% Al_2O_3 for wear resistance, and 9% carbon for lubricity. The composite was integrally cast with the engine block, had improved cooling efficiency, and exhibited improved wear and a 50% weight savings over cast iron, without increasing the engine package size. While this concept was initially implemented in the Honda Prelude 2.3L engine, it has also been used in the Honda S2000, Toyota Celica, and Porsche Boxtser engines (Hunt and Miracle, 2001).

An important application of MMCs in the automotive sector is as a driveshaft. (Hoover, 1991) The key limitation in this case arises from the critical rotational speed at which the material becomes dynamically unstable. The critical rotational speed (N_c) is given by (Allison et al., 1997):

$$N_c = \frac{15\pi}{L^2}\left[\frac{E}{\rho}(R_o + R_i)^2\right]^{\frac{1}{2}}$$

where L is the length of the driveshaft, E is the Young's modulus, ρ is the density, and R_o and R_i are the outer and inner radii of the shaft, respectively.

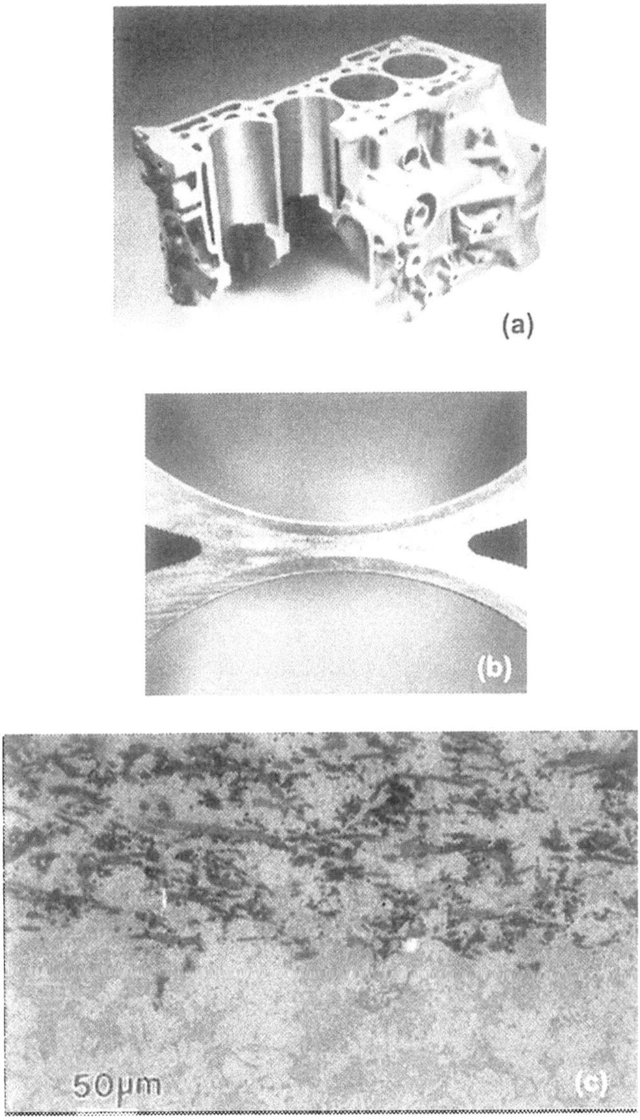

Fig. 11.12 Hybrid particulate reinforced Al matrix composite used as a cylinder liner in the Honda Prelude (parts (a) and (b) courtesy of D. Miracle). The composite consisted of an Al-Si matrix with 12% Al_2O_3 for wear resistance, and 9% carbon for lubricity. (a) Prelude engine block, (b) magnified view of cylinder liner, and (c) microstructure of composite showing carbon short fibers (black) and Al_2O_3 fibers (dark gray).

The important point to be made from this equation is that the material parameter that controls critical speed is the specific modulus, E/ρ. One of the requirements of the driveshaft is that it be welded to a yoke. Thus, SiC reinforcement is precluded from material selection, because of the harmful reaction products formed between Al and SiC in liquid-phase processes (see chapter 4). Duralcan has used a $6061/Al_2O_3/20_p$ composite, which exhibits a 36% increase in specific modulus over steel, Fig. 11.13 (Koczak et al., 1993).

Another important potential replacement of steel by SiC particle reinforced Al matrix composite is in the connecting rod, Fig. 11.14. The connecting rod requires high fatigue resistance at temperatures as high as 150°C. A lighter connecting rod would result in: (a) 12-20% reduction in secondary shaking force, (b) 0.5-1% improvement in fuel economy (with lightweight piston and pin), (c) 15-20% increase in peak RPM, (d) decreased bearing width (package improvement), and (e) increased bearing and crankshaft durability. Initial attempts at developing an MMC connecting rod used hot-pressing followed by extrusion, commonly used to fabricate aerospace components. This technique proved to be too costly in the automotive sector (where production volume is much larger than in the aerospace industry) because of the large amount of wasted material. Near-net shape sinter forging (see chapter 3) was used to fabricate MMC connecting rods with tensile and fatigue properties comparable to those of extruded materials (Chawla et al., 2002). A prototype model of this rod is shown in Fig. 11.15.

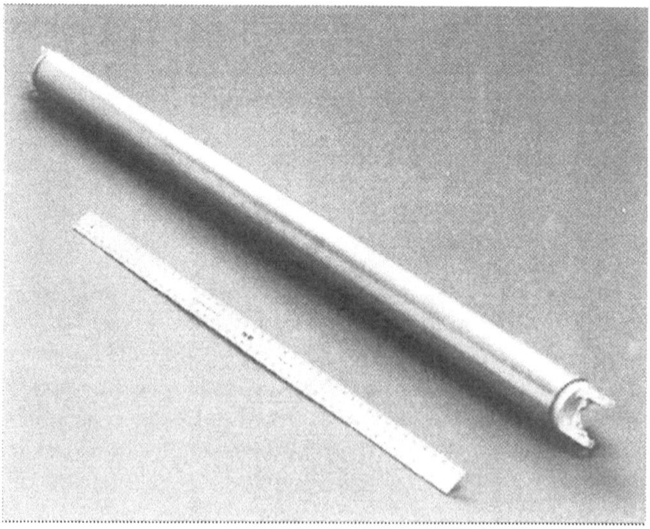

Fig. 11.13 $6061/Al_2O_3/20_p$ composite used as driveshaft in the Corvette (courtesy of D. Miracle). The composite exhibits a 36% increase in specific modulus over steel.

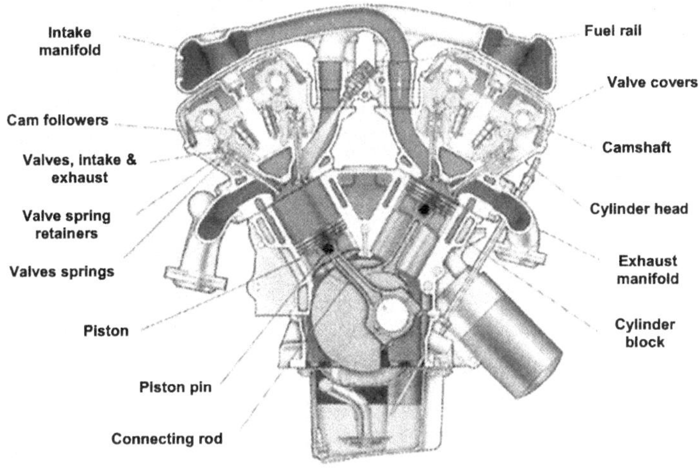

Fig. 11.14 Cross-section of a passenger car engine showing the location of the connecting rod (courtesy of J. Allison).

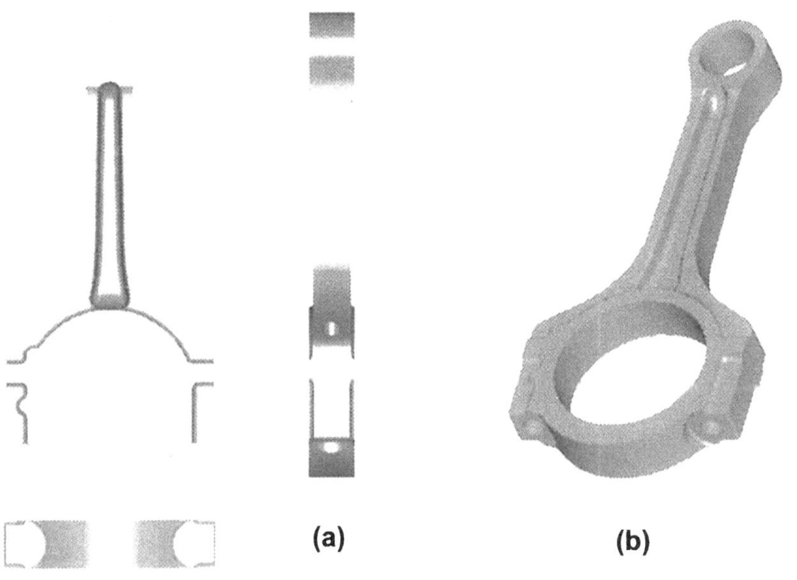

Fig. 11.15 Models of sinter-forged MMC connecting rods: (a) two-dimensional view and (c) three-dimensional view (courtesy of F. Liu).

In Table 11.2 we compare the weight of the MMC rod vis-à-vis that of the steel rod (Chawla and Chawla, 2004). A 57% weight savings was achieved with the MMC rod, with a moderate increase in cost. In addition, it is estimated that for every 1 kg of weight removed from connecting rods, 7 kg of supporting and counterbalancing structure can be eliminated (Hunt and Miracle, 2001). Other demanding powertrain applications are intake and exhaust valves. These components must have good high cycle fatigue performance at elevated temperature, good sliding wear resistance, and creep resistance. Austenitic stainless steel was replaced by a TiB_2 reinforced Ti matrix composite in the Toyota Altezza (Hunt and Miracle, 2001).

Table 11.2 Weight comparison of MMC and steel connecting rods.

	2080/SiC/20$_p$	**Steel**
Pin weight (g)	65.2	144.7
Crank weight (g)	184.0	437.7
Total weight (g)	249.2	582.4

The use of particulate MMCs, particularly Al-based MMCs, has been made in brake drums and brake rotors as a replacement for cast iron, Fig. 11.16. The high wear resistance and thermal conductivity coupled with 50-60% weight savings, make MMCs quite attractive for this application. An

Fig. 11.16 Particulate MMCs for use in brake drums and brake rotors, as a replacement for cast iron (courtesy of D. Miracle). The high wear resistance and thermal conductivity coupled with 50-60% weight savings, make MMCs quite attractive for this application.

intensive development effort was carried out using cast 359Al/SiC/20_p composite. While the costs for this rotor were somewhat higher than those for cast iron, the benefits were justified in a number of specialty vehicles, such as the Plymouth Prowler, Lotus Elise, and others (Hunt and Miracle, 2001). Use of "selective reinforcement" whereby the hard SiC particles are used where high strength and wear resistance are required would appear to be quite attractive. This also minimizes the machining costs since the SiC-free regions are easier to machine.

Metal matrix composite foams have also been developed for damping/energy-absorbant applications in automobiles (Leitlmeier et al., 2002). A 6061Al/Al_2O_3/22_p composite foam was has been used in Ferrari cars. Figure 11.17(a) shows the three-dimensional microstructure of the foam obtained by computed x-ray tomography. The pores are a few millimeters in diameter and are homogeneously distributed. Figure 11.17(b) shows the microstructure of the composite (Babcsan et al., 2004). The SiC particles stabilize the wall thickness, Fig. 11.17(b), and segregate to the cell wall Fig. 11.17(c). The cell walls are about twice the SiC particle diameter.

11.2.1 Railroad Brakes

The driving force for lightweight railway vehicles has also prompted the use of high performance MMCs (Zeuner et al., 1998). A conventional brake system for a railway vehicle, shown in Fig. 11.18, consists of four brake disks, calipers, hand brake, and electromagnetic track brake. This makes up about 20% of the total weight of the bogey. Zeuner et al. (1998) used a selectively cast Al matrix composite with SiC particles for this application. The composites were fabricated by a multipouring process, where alternating layers of the unreinforced alloy and MMC were cast in successive layers. This contributed to reduced cost by using less MMC, and placing the composite in the strategically important region (i.e., in contact with the wearing surface). In this particular application, the weight was reduced from 115 kg for a spheroidal graphite iron disk to 65 kg for the MMC disk, a weight savings of 43%. Figure 11.19 shows photographs of MMC and steel brakes after wear testing. One can see that the steel brake shows a significant amount of cracking, while the MMC brake is in relatively good condition.

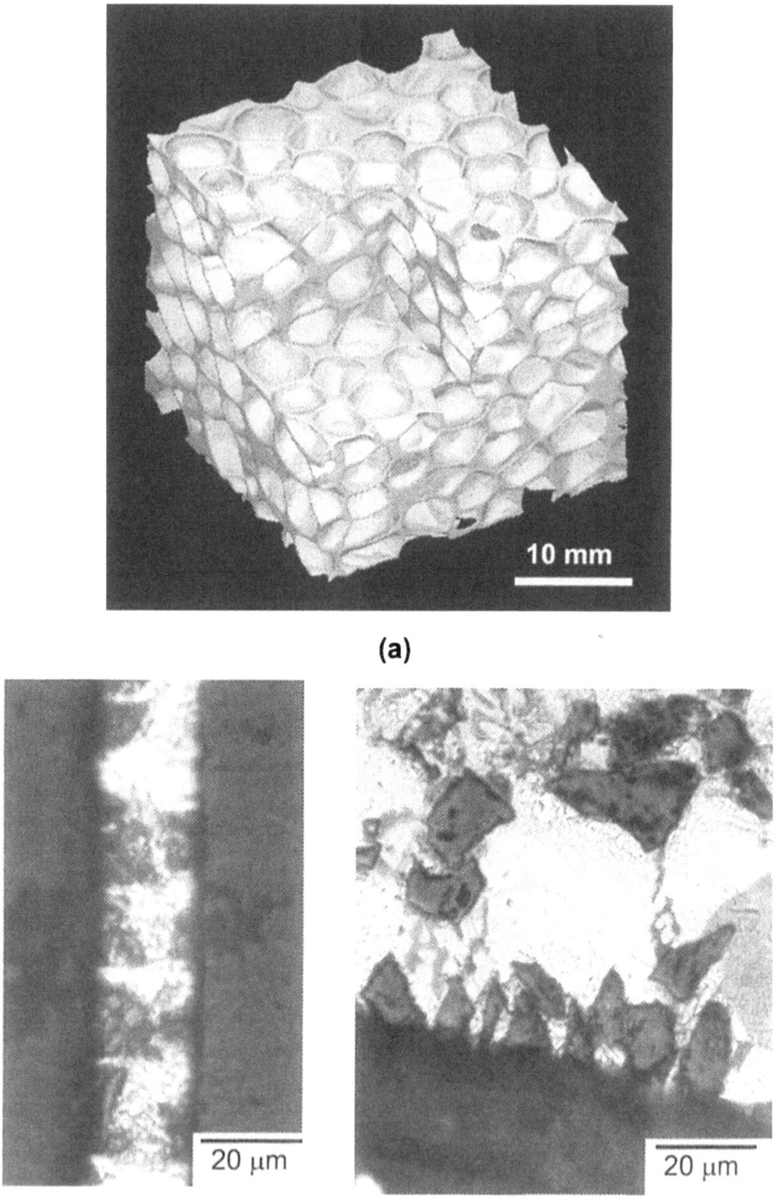

(a)

(b) (c)

Fig. 11.17 $6061/Al_2O_3/22_p$ composite foam used in Ferrari cars (courtesy of H.P. Deigischer): (a) three-dimensional microstructure obtained by computed x-ray tomography, (b) microstructure in the cell wall, which is about twice the SiC particle diameter, and (c) segregation of SiC particles to the cell wall.

Fig. 11.18 High speedway railroad coach bogey with four brake disks produced by SAB Wabco (courtesy of H. Ruppert).

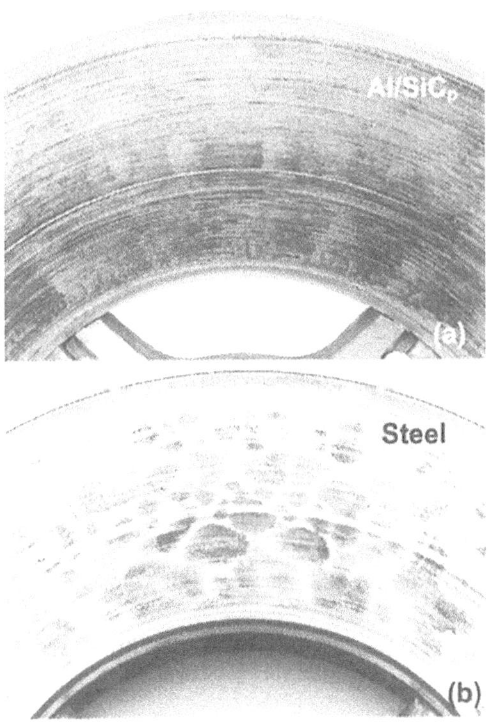

Fig. 11.19 Thermal cracking behavior in (a) Al/SiC$_p$ and (b) steel after wear testing (courtesy of H. Ruppert). The steel brake shows significant amount of cracking (white lines) while the MMC is in relatively good condition.

11.3 ELECTRONICS AND THERMAL MANAGEMENT

A very important market area for aluminum and to a lesser degree copper matrix MMCs is in electronic packaging and thermal management. Metal-matrix composites can be tailored to have optimal thermal and physical properties to meet requirements of electronic packaging systems, e.g., cores, substrates, carriers, and housings. The main attraction of MMCs for these applications is controlled thermal expansion with a negligible penalty in thermal conductivity. Figure 11.20 shows the decrease in coefficient of thermal expansion with increasing volume fraction of SiC in a 2080/SiC_p composite (Deng et al., 2005).

In a multilayered package, a solder joint bonds the substrate (such as AlN) to the baseplate, usually made of Cu. The thermal expansion mismatch between Cu and AlN results in thermal stresses in the solder that may lead to failure of the package. By making the baseplate of an MMC, the CTE may be tailored to that of AlN (or any other suitable substrate). From a heat dissipation point of view, a metallic matrix (such as Cu or Al) reinforced with SiC will not lose much in the way of thermal conductivity, because the conductivity of SiC is similar to that of Al. In order to attain the necessary requirements in thermal expansion and thermal conductivity, composite with

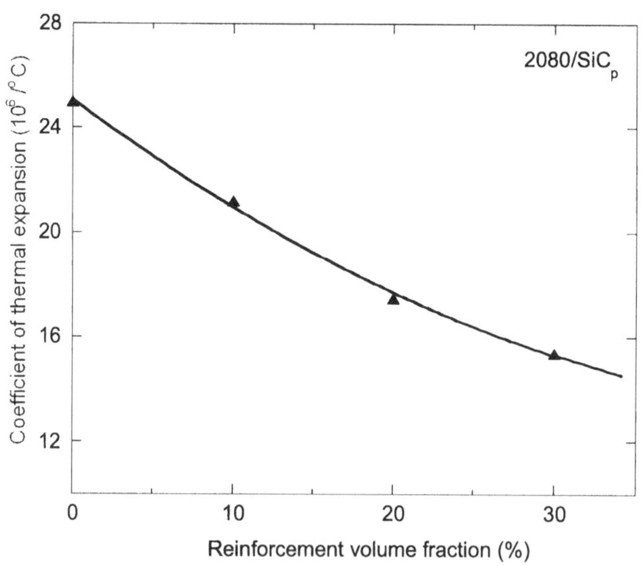

Fig. 11.20 Coefficient of thermal expansion (CTE) of a SiC particle reinforced Al 2080 matrix composite (after Deng et al., 2005). With increasing volume fraction of the low expansion ceramic particle reinforcement, the composite CTE decreases.

55-65% reinforcement particles are typically used. Such high volume fraction of reinforcement is obtained by innovative techniques for enhancing the degree of packing of particles. In particular, a bimodal distribution of particles has been used to obtain the desired packing because very small particles can fit in the space between large reinforcement particles, Fig. 11.21. Examples of SiC particle reinforced aluminum used in microprocessor and optoelectronic packaging applications are shown in Fig. 11.22.

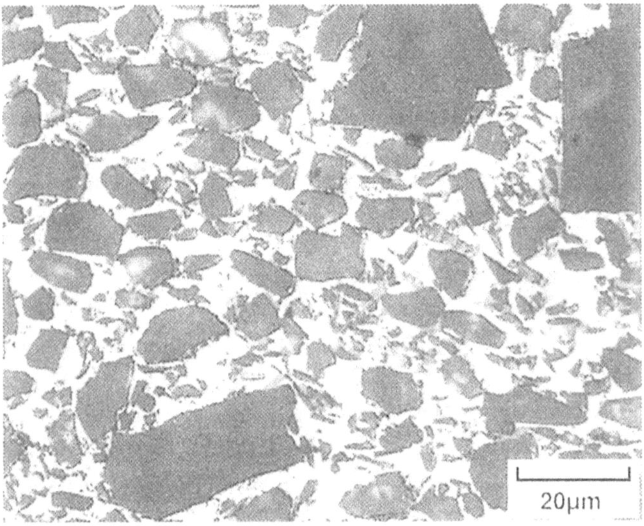

Fig. 11.21 Microstructure of SiC particle reinforced Al matrix composite used for thermal applications (courtesy of M. Occhionero, Ceramic Process Systems). The bimodal particle size distribution enables a large volume fraction of particles (> 50%) to be used.

Fig. 11.22 Examples of SiC particle reinforced Al matrix composites used in microprocessor and optoelectronic packaging applications (courtesy of courtesy of M. Occhionero, Ceramic Process Systems).

An early application was in microwave packaging where a 40 vol.% SiC particle reinforced aluminum MMC replaced Kovar, a heavier Ni-Co-Fe alloy. The major drivers here were weight savings, with a 65% cost reduction realized, along with improved thermal conductivity over the baseline alloy. Increased thermal conductivity and lightweight in a CTE-matched material are also the features being exploited in the use of both Al/SiC and Al/carbon MMCs in Insulated Gate Bipolar Transistors (IGBT) power module base plates. Printed wiring board cores are also being made of Al-SiC materials in replacement of conventional copper or aluminum cores. In addition to the CTE matching, the increased specific stiffness of these MMCs reduces the thermal cycling and vibration-induced fatigue. Yet another application in which the thermophysical properties of Al/SiC are important is in carriers for hybrid circuits for power amplifiers in cellular phone base stations. Continuous boron fiber reinforced aluminum composites made by diffusion bonding have also been used as heat sinks in chip carrier multilayer boards.

Unidirectionally aligned, pitch-based carbon fibers in an aluminum matrix can have high thermal conductivity along the fiber direction. The conductivity transverse to the fibers is about two-thirds that of aluminum. Such a C/Al composite is useful in heat-transfer applications where weight reduction is an important consideration, e.g., in high density, high speed integrated circuit packages for computers and in base plates for electronic equipment. Another possible use of this composite is to dissipate heat from the leading edges of wings in high speed airplanes.

An interesting application involved MMCs for covers of inertial guidance system for a missile system. The requirement here was that the MMC should match the coefficient of thermal expansion of beryllium covers. The ability to tailor the CTE through reinforcement content and matrix selection proved invaluable, and a 6061/SiC/40_p material was used. It should be noted that beryllium, used in this application, is very expensive to machine and has significant toxicity problems. In contrast, the MMC covers were precision forged to near net shape, with some minimal final machining required.

11.4 FILAMENTARY SUPERCONDUCTING MAGNETS

Filamentary superconducting composites have some very important applications. Examples of applications of metal matrix composite superconducting coils include (Cyrot and Pavuna, 1992):

- High-field magnets for research in high energy and in condensed matter physics.

- Magnetic resonance imaging (MRI), which requires extremely uniform magnetic fields of ~1–2 Tesla).
- Coils for windings in motors and generators.
- Magnetic levitating (MAGLEV) coils for high-speed trains.
- Magnetohydrodynamic and electromagnetic thrust systems for propulsion in ships and submarines.

Since the discovery of the so-called high-T_c superconductors, much emphasis has been placed on the high T_c of these superconductors. As pointed out in chapter 3, one should be mindful of not only just T_c but also J_c (critical current density) and H_c (critical magnetic field). As a rule of thumb for most superconductor applications, the critical temperature, T_c, of a technological superconductor should be about twice the use temperature (Cyrot and Pavuna, 1992). This implies that superconductors that are used with liquid helium (4.2 K) as a coolant should have a $T_c > 8$ K. In fact, due to heating of the magnet, the actual operational temperature is closer to ~ 7 K, so one needs a $T_c \sim 15$ K. Thus, to use the oxide high-temperature superconductors (HTS) in applications at 77 K one needs a superconductor with a $T_c \sim 150$ K. Such a material has not yet been synthesized; the thallium-compound with $T_c \sim 125$ K is the best candidate for the "true" technological material at liquid nitrogen temperature.

Nb_3Sn/Cu superconducting composites are used for magnetic fields greater than 12T. Such high fields are encountered in thermonuclear fusion reactors, and superconducting composite magnets would represent a sizable fraction of the capital cost of such a fusion power plant. The main difference between the magnets used in a fusion reactor and in power transmission is that the former use superconductors at very high magnetic fields while the latter use them at low fields.

A large-scale application of Nb-Ti/Cu superconducting magnets is in magnetically levitated trains. Japan National Railways has tested such trains over a small stretch at speeds over 500 km/h. Figure 11.23(a) shows a superconductor cable with 15 strands. A higher magnification micrograph of an individual strand, Fig. 11.23(b), shows 1060 superconducting filaments. Nb-Ti/Cu superconductor composites are also employed in pulsed magnets for particle accelerators in high-energy physics.

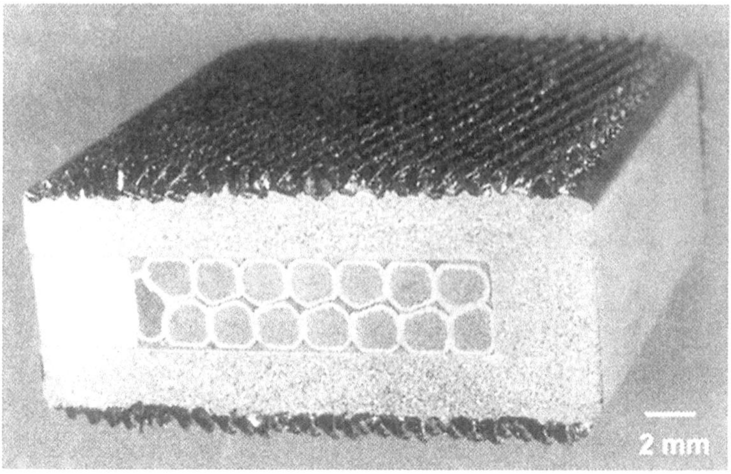

(a)

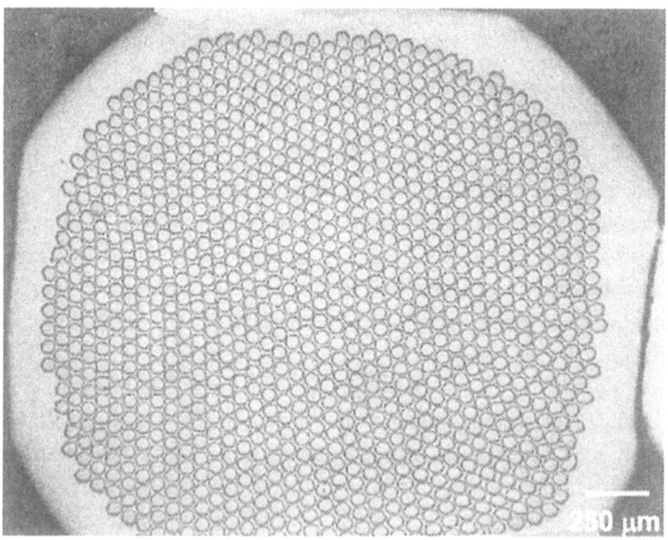

(b)

Fig. 11.23 (a) Compacted strand superconducting cable made of 15 multifilamentary strands and (b) magnified view of one of the strands containing 1060 filaments (courtesy of Hitachi Cable Co.).

Some use of superconductors in power generation has been made by companies such as GE and Siemens Westinghouse. In some cases, by using superconducting materials, the designers could make a generator that developed a much stronger magnetic field than a conventional generator, permitting a significant reduction in the size of the generator for the same power output. Superconducting magnets do require cryogenic temperatures to operate, but the cost of this refrigeration is more than compensated by the energy savings. One has only to remember that, in a superconducting machine working with almost zero resistance, the normal losses associated with the flow of electricity in rotor windings of a conventional machine are absent, resulting in a higher efficiency and reduced operating costs. It is worth pointing out that a big problem in one such project by GE was to prevent the movement of rotor windings under the intense centrifugal and magnetic forces exerted on them. The rotor spins at a speed of 360 rpm. Thus, even an infinitesimally small movement of these components would generate enough heat by friction to quench the superconductors. The GE researchers used a special vacuum epoxy-impregnation process to bond the Nb-Ti superconductors into rock solid modules and strong aluminum supports to hold the windings rigidly.

An important landmark in the development of high-temperature oxide superconductors occurred in 1997, when Geneva's electric utility, SIG, put an electrical transformer using HTS wires into operation. This transformer was built by ABB; it used the flexible HTS wires made by American Superconductor by the process involving packing of the raw material into hollow silver tubes, drawing into fine filaments, grouping the multifilaments in another metal jacket, further drawing and heat treating to convert the raw material into the oxide superconductor (see Chapter 4). This transformer loses only about one fifth of the AC power losses of the conventional ones. Because HTS wires can carry a higher current density, this new transformer is more compact and lighter than a conventional transformer. Liquid nitrogen used as a coolant in the HTS transformer is safer than oils used as insulators in conventional transformers. These superconductors are termed "first-generation" superconductors, Fig. 11.24(a). Newer, second-generation coated superconductors, made by the RABiTS process (see chapter 4), consist of a superconductor layer, coated with a noble metal on an alloy substrate, Fig. 11.24(b). Fig. 11.25 shows the second generation superconductors in tape form.

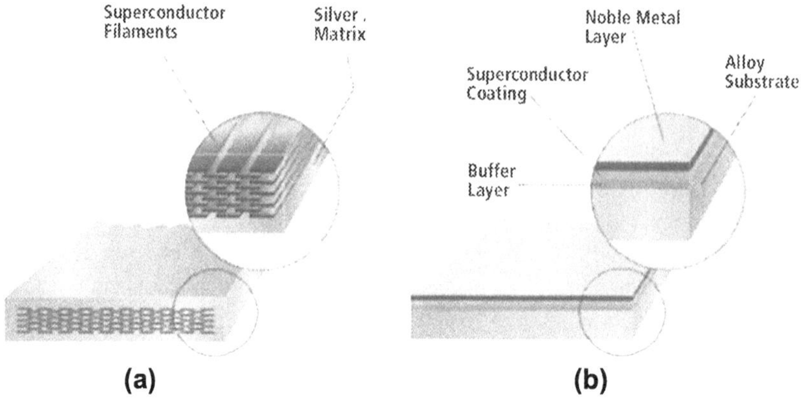

Fig. 11.24 Superconducting metal matrix composites: (a) first generation high temperature superconductor (HTS) filaments in a silver matrix (in commercial production), (b) second generation coated conductor composite, consisting of a superconductor layer, coated with a noble metal, on an alloy substrate (courtesy of J. Jackson, American Superconductor).

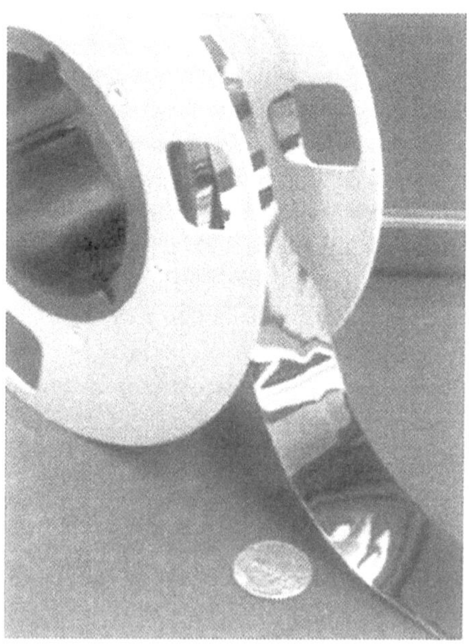

Fig. 11.25 Second generation superconductors, processed by RABiTS process, in tape form (courtesy of J. Jackson, American Superconductor).

We now describe a very important application of Nb-based superconducting composites that has become quite commonplace – magnetic resonance imaging. The phenomenon of nuclear magnetic resonance (NMR) is exploited in the technique of magnetic resonance imaging (MRI). This technique has a major advantage over X-ray radiography in that it is a noninvasive diagnostic technique, and thus the human body is not exposed to an ionizing radiation. Use is made of the electromagnetic characteristics of the nuclei of elements such as hydrogen, carbon, and phosphorus that are present in the human body. The nuclei of these elements act as bar magnets when placed in a strong magnetic field. The patient is placed in the center of a very powerful magnet. When the magnetic field is turned on, the nuclei of the elements in the patient's body part under examination realign along the magnetic field direction. If we apply a radio frequency field, the nuclei will reorient. And if we repeat this process over and over again, the nuclei will resonate. The resonance frequency is picked up by a sensitive antenna, amplified, and processed by a computer into an image. The MRI images, obtained from the resonance patterns are more detailed and have a higher resolution than traditional techniques of visualization of soft tissues. And best of all, these images are obtained without exposing the patient to a radiation or performing biopsy.

The superconducting solenoid, made from Nb-Ti/Cu composite wire, is immersed in a liquid helium cryogenic Dewar. The liquid helium is consumed at about 4 ml per hour, and each refill of the Dewar lasts about three months. Commercially manufactured NMR (nuclear magnetic resonance) spectrometer systems, also called magnetic resonance imaging systems, for medical diagnostics became available in the 1980s. As mentioned above, the big advantage of magnetic resonance imaging in clinical diagnostics is that it does not expose the patient to ionizing radiation and its possible harmful side effects. Of course, MRI techniques do not have to use superconducting magnets but certain advantages exist with superconductors, for example, better homogeneity and resolution and higher field strengths than are available with conventional magnets. The disadvantage is that higher fields with superconductors lead to greater shielding problems.

11.5 POWER CONDUCTORS

A relatively new application of MMCs is in the field of power transmission cable (3M, 2003). The cable consists of a composite core, consisting of Al_2O_3 continuous fibers (Nextel 610) in an Al matrix, and is wrapped with Al-Zr wires. Fig. 11.26(a) shows a power line with composite conductors installed in Buckeye, Arizona. The composite core bears most of the load,

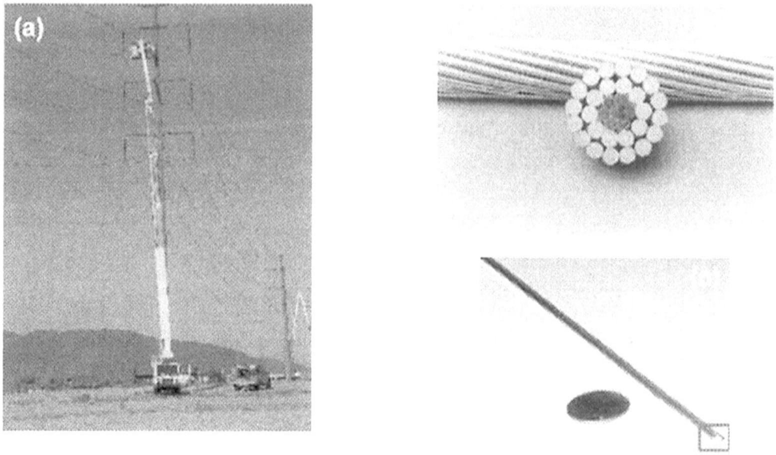

Fig. 11.26 Continuous fiber aluminum composite conductors: (a) 230 kV power line in Buckeye, AZ, (b) stranded composite conductor with aluminum alloy cables surrounding the composite core, and (c) individual composite filament (courtesy of H. Devé, 3M Co.)

due to its much higher stiffness and strength. The maximum allowable temperature of the core (300°C) is also higher than that of the surrounding wires (240°C). The properties of the composite core and surrounding Al-Zr strands are shown in Table 11.3 (Kawakami et al., 1991).

Table 11.3 Properties Al/Al_2O_{3f} composite core, and Al-Zr wire (Kawakami et al., 1991).

Property	Composite Core	Al-Zr Wire
Tensile strength	1380 MPa	162 MPa
Density	3.3 g/cm^3	2.7 g/cm^3
Young's modulus	215-230 GPa	–
Conductivity	24 % IACS*	> 60% IACS
Maximum allowable temperature	300°C	240°C
Fatigue endurance limit (10^7 cycles)	690 MPa	–

* IACS stands for International Annealed Copper Standard. Commonly, the electrical conductivity in power industry is expressed in terms of a percentage of this standard.

Several tests were conducted on the actual conductor, including sag-tension behavior, vibration fatigue, thermal expansion, lightning resistance, and electrical resistance. These were conducted on a 235 m span installed in Les Renardieres, France. Lightning resistance was conducted by applying lightning arcs to the wires and monitoring damage, Fig. 11.27. The level of damage between the Aluminum Conductor Composite Reinforced (ACCR)

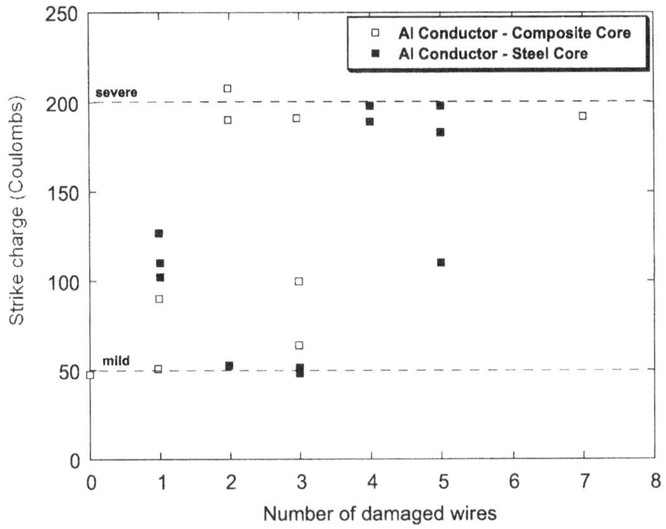

Fig. 11.27 25 Coulomb (C) charge vs. number of damaged wires (after 3M Aluminum Conductor Composite Reinforced Technical Notebook, 2003). The number of damaged wires increases with charge level, with little difference between conductor types.

and the aluminum-steel reinforced conductor (ACSR), was similar. Damage was typically restricted to the outer aluminum layers and consisted of individual strand fracture, "splattering," and/or localized melting of the strand (3M).

Another important aspect of the testing of the conductors was vibration fatigue resistance. Long conductors were tested with a 7 m long gage length, Fig. 11.28(a). The applied peak-to-peak amplitude was varied between 0.7-1.4 mm, resulting in stresses approximately 25% of the fracture strength, and a turning angle of 5°. An endurance limit (taken here as 100 million cycles) was observed at an amplitude of 0.03 in. Figure 11.28(b) shows the onset of aluminum strand failures as a function of amplitude and vibration cycles. The aluminum wires always failed before the composite wires, since the composite core wires had a higher fatigue resistance.

11.6 RECREATIONAL AND SPORTING GOODS

Particulate metal matrix composites, especially with light metal-matrix composites such as aluminum and magnesium, also find applications in automotive and sporting goods. In this regard, the price per kg becomes the

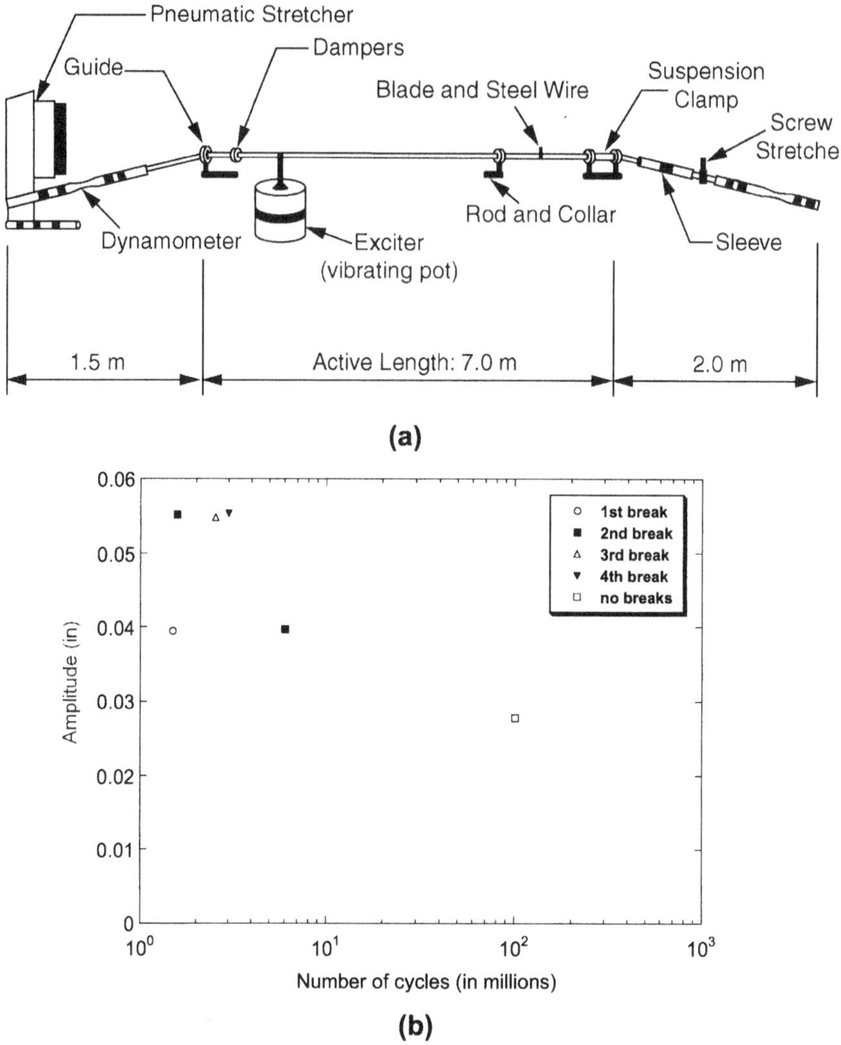

Fig. 11.28 (a) Schematic of set-up for vibration fatigue tests conducted on a 284-kcmil conductor and (b) displacement amplitude versus fatigue cycles curve. Below an amplitude of 0.03 in, the conductor did not fail (after 3M Aluminum Conductor Composite Reinforced Technical Notebook, 2003).

driving force for application. An excellent example involves the use of Duralcan particulate MMCs to make mountain bicycles. The Specialized Bicycle Co. in the Unites States sells these bicycles with the frame made from extruded tubes of 6061 aluminum containing about 10% alumina particles. The primary advantage is the gain in stiffness.

The search for improved performance in recreational products often leads to new and exciting materials options. Metal matrix composites have been applied in both prototype and production applications of recreational products. The primary property advantage that has been utilized is the higher specific stiffness, especially in the golf club and bike frame applications.

An interesting recreational application of MMCs is in track shoe spikes (Grant, 1999), shown in Fig. 11.29. The composite consists of an aluminum alloy matrix reinforced with particulate reinforcement of aluminum oxide;

Fig. 11.29 (a) Particle reinforced MMCs used in track shoe spikes and (b) track shoe with MMC spikes (courtesy of T. Wang, Omni-Lite Corp.).

silicon carbide, boron carbide, or titanium carbide. The volume fraction of particles can range between 5-30%. The composites are processed using a technique called "progressive cold forging," where about 300 parts/minute can be produced. The unique shapes of the spikes are designed to compress the track without providing unwanted impact and stress to the athlete's feet and legs.

11.7 WEAR-RESISTANT MATERIALS

Extensive use if made of metal matrix composites in applications requiring wear resistance. Carbides, in general, and tungsten carbide, in particular, are very hard materials. Combined with a suitable ductile metal matrix we can get a composite that is very useful for cutting, grinding, drilling, roll-surfaces for rolling mills and nibs of dies for wire drawing and such operations. One can get, by suitable microstructural design techniques, a wide range of hardness (which relates to wear resistance) and toughness combinations (Deng et al., 2001; Deng et al., 2002). Commonly, the volume fraction of WC particles varies between 0.70 and 0.90. The particle size typically ranges between 0.2 and 15 μm. Besides, the straight tungsten carbide/cobalt composite, one may incorporate different amounts of other carbides such as titanium carbide (TiC), tantalum carbide (TaC) and niobium carbide (NbC).

Microstructurally, tungsten carbide based composites are isotropic and quite homogenous. They have provided a technically very effective and reliable and economically very reasonable product. Figure 11.30 shows a roller-cone bit used for oil well drilling; the rock cutting inserts, made of WC/Co composites, on the cones.

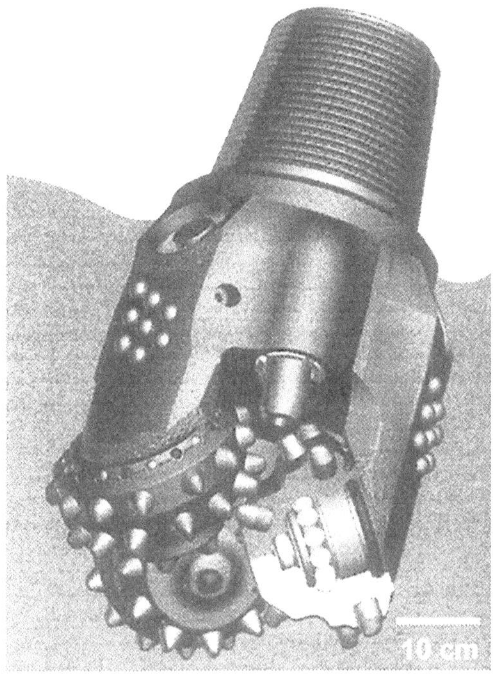

Fig. 11.30 Roller cone bit used for oil well drilling. The rock cutting inserts are made of WC/Co metal matrix composites. (Courtesy of A. Griffo, Smith International, Inc.)

References

3M Aluminum Conductor Composite Reinforced Technical Notebook, (2003) 3M, Minneapolis, MN, pp. 1-28.

Allison, J.E., J.W. Jones, and L.C. Davis, (1997) in *Composites Engineering Handbook*, (P.K. Mallick, ed.), Marcel Dekker, New York, pp. 975-980.

Babcsan, N., D. Leitlmeier, H.P. Degischer, J. Banhart, (2004) *Adv. Eng. Mater.*, **6**, 421-428.

Chawla, K.K., (1998) *Fibrous Materials*, Cambridge University Press.

Chawla, K.K., and N. Chawla, (2004) in *Kirk-Othmer Encyclopedia*, John-Wiley and Sons.

Chawla, N., J.J. Williams, and R. Saha, (2002) *J. Light Metals*, **2**, 215-227.

Deng, X., B.R. Patterson, K.K. Chawla, M.C. Koopmnan, C. Mackin, Z. Fang, G. Lockwood, and A. Griffo, (2001) *Int. J. Refrac. Met. Hard Mater.*, **19**, 547-552.

Deng, X., B.R. Patterson, K.K. Chawla, M.C. Koopmnan, C. Mackin, Z. Fang, G. Lockwood, and A. Griffo, (2002) *J. Mater. Sci. Lett.*, **21**, 707-709.

Deng, X., D.R.M. Schnell, and N. Chawla, (2005) *Mater. Sci. Eng..*, submitted.

Donomoto, T., N. Miura, K. Funatani, and N. Miyake, (1983) *SAE Tech. Paper no. 83052.*

Evans, A.,C.S. Marchi, and A. Mortensen, (2003) *Metal Matrix Composites in Industry*, Kluwer Academic Publishers, Dordrecht.

Grant, D., (1999) *United States Patent*, # 5,979,084.

Hoover, W., in *12th Risø International Symposium*, (1991) (N. Hansen et al. eds.), Roskilde, Denmark, Risø National Laboratory, pp. 387-392.

Hunt, W.H., and D.B. Miracle, (2001) in *ASM Handbook – Composites*, vol. 21, 1029-1032.

Koczak, M.J., S.C. Khatri, J.E. Allison, and M. Bader, (1993) in *Fundamentals of Metal Matrix Composites*, (S. Suresh, A. Mortensen, and A. Needleman, eds.), Butterworth-Heinemann, Stoneham, MA.

Kawakami, K., M. Okuno, K. Ogawa, M. Miyauchi, and K. Yoshida, (1991) *Furukawa Rev.*, **9**, 81-85.

Leitlmeier, D., H.P. Degischer, H. Flankl, (2002) *Adv. Eng. Mater.*, **4**, 735-740.

Miracle, D.B., (2001) in *ASM Handbook – Composites*, vol.21, 1043-1049.

Rawal, S.P., (April 2001) *JOM*, **53**, 14.

Shakesheff, A.J., and G. Purdue, (1998) *Mater. Sci. Tech.*, **14**, 851.

Vlot, A., and J.W. Gunnick., (2001) *Fibre Metal Laminates: An Introduction*. Kluwer Academic Publishers, Netherlands.

Vogelesang, L.B., J. Schijve, and R. Fredell, "Fiber Metal Laminates: Damage Tolerant Aerospace Materials," *Case Studies in Manufacturing with Advanced Materials*, vol. 2, (1995) (A. Demaid and J.H.W. de Wit, eds.), Amsterdam: Elsevier, pp. 253–271.

Zeuner, T., P. Stojanov, P.R. Sahm, H. Ruppert, and A. Engels, (1998) *Mater. Sci. Tech.*, **14**, 857-863.

AUTHOR INDEX

C

D

E

F

L

M

N

O

P

R

S

T

Y

Z

SUBJECT INDEX

W

Y

ABOUT THE AUTHORS

Nikhilesh Chawla is Associate Professor of Materials Engineering at Arizona State University. He received his Ph.D. from the University of Michigan. Prior to ASU, he was a postdoctoral fellow jointly at Ford Motor Company and the University of Michigan, and a senior development engineer at Hoeganaes Corporation. His research interests encompass the mechanical behavior and modeling of advanced materials at bulk and small length scales. Professor Chawla is the recipient of the National Science Foundation Early Career Development Award and the Office of Naval Research Young Investigator Award. He's also the recipient of the 2004 Bradley Stoughton Award for Young Teachers, given by ASM International. During 2004-06, he was Chair of the Composite Materials Committee of The Minerals, Metals, and Materials Society, and serves on the Editorial Board for *Metallurgical and Materials Transactions*.

Krishan Chawla is Professor of Materials Science and Engineering at the University of Alabama at Birmingham, USA. He received his Ph.D. from the University of Illinois *at* Urbana-Champaign. His research interests encompass processing, microstructure, and mechanical behavior of materials. He has taught and/or done research at several universities around the world. Professor Chawla has served as a Program Director for metals and ceramics in the Division of Materials Research, National Science Foundation. He is a fellow of ASM international. Among his other awards are: Distinguished Researcher Award at New Mexico Tech, Distinguished Alumnus Award from Banaras Hindu University, Eshbach Society Distinguished Visiting Scholar award at Northwestern University, Faculty Fellow award at Oak Ridge National Laboratory. Professor Chawla is the author or coauthor of various textbooks in the area of materials and serves on the editorial boards of a number of journals.

FSC
www.fsc.org
MIX
Papier | Fördert gute Waldnutzung
FSC® C083411

Kolibri 360 GmbH
Ferdinand-Jühlke-Straße 7
99095 Erfurt, Deutschland
produktsicherheit@kolibri360.de